T0183747

Lecture Notes in Computer Science   **11410**

Commenced Publication in 1973
Founding and Former Series Editors:
Gerhard Goos, Juris Hartmanis, and Jan van Leeuwen

Editorial Board

David Hutchison
  Lancaster University, Lancaster, UK
Takeo Kanade
  Carnegie Mellon University, Pittsburgh, PA, USA
Josef Kittler
  University of Surrey, Guildford, UK
Jon M. Kleinberg
  Cornell University, Ithaca, NY, USA
Friedemann Mattern
  ETH Zurich, Zurich, Switzerland
John C. Mitchell
  Stanford University, Stanford, CA, USA
Moni Naor
  Weizmann Institute of Science, Rehovot, Israel
C. Pandu Rangan
  Indian Institute of Technology Madras, Chennai, India
Bernhard Steffen
  TU Dortmund University, Dortmund, Germany
Demetri Terzopoulos
  University of California, Los Angeles, CA, USA
Doug Tygar
  University of California, Berkeley, CA, USA

More information about this series at http://www.springer.com/series/7411

Seth Gilbert · Danny Hughes
Bhaskar Krishnamachari (Eds.)

Algorithms for Sensor Systems

14th International Symposium on Algorithms and Experiments
for Wireless Sensor Networks, ALGOSENSORS 2018
Helsinki, Finland, August 23–24, 2018
Revised Selected Papers

 Springer

Editors
Seth Gilbert (ID)
National University of Singapore
Singapore, Singapore

Bhaskar Krishnamachari (ID)
University of Southern California
Los Angeles, CA, USA

Danny Hughes (ID)
KU Leuven
Heverlee, Belgium

ISSN 0302-9743   ISSN 1611-3349   (electronic)
Lecture Notes in Computer Science
ISBN 978-3-030-14093-9   ISBN 978-3-030-14094-6   (eBook)
https://doi.org/10.1007/978-3-030-14094-6

Library of Congress Control Number: 2019932168

LNCS Sublibrary: SL5 – Computer Communication Networks and Telecommunications

© Springer Nature Switzerland AG 2019
This work is subject to copyright. All rights are reserved by the Publisher, whether the whole or part of the material is concerned, specifically the rights of translation, reprinting, reuse of illustrations, recitation, broadcasting, reproduction on microfilms or in any other physical way, and transmission or information storage and retrieval, electronic adaptation, computer software, or by similar or dissimilar methodology now known or hereafter developed.
The use of general descriptive names, registered names, trademarks, service marks, etc. in this publication does not imply, even in the absence of a specific statement, that such names are exempt from the relevant protective laws and regulations and therefore free for general use.
The publisher, the authors and the editors are safe to assume that the advice and information in this book are believed to be true and accurate at the date of publication. Neither the publisher nor the authors or the editors give a warranty, express or implied, with respect to the material contained herein or for any errors or omissions that may have been made. The publisher remains neutral with regard to jurisdictional claims in published maps and institutional affiliations.

This Springer imprint is published by the registered company Springer Nature Switzerland AG
The registered company address is: Gewerbestrasse 11, 6330 Cham, Switzerland

Preface

The 14th International Symposium on Algorithms and Experiments for Wireless Networks was held during August 23–24, 2018 at Helsinki, Finland.

AlgoSensors is an international symposium dedicated to the algorithmic aspects of wireless networks. Originally focused on sensor networks, it now covers algorithmic issues arising in wireless networks of all types of computational entities, static or mobile, including sensor networks, sensor-actuator networks, and autonomous robots. The focus is on the design and analysis of algorithms, models of computation, and experimental analysis.

AlgoSensors 2018 was organized as part of ALGO 2018, which also hosts Algo-Cloud, ATMOS, ESA, IPEC, WABI, and WAOA. Submissions were sought in all areas of wireless networking, broadly construed, including both theoretical and experimental perspectives.

AlgoSensors 2018 received a total of 39 submissions. Although they were treated in a unified manner, there were two tracks for submissions: algorithms (30 submissions) and experiments (nine submissions). After a rigorous and thorough review process by our Program Committee of international experts, 15 papers were accepted for the final program (12 from algorithms submissions and three from the experimental submissions).

We would like to thank all the TPC members for their valuable time in reviewing the papers, our publicity chair, Theofanis Raptis, for bringing the venue to the attention of researchers around the world, our local co-chairs, Parinya Chalermsook, Petteri Kaski, and Jukka Suomela, for their assistance in organizing the event, and the AlgoSensors Steering Committee, particularly Sotiris Nikoletseas (Chair) for his guidance and assistance in every step of the process.

January 2019

Seth Gilbert
Danny Hughes
Bhaskar Krishnamachari

Organization

General Chair

Robert Bai   University of Cambridge, UK

Program Committee Chairs

Seth Gilbert   National University of Singapore, Singapore
Danny Hughes   KU Leuven, Belgium
Bhaskar Krishnamachari   University of Southern California, USA

Steering Committee

Josep Diaz   U.P. Catalunya, Spain
Magnus M. Halldorsson   Reykjavik University, Iceland
Bhaskar Krishnamachari   University of Southern California, USA
P. R. Kumar   Texas A&M University, USA
Sotiris Nikoletseas   University of Patras and CTI, Greece
  (Chair)
Jose Rolim   University of Geneva, Switzerland
Paul Spirakis   University of Patras and CTI, Greece
Adam Wolisz   T.U. Berlin, Germany

Program Committee

Chaodong Zheng   Nanjing University, China
Stephan Holzer   Massachusetts Institute of Technology, USA
Bogdan Chlebus   University of Colorado Denver, USA
Jó Ueyama   University of São Paulo, Brazil
Christian Schindelhauer   University of Freiburg, Germany
Patrick Thiran   Ecole Polytechnique Fédérale de Lausanne, Switzerland
Magnus M. Halldorsson   Reykjavik University, Iceland
Gowri Sankar   University of Southern California, USA
  Ramachandran
Raphael Eidenbenz   ABB Corporate Research, Switzerland
Qiang-Sheng Hua   Huazhong University of Science and Technology, China
Kevin Lee   Nottingham Trent University, UK
Chantal Taconet   Télécom SudParis, France
Jose Proenca   HASLab - INESC TEC/University of Minho, Portugal
Dariusz Kowalski   University of Liverpool, UK
Utz Roedig   Lancaster University, UK
Kristof Van Laerhoven   University of Siegen, Germany

Fabian Kuhn   University of Freiburg, Germany
Philipp Sommer   ABB Corporate Research, Switzerland
Philipp M. Scholl   Universität Freiburg, Germany
Davy Preuveneers   K.U. Leuven, Belgium
Artur Czumaj   The University of Warwick, UK

Contents

Local Gossip and Neighbour Discovery in Mobile Ad Hoc Radio Networks

Avery Miller[✉][iD]

University of Manitoba, Winnipeg, MB R3T 2N2, Canada
avery.miller@umanitoba.ca

Abstract. We propose a new task called δ-local gossip, which can be viewed as a variant of both gossiping and geocasting. We motivate its study by showing how the tasks of discovering and maintaining neighbourhood information reduce to solving δ-local gossip. We then provide a deterministic algorithm that solves δ-local gossip when nodes travel on a line along arbitrary continuous trajectories with bounded speed.

Keywords: Mobile radio networks · Information dissemination · Neighbour discovery

1 Introduction

Designing deterministic algorithms for communication tasks in mobile radio networks is difficult due to the fact that the neighbourhoods of nodes can change frequently and unpredictably while an algorithm is executing. Overcoming such difficulties is an important step towards designing ad hoc networks of devices that can cooperate to perform tasks with little to no human supervision, such as self-driving cars or aerial robotic drones. When a radio-equipped device is turned on, it may know some initial information about itself and possibly its location, but it does not have any information about other nearby devices. To form an ad hoc network, devices need to coordinate communication amongst themselves, despite wireless signal interference and changing neighbourhoods, so that they can perform more complex and interesting tasks. Solutions to such tasks often assume that every device initially has information about all other devices within its communication range (or even slightly beyond). We are interested in filling the current gaps in what is known about collecting this initial information without making simplifying assumptions, e.g., without assuming that there is an initial period during which devices are stationary.

To formalize the notion of sharing and collecting information among nearby mobile nodes, we define a new task called δ-local gossip. This task can be viewed as a variant of both gossiping [16] and geocasting [3]: each node initially has a piece of information that it needs to share, but rather than sharing it with all other nodes (as in gossiping), it shares this information with all nodes that enter a certain region (as in geocasting). However, the region is defined as all

© Springer Nature Switzerland AG 2019
S. Gilbert et al. (Eds.): ALGOSENSORS 2018, LNCS 11410, pp. 1–14, 2019.
https://doi.org/10.1007/978-3-030-14094-6_1

points that are within a δ-multiple of p's transmission range, so it depends on p's movement (unlike in geocasting where the region is fixed).

In this paper, we restrict our study to deterministic solutions for δ-local gossip and neighbour discovery. Our motivation stems from the fact that the collection of information could be an initial step or a subroutine of an algorithm that performs a more complex task. Randomized solutions that allow small amounts of error may cause subsequent algorithms that depend on accurate initial information to fail. For example, algorithms for calculating transmission schedules, for determining routing paths, or for avoiding physical collisions, could all be sensitive to errors in information about nearby nodes. This is particularly important in applications where human life is at risk. Further, one cannot amplify the success probability of a randomized algorithm for information collection by using repetition, since the neighbourhoods of mobile nodes can change across different executions.

The organization of the paper is as follows. In Sect. 1.1, we define our models of mobile radio networks and define δ-local gossip. In Sect. 2, we motivate the study of δ-local gossip by showing its connections to neighbour discovery. In Sect. 3, we give a solution to δ-local gossip in a specific model that involves nodes moving on a line along arbitrary continuous trajectories with bounded speed. Due to limited space, the formal proofs have been omitted, but can be found in [17].

1.1   Models and Definitions

We formally define a mobile radio network model called *Mobile-Rcv*, which originally appeared in [8]. The network consists of a fixed set $\mathcal{V}$ of n nodes with distinct identifiers. With each node p, we associate a *trajectory function* p: given a time τ as input, $p[\tau]$ is p's location in the environment at time τ. Each node moves along a continuous trajectory, and, at all times, its speed is bounded above by some known constant σ. Each node can accurately determine its location relative to a global origin. Unless specified otherwise, we assume that every node knows its entire trajectory function. We assume that nodes have synchronized clocks. Time is partitioned into slots, where each slot is an interval of fixed length. For all $i \geq 1$, all nodes execute slot i of their algorithm at the same time.

Each node in $\mathcal{V}$ possesses a radio with which it can perform a transmission on the shared channel in any slot t. A node can only start transmissions at slot boundaries, and the length of each transmission is exactly the length of one slot. If a node is not transmitting in a slot, we say that it is *listening*. Each radio has *transmission radius* R and *interference radius* R'. A node p receives a message from a node q if and only if all of the following hold: (1) p does not transmit during slot t; (2) q transmits during slot t, and, at all times τ in slot t, the distance between p and q at time τ is at most R; (3) for all nodes $q' \neq q$ that transmit during slot t, at all times τ in slot t, the distance between q' and p at time τ is greater than R'. If the first two of the above conditions are satisfied, but the third is not, then we say that a *transmission collision* occurs due to interference. We assume that nodes do not possess collision detectors, so, for each node p and each slot t, either p receives a message in $\{0,1\}^*$ from a node q during slot t, or, p receives $\bot$ during slot t (which represents "no message"). We assume that all nodes know the values of R, R', and σ. For two nodes u, v, we say that v *is a neighbour of u for slot t*, if, when u is the only transmitting node in slot t, v receives u's transmitted message in slot t. In particular, u and v are neighbours for slot t if the distance between them for the entirety of slot t is at most R. The *neighbour graph G_{nbr}* is a sequence of graphs $\{G_t\}_{t \geq 1}$ where, for each $t \geq 1$, we have $V(G_t) = \mathcal{V}$ and $\{u, v\} \in G_t$ if and only if u and v are neighbours for slot t.

Local Gossip. Informally, the main task considered in this paper is to guarantee that every node obtains a piece of information from each node that is ever located within a δ-multiple of its transmission radius. For example, if $\delta > 1$, this could allow nodes to get advance warning before another node becomes its neighbour.

Definition 1 (δ-local gossip). *Suppose that each node p in the network initially possesses a piece of information I_j. Given a constant $\delta > 0$, there is a known $T_\delta > 0$ such that each node terminates at the end of slot T_δ and, at termination, for all nodes p_j and p_k such that $d(p_j, p_k) \leq \delta R$ at some time before or at the end of slot T_δ, p_j knows I_k.*

The condition that $d(p_j, p_k) \leq \delta R$ at some time during the execution is quite weak. For example, the distance between p_j and p_k could be greater than δR up until the exact moment that they both terminate. Also, if $\delta > 1$, it is possible that nodes p_j and p_k are never within transmission range of one another. Further, the task does not even require that p_j and p_k receive at least one transmission from one another, even if they are neighbours for the entire execution. This could happen due to signal interference from simultaneous transmissions.

At a high level, δ-local gossip is similar to geocasting [3], in that information must be disseminated to all nodes within a specified region. However, there are several significant differences. First, the geocasting task specifies a single source with a single piece of information, whereas, in δ-local gossip, each node has a distinct piece of information that it must share. In this sense, δ-local gossip can be considered a 'multi-source' version of geocasting. Also, in geocasting, the region within which the source's information must be shared is fixed relative to the starting point of the source, whereas, in δ-local gossip, the region follows the node's trajectory. Finally, in geocasting, only nodes that stay within the specified region for a long time must receive the source's message, whereas, in δ-local gossip, for any node q that is within distance δR from a node p at *some* time during the execution, p must receive q's piece of information.

1.2   Related Work

Motived by real-world networks, there has been increased activity in the theoretical study of distributed algorithms for mobile and dynamic networks [2, 15]. Much of this study has focused on dynamic graph models, while some models opt to explicitly include device movement along continuous trajectories and must deal with the resulting geometric considerations. In this latter category, the existing work about information dissemination (i.e., broadcasting, gossiping, geocasting) is most closely related to our work [1,3,16].

A survey of results regarding neighbour discovery in mobile networks can be found in [12]. In the model we consider in this paper, some previous work solves the *neighbourhood maintenance* task, in which it is assumed that each node initially knows its neighbourhood exactly, and, at all times before termination, each node must maintain an accurate list of its neighbours. Two solutions are known for this task in the *Mobile-Rcv* model: one for the line network by Ellen, Welch and Subramanian [8], which we refer to as the EWS algorithm, and one for road networks by Chung, Viqar and Welch [4], which we refer to as the CVW algorithm. These algorithms work if a certain amount of neighbourhood information is initially known by each node. In [4], they state that this information can be acquired by running a gossiping algorithm for static networks, under the assumption that all nodes stay close to their starting position for the entire execution. In Sect. 2, we will show how a δ-local gossip algorithm can be used to ensure that each node acquires this initial information without making such an assumption. In the *Mobile-Rcv* model with $R' = R$, Cornejo et al. [6] provide an algorithm for a neighbour discovery task that we call "continual stable-neighbour discovery" in Sect. 2. Their solution assumes the existence of a MAC layer, as defined in [14] (without aborts), so they do not need to consider transmission collisions. There are currently no deterministic implementations of the MAC layer that they use, and the probabilistic implementations that they cite do not guarantee message delivery.

2   Neighbour Discovery and Local Gossip

Neighbour discovery has been well-studied in the case of static networks, i.e., where the devices do not move or fail [5,10,13,18,19]. However, in mobile networks, it is not even clear how to define the task: since the neighbour graph G_{nbr} is a dynamic graph that can change often during an algorithm's execution, which nodes should a node v consider as its neighbours? In this section, we formalize neighbour discovery tasks and prove reductions that show the natural connections between neighbour discovery and δ-local gossip.

To solve a neighbour discovery task, each node p must calculate, at the start of each slot t, a *neighbour list for slot t*, which we denote by $\text{List}(p, t)$. To define a neighbour discovery task, we will specify conditions on the relationship between $\text{List}(p, t)$ and p's actual set of neighbours in G_t, denoted by $NBRS(p, t)$. In all cases, we assume that nodes have no initial information about their neighbours.

We distinguish between neighbourhood discovery tasks in several ways. First, we consider the *permanence* of the task: do nodes determine their neighbourhoods for a specific slot, i.e., *one-time* discovery, or do nodes keep learning about their neighbourhoods as the neighbourhoods change, i.e., *continual* discovery? One-time algorithms are useful in order to satisfy the initial conditions of a subsequent algorithm. In this case, we want all nodes to terminate the neighbour discovery algorithm at the same time, and we want all nodes to determine who their neighbours are at termination. In contrast, when discovering neighbours to facilitate on-going communication tasks such as routing, we want nodes to continually learn about changes in their neighbourhood. Next, we consider the *accuracy* of the task. In some problems, each node's neighbour list must match its actual neighbourhood exactly. For other problems, each node's neighbour list will be a subset or superset of its neighbourhood. Depending on the application, algorithms that calculate inexact neighbour lists might be sufficient. For example, if nodes are keeping a list of neighbours in order to avoid physical collisions, then, as long as each node's neighbour list is always a superset of its actual neighbourhood, the system can accomplish its goal. However, such a solution might be less efficient than in the case where exact neighbourhoods are known.

Using the above distinctions, we give formal definitions for six neighbour discovery tasks and provide reductions from each task to δ-local gossip. The reductions are general, i.e., they do not depend on our specific communication model. In what follows, suppose that we have an algorithm $\texttt{LG}(\delta)$ that solves δ-local gossip, and we denote its running time by $|\texttt{LG}(\delta)|$.

One-Time Exact Neighbour Discovery. A solution to this task guarantees that, at termination, each node knows the ID of each of its neighbours for the next slot. Such an algorithm is useful as an initialization step before executing an algorithm that assumes that each node initially knows its exact neighbourhood. This is equivalent to how neighbour discovery is defined for static networks. A parameter T_{Init} specifies how long the discovery process takes.

Definition 2. *An algorithm solves* one-time exact neighbourhood discovery *if, for some known $T_{Init} > 0$, every execution terminates at the end of slot T_{Init}, and, at termination, each node p has* $\texttt{List}(p, T_{Init} + 1) = NBRS(p, T_{Init} + 1)$.

Define algorithm $\texttt{OENL}$ as the execution of $\texttt{LG}(1)$, where, for each p_j, the value of I_j is p_j's trajectory for slot $|\texttt{LG}(1)| + 1$.

Lemma 1. $\texttt{OENL}$ *solves one-time exact neighbour discovery with* $T_{Init} = |\texttt{LG}(1)| + 1$.

Continual Exact Neighbour Discovery. This is the strongest possible version of neighbour discovery: after some fixed number of slots, all nodes know the identity of all neighbours at all times.

Definition 3. *An algorithm solves* continual exact neighbour discovery *if, for some known $T_{Init} > 0$, during every execution,* $\texttt{List}(p, t) = NBRS(p, t)$ *for every node p and every slot $t \geq T_{Init} + 1$.*

First, we consider models where each node knows its entire future trajectory. One approach is to use a known neighbourhood maintenance algorithm. For example, in the *Mobile-Rcv* model where nodes travel along arbitrary, continuous trajectories on a line, we could run the $\texttt{EWS}$ algorithm [8]. Similarly, in the

Mobile-Rcv model where nodes travel along arbitrary, continuous trajectories in a road network, we could run the CVW algorithm [4]. However, both of these algorithms make strong assumptions about each node initially knowing the entire future trajectories of all other nodes within a certain distance. Let EWSINIT be the algorithm that consists of executing a $\frac{4}{3}$-local gossip algorithm where, for each node p_j, we take I_j to be p_j's entire future trajectory. Let CVWINIT be the algorithm that consists of executing a $\frac{13}{11}$-local gossip algorithm where, for each node p_j, we take I_j to be p_j's entire future trajectory. The following results can be verified using the model constraints from [8] and [4], respectively.

Lemma 2. *In the Mobile-Rcv model where nodes travel along arbitrary continuous trajectories on the line, executing EWSINIT followed by EWS solves continual exact neighbour discovery with $T_{Init} = |\mathtt{LG}(\frac{4}{3})|$.*

Lemma 3. *In the Mobile-Rcv model where nodes travel along arbitrary continuous trajectories in road networks, executing CVWINIT followed by CVW solves continual exact neighbour discovery with $T_{Init} = |\mathtt{LG}(\frac{13}{11})|$.*

We can weaken the assumption about entire trajectory knowledge by using δ-local gossip to communicate trajectory updates. In each phase of our algorithm, called CENL, each node shares enough future trajectory information to ensure that all nodes can correctly calculate their neighbourhoods for all slots up until the completion of the next phase. More specifically, for some fixed δ (to be specified later), let phase $i \geq 0$ consist of the slots $i|\mathtt{LG}(\delta)|+1, \ldots, (i+1)|\mathtt{LG}(\delta)|$. During each phase i, all nodes execute $\mathtt{LG}(\delta)$, where each p_j sets I_j to be its trajectory for slots $(i+1)|\mathtt{LG}(\delta)|+1, \ldots, (i+2)|\mathtt{LG}(\delta)|$. So, by the end of phase i, each neighbour of p_j during phase $i+1$ receives p_j's trajectory for all times in phase $i+1$. The value of δ is chosen so that, for any node p_j, every node that is a neighbour of p_j during at least one slot in phase $i+1$ is located within distance δR of p_j at some time during phase i. It is sufficient to choose δ such that $\delta R \geq R + 2\sigma|\mathtt{LG}(\delta)|$.

Lemma 4. *CENL solves continual exact neighbour discovery with $T_{Init} = |\mathtt{LG}(\delta)|$, where δ satisfies $\delta R \geq R + 2\sigma|\mathtt{LG}(\delta)|$.*

One-Time Stable-Neighbour Discovery. In some applications, it might be useful to learn about neighbours that, in a sense, can be considered to be more 'dependable'. That is, node p cares only about neighbours that will stay nearby for a while, and ignores nodes that enter its neighbourhood and then leave shortly afterward. For example, in the context of routing algorithms, such a neighbour discovery algorithm could be used to periodically update routing tables at every node.

Definition 4. *An algorithm solves* one-time stable-neighbour discovery *if, for some fixed $T_{Init} > 0$ and some fixed $T_{Stable} > 0$, every execution terminates at the end of slot T_{Init}, and, for all nodes p, $q \in \mathtt{List}(p, T_{Init} + 1)$ if and only if $q \in NBRS(p, t)$ for all $t \in \{T_{Init} + 1, \ldots, T_{Init} + T_{Stable}\}$.*

Suppose that, for some $F_{\text{traj}} > |\text{LG}(1)|$, each node initially knows its trajectory for slots $|\text{LG}(1)| + 1, \ldots, F_{\text{traj}}$. Define algorithm OSNL as the execution of $\text{LG}(1)$ where, for each node p_j, the value of I_j is defined as its trajectory for slots $|\text{LG}(1)| + 1, \ldots, F_{\text{traj}}$. At the end of slot $|\text{LG}(1)|$, let $\text{List}(p, |\text{LG}(1)| + 1) = \{p_k \mid I_k$ has been received, and, p_k will be a neighbour for all slots $|\text{LG}(1)| + 1, \ldots, F_{\text{traj}}\}$.

Lemma 5. OSNL *solves one-time stable-neighbour discovery with* $T_{Init} = |\text{LG}(1)|$ *and* $T_{Stable} = F_{\text{traj}} - |\text{LG}(1)|$.

Continual Stable-Neighbour Discovery. In the continual version of stable-neighbour discovery, we want node p to consider a node q as a neighbour if and only if q enters p's neighbourhood and stays for a while. Informally, if q is a neighbour of p for all slots during some sufficiently long interval, then q must be contained in p's neighbour list for some suffix of the interval. Conversely, if p includes q in its neighbour list for some slot t, it must be the case q is a neighbour of p for a sufficiently long interval that includes t. A parameter T_{Stable} specifies how many slots must elapse before two neighbours are considered "stable", while a parameter T_{Delay} specifies an upper bound on the amount of delay before a node adds a stable neighbour to its neighbour list.

Definition 5. *An algorithm solves* continual stable-neighbour discovery *if there exist* $T_{Stable} > 0$ *and* $T_{Delay} \geq 0$ *such that, for every node p, every slot t, and every* $T \geq t + T_{Stable} - 1$: *if* $q \in NBRS(p, t')$ *for all* $t' \in \{t, \ldots, T\}$, *then* $q \in \text{List}(p, t'')$ *for all* $t'' \in \{t + T_{Delay}, \ldots, T\}$, *and, if* $q \in \text{List}(p, t)$, *then there exists* $t' \in \{t - T_{Stable} + 1, \ldots, t\}$ *such that* $q \in NBRS(p, t')$ *for all* $t'' \in \{t', \ldots, t' + T_{Stable} - 1\}$.

We now describe an algorithm, CSNL, for continual stable-neighbour discovery. At a high level, our solution proceeds in phases, each consisting of an execution of $\text{LG}(1)$. Initially, each node's neighbour list is empty for all slots. When a node p receives trajectory information about a node q that will be its neighbour for a while, then p adds q to its neighbour list for the appropriate future slots. More specifically, we ensure that, by the end of each phase i, each node p receives q's trajectory information for phases i and $i + 1$, for each node q within distance R of p at some time during phase i. Using this trajectory information, p can determine if q is a neighbour for a suffix of phase i and a prefix of phase $i + 1$ whose lengths a and b, respectively, add up to at least $|\text{LG}(1)| + 1$ slots. In this case, p adds q to its neighbour list for the first b slots of phase $i + 1$. Our algorithm proceeds as follows at each node p_j and for each phase $i \geq 0$ consisting of slots $i|\text{LG}(1)| + 1, \ldots, (i + 1)|\text{LG}(1)|$:

- Set I_j to be p_j's trajectory for slots $i|\text{LG}(1)| + 1, \ldots, (i + 2)|\text{LG}(1)|$. Run $\text{LG}(1)$ at the start of slot $i|\text{LG}(1)| + 1$.
- At the end of each slot during which a message containing some I_k is received: find the largest positive integer $a \leq |\text{LG}(1)|$ such that p_j and p_k are neighbours for every slot in $\{(i+1)|\text{LG}(1)| + 1 - a, \ldots, (i+1)|\text{LG}(1)|\}$, and find the largest positive integer $b \leq |\text{LG}(1)|$ such that p_j and p_k are neighbours for every slot

in $\{(i+1)|\mathtt{LG}(1)|+1,\ldots,(i+1)|\mathtt{LG}(1)|+b\}$. If $a+b \geq |\mathtt{LG}(1)|+1$, put p_k in $\mathtt{List}(p_j,t)$ for each $t \in \{(i+1)|\mathtt{LG}(1)|+1,\ldots,(i+1)|\mathtt{LG}(1)|+b\}$.

Lemma 6. $\mathtt{CSNL}$ *solves continual stable-neighbour discovery with* $T_{Stable} = |\mathtt{LG}(1)|+1$ *and* $T_{Delay} = |\mathtt{LG}(1)|$.

One-Time Delayed Neighbour Discovery. Another version of neighbourhood discovery involves each node discovering which nodes are its neighbours, but allowing a certain amount of delay before this occurs. An upper bound on the amount of delay is specified by a T_{Del} parameter. A node's neighbour list at termination could be a subset or a superset of its actual neighbourhood, but the learned neighbourhood information can still provide useful estimates. For example, if there is a known upper bound σ on the speed of nodes, then, knowing that a node q was a neighbour t' slots ago can help provide a range of possible locations for q in the current slot.

Definition 6. *An algorithm solves* one-time delayed neighbour discovery *if, for some fixed* $T_{Init} > 0$ *and some fixed* $T_{Del} > 0$, *every execution terminates at the end of slot* T_{Init}, *and, for all nodes* p, *node* $q \in \mathtt{List}(p,T_{Init}+1)$ *if and only if* $q \in NBRS(p,t)$ *for some* $t \in \{T_{Init}-T_{Del}+1,\ldots,T_{Init}\}$.

Define algorithm $\mathtt{ODNL}$ as the execution of $\mathtt{LG}(1)$, where, for each p_j, the value of I_j is defined as its trajectory for slots $1,\ldots,|\mathtt{LG}(1)|$. For each I_j received, p can determine whether or not p_j was a neighbour at some time during the previous $|\mathtt{LG}(1)|$ slots.

Lemma 7. $\mathtt{ODNL}$ *solves one-time delayed neighbour discovery with* $T_{Init} = T_{Del} = |\mathtt{LG}(1)|$.

Continual Delayed Neighbour Discovery. In the continual version of delayed neighbour discovery, we want each node to discover its neighbours within a bounded number of slots, and, if a node p has a node q in its neighbour list, then q must have been a neighbour of p in the recent past.

Definition 7. *An algorithm solves* continual delayed neighbour discovery *if, for some fixed* $T_{Del} > 0$, *for every node* p *and every slot* t, *if* $q \in NBRS(p,t)$, *then* $q \in \mathtt{List}(p,t')$ *for some* $t' \in \{t+1,\ldots,t+T_{Del}\}$. *Also, if* $q \in \mathtt{List}(p,t)$, *then* $q \in NBRS(p,t')$ *for some* $t' \in \{t-T_{Del},\ldots,t-1\}$.

Define algorithm $\mathtt{CDNL}$ as a sequence of phases, each consisting of an execution of $\mathtt{LG}(1)$. In each phase $i \geq 0$, each node p_j sets I_j to be its trajectory for slots $i|\mathtt{LG}(1)|+1,\ldots,(i+1)|\mathtt{LG}(1)|$. If a node p_j receives a message containing I_k during some slot t in phase i, it compares I_k with its own trajectory for all slots that occur during phase i. If it determines that, for some slot t' of phase i, p_k is its neighbour, it adds p_k to $\mathtt{List}(p_j,t'+1),\ldots,\mathtt{List}(p_j,(i+1)|\mathtt{LG}(1)|+1)$.

Lemma 8. $\mathtt{CDNL}$ *solves continual delayed neighbour discovery with* $T_{Del} = |\mathtt{LG}(1)|$.

3   Solving Local Gossip

For the *Mobile-Rcv* model where nodes travel along continuous trajectories on a line with speed bounded above by σ, we describe a region-based transmission schedule RBSched and use it to solve δ-local gossip. At a high level, each phase of RBSched consists of a subset of the nodes running a particular schedule that is based on cover-free families of sets. By carefully specifying which nodes participate in which phases, we ensure that each node transmits frequently and that the contents of its transmitted message gets propagated quickly in both directions.

The Phase Schedule. In each phase of RBSched, we use a schedule called PS based on cover-free families of sets [9,11]. For any set S, an *r-cover for S* is a set family of size r that does not contain S and whose union does contain S. For $r \geq 1$, an *r-cover-free family* $\mathscr{F}$ is a family of sets such that, for each $S \in \mathscr{F}$, there is no r-cover for S consisting of sets from $\mathscr{F} - \{S\}$. The number of sets in a family $\mathscr{F}$ is called the *size* of $\mathscr{F}$, denoted by $|\mathscr{F}|$. The *length* T of a family $\mathscr{F}$ is the largest integer contained in at least one set in $\mathscr{F}$. It is known that there exists r-cover-free families with $T \in O(r^2 \log |\mathscr{F}|)$ [9]. It was shown in [5] that an r-cover-free family with size $|\mathscr{F}|$ and length T can be converted into a transmission schedule with T slots for nodes with IDs in the range $\{1, \ldots, |\mathscr{F}|\}$ such that, for each set X of at most $r + 1$ nodes, for each node $p \in X$, there is a slot in which p is scheduled to transmit and all nodes in $X - \{p\}$ listen.

In what follows, we denote by Δ a known upper bound on the maximum degree of any vertex in the neighbour graph, taken over all slots t, i.e., $\Delta \geq \max_{v,t}\{|NBRS(v,t)|\}$. Let PS denote the schedule obtained from a $((3+2\rho)(\Delta + 1))$-cover-free family of size U, where $\rho = \lceil(R'/R)\rceil$. Let $|PS|$ denote the number of slots in the resulting schedule. From the discussion above, we can assume that $|PS| \in O(\rho^2 \Delta^2 \log U)$, and that the following observation holds.

Observation 1. *Consider any set X of at most $(3+2\rho)(\Delta+1)$ nodes. For each node $p \in X$, there is a slot in PS during which p is scheduled to transmit and all nodes in $X - \{p\}$ listen.*

Different implementations for PS could increase the algorithm's robustness. For example, the $(r; d)$-cover-free families from Dyachkov et al. [7], with $r = (3 + 2\rho)(\Delta + 1)$, could be used to construct a schedule that gives a stronger property than Observation 1: for each node $p \in X$, there are at least d slots during which p is scheduled to transmit and all nodes in $X - \{p\}$ listen. This could help in models which include some amount of unpredictable interference or some number of transient failures.

The Full Schedule. Our model assumes that each node can accurately determine its location relative to a global origin. Using this origin, we divide the line into regions of length R, each overlapping its neighbouring regions by $5\sigma|PS|$ units. Specifically, for all $z \in \mathbb{Z}$, *region* ψ_z is the set of points $[z(R - 5\sigma|PS|), z(R - 5\sigma|PS|) + R)$. We partition the set of time slots into phases of $|\mathsf{PS}|$ slots each, i.e., for all $a \in \mathbb{N}$, *phase* π_a is the set of slots $a|\mathsf{PS}| + 1, \dots, (a + 1)|\mathsf{PS}|$. In each phase, a set of *participants* will start executing PS, and those that stay within the same region for the entire phase are *survivors*. More formally, a node p is a *participant in phase* π_a *for region* ψ_z if p is located in region ψ_z at the start of phase π_a and $z \equiv a \mod 2$. Denote by $\mathcal{P}_{a,z}$ the set of such participants. A node p is a *survivor in phase* π_a *for region* ψ_z if p is located in region ψ_z at all times during phase π_a and $z \equiv a \mod 2$. Denote by $\mathcal{S}_{a,z}$ the set of such survivors.

We now define the schedule RBSched. At the beginning of each phase π_a, each node p checks if its current location is in a region ψ_z such that $a \equiv z \mod 2$. If this is not the case, then node p listens for the entire phase. If located in such a region ψ_z, then $p \in \mathcal{P}_{a,z}$ and so it starts running PS at the start of phase. If p ever leaves region ψ_z during phase π_a, it immediately stops executing PS and listens for the remainder of the phase. When a node p_j transmits, its message contains all of the information that it knows, i.e., $I_j \cup \{I_k \mid \text{node } p_j \text{ has previously received } I_k\}$. The total number of phases will be specified later.

Analyzing the Schedule. From our definitions of regions and phases in the previous section, we notice that: each region is of size R, so that two nodes in the same region are within communication range; the overlap between regions is large enough so that, for example, if a node crosses the rightmost border of region ψ_z frequently within some phase, then the node is actually found in region ψ_{z+1} for the entire phase; and, it is also the case that even if a node travels at maximum speed σ, it cannot pass through an entire region during a single phase. We depend on two assumptions that relate the various model parameters. Assumption (A1) is about density: between the leftmost and rightmost nodes, there is never an empty line segment of length $\frac{R}{2}$. Assumption (A2) is that $R \geq 10\sigma|\mathsf{PS}|$, which ensures that each region ψ_z only overlaps with regions ψ_{z-1} and ψ_{z+1}.

For our analysis, we define two types of "windows". At a high level, these windows represent how fast information propagates through the network. When a transmission containing some information I occurs during a phase π_a, we consider windows that move leftward and rightward, once every 3 phases, from a region ψ_z where the transmission originated. Formally, we say that, for phase $\pi_{a+3\varphi}$, a node q is *located in a rightward-moving window* if it is located in region $\pi_{z+\varphi}$ for the entire phase. We make analogous definitions for leftward-moving windows. We will determine an upper bound on the elapsed time before an arbitrary node q is found within a window, which gives a bound on how soon q will receive I.

Node Location Bounds. Based on the upper bound σ on node speed and the length $|\mathsf{PS}|$ of each phase, we get the following upper bound on how far a node can travel from a known location within a bounded amount of time.

Observation 2. *If, at some time during phase* $\pi_{a'}$, *node p is located at a point x, then, at all times from the beginning of phase* $\pi_{a'-k}$ *until the end of phase* $\pi_{a'+k}$, *p is located in* $[x - (k + 1)\sigma|\mathsf{PS}|, x + (k + 1)\sigma|\mathsf{PS}|]$.

Next, we use Observation 2 to prove that each node is a survivor at least once in every two consecutive phases, which implies that each node transmits often.

Lemma 9. *Suppose that, at the beginning of phase $\pi_{a'}$, node p is located in segment $SEG_{z'}$. Then, $p \in \mathcal{S}_{a',z'} \cup \mathcal{S}_{a'+1,z'}$.*

Using the definitions of windows and the limit on node speed, we can prove the following result, which says that all nodes found within a certain area of the environment have all been located within a window at some time.

Theorem 1. *Suppose that $a \equiv z \mod 2$ and $\gamma \geq 0$. If node q is located in $((z - \gamma)(R - 5\sigma|\mathsf{PS}|) + \sigma|\mathsf{PS}|, (z + \gamma)(R - 5\sigma|\mathsf{PS}|) + R - \sigma|\mathsf{PS}|)$ at the end of phase $\pi_{a+3\gamma}$, then, for some $\varphi \in \{-\gamma, \ldots, \gamma\}$, $q \in \mathcal{S}_{a+3|\varphi|,z+\varphi}$.*

Message Propagation Bounds. Suppose that p is a survivor in phase π_a for region ψ_z, and that p transmits a message containing I during phase π_a. Our goal is to show that I soon gets re-transmitted by nodes in ψ_z and nearby regions. First, we show that all survivors in phase π_a for region ψ_z, other than p, receive I during phase π_a.

Lemma 10. *Every $q \in \mathcal{S}_{a,z}$ receives a message from each $p \in \mathcal{S}_{a,z} - \{q\}$ during phase π_a.*

To prove the above, we determine which nearby regions might contain nodes that cause transmission collisions at nodes in ψ_z during phase π_a, set X to be all nodes that are located in these regions, and then bound $|X|$ from above by $(3 + 2\rho)(\Delta + 1)$. Then, by Observation 1, we note that PS is a schedule that ensures that, for each node $p' \in X$, there is a slot in phase π_a such that p' transmits and all nodes in $X - \{p'\}$ listen. So, for each $p \in \mathcal{S}_{a,z}$, there is a slot during which p transmits and that p's transmission is received by all other nodes that are within radius R of p for the entire slot (which includes all of $\mathcal{S}_{a,z} - \{p\}$).

A careful analysis shows that I gets transmitted again three phases later by at least one node in each region that overlaps with ψ_z. An induction argument shows that this propagation of I continues in both directions, essentially at a rate of one region per every three phases, as stated in the following result.

Theorem 2. *Suppose that $p \in \mathcal{S}_{a,z}$ and p transmits I during phase π_a. For every $\ell \in \mathbb{Z}$, if $\mathcal{S}_{a+3|\ell|,z+\ell} \neq \emptyset$, then there exists a node in $\mathcal{S}_{a+3|\ell|,z+\ell}$ that transmits I during phase $\pi_{a+3|\ell|}$.*

Theorems 1 and 2 are combined to provide the following guarantee about the speed of information dissemination amongst nodes following $\mathsf{RBSched}$.

Theorem 3. *Suppose that there is a node $p \in \mathcal{S}_{a,z}$ for some phase π_a and region ψ_z. During phase π_a, suppose that p transmits a message containing information I. If, for some $\gamma \geq 0$, a node q is located in $((z - \gamma)(R - 5\sigma|\mathsf{PS}|) + \sigma|\mathsf{PS}|, (z + \gamma)(R - 5\sigma|\mathsf{PS}|) + R - \sigma|\mathsf{PS}|)$ at the end of phase $\pi_{a+3\gamma}$, then q receives I by the end of phase $\pi_{a+3\gamma}$.*

Our Local Gossip Algorithm. Let $\alpha = \left\lceil \frac{\delta R + 2\sigma|\mathsf{PS}| + 1}{R - 8\sigma|\mathsf{PS}|} \right\rceil$. Given $\delta > 0$, our algorithm for δ-local gossip, denoted by $\mathsf{LG}(\delta)$, consists of each node running the schedule $\mathsf{RBSched}$ for exactly $3\alpha + 2$ phases. Since δ, R, σ and $|\mathsf{PS}|$ are known values, it is clear that all nodes terminate the algorithm at the same time.

Suppose that, at the beginning of some phase π_a, a node p is located in some region ψ_z. Let q be a node within distance δR from p during a phase π_b with $b \geq a$. The following result limits the number of regions that lie between region ψ_z and q's location during phases after π_b. In particular, even at maximum speed, it takes 3 phases for q to move a distance of one region away from region ψ_z.

Lemma 11. *Suppose that, at the beginning of phase π_a, an arbitrary node p is located in region ψ_z. For any $k \geq \alpha$, suppose that q is within distance δR of p at some time between the beginning of phase π_a and the end of phase π_{a+3k}. At the end of phase π_{a+3k}, q is located in $((z - k)(R - 5\sigma|\mathsf{PS}|) + \sigma|\mathsf{PS}|, (z + k)(R - 5\sigma|\mathsf{PS}|) + R - \sigma|\mathsf{PS}|)$.*

Suppose p is a survivor for some region ψ_z in phase π_a when it first transmits its initial information I. If $\mathsf{RBSched}$ is executed for the next $3\alpha + 1$ phases, it follows from Lemma 11 and Theorem 3 that I has been received by each node q that is found within distance δR from p at some time during the execution. By Lemma 9, node p is a survivor at least once within two consecutive phases, which proves that running $\mathsf{RBSched}$ for $3\alpha + 2$ phases suffices.

Theorem 4. *Consider any network node p_j with initial information I_j. For any $\delta > 0$, after executing $\mathsf{LG}(\delta)$, each node that is within distance δR from p_j at some time during the execution has received I_j before termination. Further, $\delta < \alpha \leq 5\delta + 3$, which implies that $\mathsf{LG}(\delta)$ uses at most $15\delta + 11$ phases, and no fewer than $3\delta + 3$ phases, where each phase consists of $O(\rho^2 \Delta^2 \log U)$ slots.*

4   Conclusion and Future Work

Local gossip is a task that captures the need for nodes in a mobile network to share information with other nearby nodes, even if they are not within communication range of one another for a long period of time. A solution to this task can be useful as a fundamental building block in algorithms for mobile networks, and we have demonstrated that this is the case with neighbour discovery via reductions that hold very generally. The resulting solutions to neighbour discovery need not make the simplifying assumptions made elsewhere in the literature, and they are deterministic, which means that they can be used as subroutines without introducing error. By solving the local gossip task in the one-dimensional *Mobile-Rcv* model, we obtain solutions to one-time exact neighbour discovery in the same model, which can be used to answer open questions about initializing the algorithms in [4,8,16].

Important directions for future work are to solve δ-local gossip in more general environments, such as road networks or the plane, and different interference models, such as SINR. An important generalization is to weaken the assumption that the transmission radius is the same for all nodes. A challenging problem to consider would be to determine the trade-off between future trajectory knowledge and feasible values of the parameters for neighbour discovery tasks. Such a trade-off would have an impact on the design of real-world systems, where there can be varying degrees of future trajectory knowledge. For example, satellites or bus routes are fully specified in advance, while a self-driving car has a planned route until it reaches its destination, while a human-controlled car has no knowledge of its future trajectory.

Acknowledgements. This work was partially supported by NSERC Discovery Grant 2017-05936. The author would like to thank Faith Ellen for her invaluable advice that substantially increased the quality of the results and presentation of this research.

References

1. Anta, A.F., Milani, A., Mosteiro, M.A., Zaks, S.: Opportunistic information dissemination in mobile ad-hoc networks: the profit of global synchrony. Distrib. Comput. **25**(4), 279–296 (2012). https://doi.org/10.1007/s00446-012-0165-9
2. Augustine, J., Pandurangan, G., Robinson, P.: Distributed algorithmic foundations of dynamic networks. SIGACT News **47**(1), 69–98 (2016). https://doi.org/10.1145/2902945.2902959
3. Baldoni, R., Anta, A.F., Ioannidou, K., Milani, A.: The impact of mobility on the geocasting problem in mobile ad-hoc networks: solvability and cost. Theor. Comput. Sci. **412**(12–14), 1066–1080 (2011). https://doi.org/10.1016/j.tcs.2010.12.006
4. Chung, H.C., Viqar, S., Welch, J.L.: Neighbor knowledge of mobile nodes in a road network. In: 2012 IEEE 32nd International Conference on Distributed Computing Systems, 18–21 June 2012, Macau, China, pp. 486–495. IEEE Computer Society (2012). https://doi.org/10.1109/ICDCS.2012.16
5. Clementi, A.E.F., Monti, A., Silvestri, R.: Selective families, superimposed codes, and broadcasting on unknown radio networks. In: Kosaraju, S.R. (ed.) Proceedings of the Twelfth Annual Symposium on Discrete Algorithms, 7–9 January 2001, Washington, DC, USA, pp. 709–718. ACM/SIAM (2001). http://dl.acm.org/citation.cfm?id=365411.365756
6. Cornejo, A., Viqar, S., Welch, J.L.: Reliable neighbor discovery for mobile ad hoc networks. Ad Hoc Netw. **12**, 259–277 (2014). https://doi.org/10.1016/j.adhoc.2012.08.009
7. Dyachkov, A.G., Rykov, V.V., Rashad, A.M.: Superimposed distance codes. Problems Control Inform. Theory/Problemy Upravlen. Teor. Inform. **18**(4), 237–250 (1989)
8. Ellen, F., Subramanian, S., Welch, J.: Maintaining information about nearby processors in a mobile environment. In: Chaudhuri, S., Das, S.R., Paul, H.S., Tirthapura, S. (eds.) ICDCN 2006. LNCS, vol. 4308, pp. 193–202. Springer, Heidelberg (2006). https://doi.org/10.1007/11947950_22
9. Erdös, P., Frankl, P., Füredi, Z.: Families of finite sets in which no set is covered by the union of r others. Israel J. Math. **51**, 79–89 (1985)

10. Gasieniec, L., Pagourtzis, A., Potapov, I., Radzik, T.: Deterministic communication in radio networks with large labels. Algorithmica **47**(1), 97–117 (2007). https://doi.org/10.1007/s00453-006-1212-3
11. Kautz, W., Singleton, R.: Nonrandom binary superimposed codes. IEEE Trans. Inf. Theory **10**(4), 363–377 (1964)
12. Khan, A.A., Rehmani, M.H., Saleem, Y.: Neighbor discovery in traditional wireless networks and cognitive radio networks: basics, taxonomy, challenges and future research directions. J. Netw. Comput. Appl. **52**, 173–190 (2015). https://doi.org/10.1016/j.jnca.2015.03.003
13. Krishnamurthy, S., et al.: Time-efficient distributed layer-2 auto-configuration for cognitive radio networks. Comput. Netw. **52**(4), 831–849 (2008). https://doi.org/10.1016/j.comnet.2007.11.013
14. Kuhn, F., Lynch, N.A., Newport, C.C.: The abstract MAC layer. Distrib. Comput. **24**(3–4), 187–206 (2011). https://doi.org/10.1007/s00446-010-0118-0
15. Kuhn, F., Oshman, R.: Dynamic networks: models and algorithms. SIGACT News **42**(1), 82–96 (2011). https://doi.org/10.1145/1959045.1959064
16. Miller, A.: Gossiping in one-dimensional synchronous ad hoc wireless radio networks. In: Blin, L., Busnel, Y. (eds.) 4th Workshop on Theoretical Aspects of Dynamic Distributed Systems, TADDS 2012, 17 December 2012, Roma, Italy, pp. 32–43. ACM (2012). https://doi.org/10.1145/2414815.2414822
17. Miller, A.: Deterministic neighbourhood learning in ad hoc wireless radio networks. Ph.D. thesis, University of Toronto (2014). http://hdl.handle.net/1807/68138
18. Mittal, N., Krishnamurthy, S., Chandrasekaran, R., Venkatesan, S., Zeng, Y.: On neighbor discovery in cognitive radio networks. J. Parallel Distrib. Comput. **69**(7), 623–637 (2009). https://doi.org/10.1016/j.jpdc.2009.03.008
19. Vaya, S.: Round complexity of leader election and gossiping in bidirectional radio networks. Inf. Process. Lett. **113**(9), 307–312 (2013). https://doi.org/10.1016/j.ipl.2013.02.001

Competitive Routing in Hybrid Communication Networks

Daniel Jung, Christina Kolb$^{(\boxtimes)}$, Christian Scheideler, and Jannik Sundermeier

Computer Science Department, Heinz Nixdorf Institute, Paderborn University,
Fürstenallee 11, 33102 Paderborn, Germany
{jungd,ckolb,scheideler,janniksu}@mail.upb.de

Abstract. Routing is a challenging problem for wireless ad hoc networks, especially when the nodes are mobile and spread so widely that in most cases multiple hops are needed to route a message from one node to another. In fact, it is known that any online routing protocol has a poor performance in the worst case, in a sense that there is a distribution of nodes resulting in bad routing paths for that protocol, even if the nodes know their geographic positions and the geographic position of the destination of a message is known. The reason for that is that radio holes in the ad hoc network may require messages to take long detours in order to get to a destination, which are hard to find in an online fashion.

In this paper, we assume that the wireless ad hoc network can make limited use of long-range links provided by a global communication infrastructure like a cellular infrastructure or a satellite in order to compute an abstraction of the wireless ad hoc network that allows the messages to be sent along near-shortest paths in the ad hoc network. We present distributed algorithms that compute an abstraction of the ad hoc network in $\mathcal{O}\left(\log^2 n\right)$ time using long-range links, which results in c-competitive routing paths between any two nodes of the ad hoc network for some constant c if the convex hulls of the radio holes do not intersect. We also show that the storage needed for the abstraction just depends on the number and size of the radio holes in the wireless ad hoc network and is independent on the total number of nodes, and this information just has to be known to a few nodes for the routing to work.

Keywords: Greedy routing · Ad hoc networks · Convex hulls · c-competitiveness

1 Introduction

Nowadays almost every person has a cell phone. Hence, in a city center the density of cell phones would, in principle, be sufficiently high to set up a well-connected wireless ad hoc network spanning the entire city center, which could

This work was partially supported by the German Research Foundation (DFG) within the Collaborative Research Center 'On-The-Fly Computing' (SFB 901).

© Springer Nature Switzerland AG 2019
S. Gilbert et al. (Eds.): ALGOSENSORS 2018, LNCS 11410, pp. 15–31, 2019.
https://doi.org/10.1007/978-3-030-14094-6_2

then be used for interesting applications in the area of social networks. Wireless ad hoc networks have the advantage that there is no limit (other than the bandwidth and battery constraints) on the amount of data that can be exchanged while the amount of data that can be transferred at a reasonable rate via long-range links using the cellular infrastructure or satellite is limited (by some data plan) or costly. Due to the nice property of being able to exchange massive amounts of data for free, wireless ad hoc networks can be used in applications requiring the transfer of huge data, for example video sharing. However, routing in a mobile ad hoc network is challenging, even if the geographic position of the destination is known, since buildings or other obstacles like rivers may create radio holes that make it non-trivial to find a near-shortest routing path. So the question we address in this paper is:

Can long-range links be used effectively to find near-shortest routing paths in the ad hoc network?

A simple solution to that problem would be that all nodes regularly post their geographic position and the nodes within their communication range to a server in the Internet. This would allow the server to compute optimal routing paths so that whenever a node wants to forward a message to a certain destination, the server can tell it which of the neighbors to send it to. An alternative approach that we are pursuing in this paper is a purely peer-to-peer based approach in which no other equipment other than the cell phones and an infrastructure for the long-range links needs to be used. To the best of our knowledge, our approach is the first one that is making use of a global communication infrastructure in a peer-to-peer manner in order to efficiently determine short routing paths for an ad hoc network. Wireless ad hoc networks have been considered before that utilize base stations in order to exchange messages more effectively, but there, messages will be sent via long-range links to bridge long distances while we will only allow messages to be sent via ad hoc links.

1.1   Model

Throughout this paper, we consider $V \subset \mathbb{R}^2$ to be a set of nodes in the Euclidean plane with unique IDs (e.g., phone numbers), where $|V| = n$. For any given pair of nodes $u = (u_x, u_y)$, $v = (v_x, v_y)$, we denote the Euclidean distance between u and v by $||uv|| = \sqrt{(u_x - v_x)^2 + (u_y - v_y)^2}$. We model our cell phone network as a hybrid directed graph $H = (V, E, E_{AH})$ where the node set V represents the set of cell phones, an edge (v, w) is in E whenever v knows the phone number (or simply *ID*) of w, and an edge $(v, w) \in E$ is also in the *ad hoc edge set* E_{AH} whenever v can send a message to w using its Wifi interface. For all edges $(v, w) \in E \setminus E_{AH}$, v can only use a long-range link to directly send a message to w. We adopt the *unit disk graph model* for the edges in E_{AH}, where for any point set $V \subseteq \mathbb{R}^2$ the *Unit Disk Graph* of V, UDG(V), is a bi-directed graph that contains all edges (u, v) with $||uv|| \leq 1$. We assume UDG(V) to be strongly connected so that a message can be sent from every node to every other node in V by just using ad hoc edges. While the ad hoc edges are fixed, the nodes can nevertheless change E over time: If a node v knows the IDs of nodes w and w', then it can send the ID of w to w', which adds (w, w') to E. This procedure is called *ID-introduction*. Alternatively, if v deletes the address of some node w with $(v, w) \in E$, then (v, w) is removed from E. There are no other means of changing E, i.e., a node v cannot learn about an ID of a node w unless w is in v's UDG-neighborhood or the ID of w is sent to v by some other node.

Moreover, we consider synchronous message passing in which time is divided into rounds. More precisely, we assume that every message initiated in round i is delivered at the beginning of round $i + 1$, and a node can process all messages in a round that have been delivered at the beginning of that round.

1.2   Objective

Our objective is to design an efficient routing algorithm for ad hoc networks, where the source s of a message knows the ID of the destination t, or in other words, $(s, t) \in E$. This is a reasonable constraint since cell phone users normally wouldn't call cell phones whose users are unknown to them. Thus, whenever a message needs to be sent from a source s to some destination t, we assume that the geographic position of t is known, since s can ask t via a long-range link for $t's$ geographic position before sending the message towards t using the ad hoc network.

Our routing algorithm consists of two parts: After determining the radio holes of the wireless ad hoc network, we compute an abstraction, i.e., a compact representation of these radio holes and use that abstraction in order to route messages along c-competitive paths. We call a routing strategy *c-competitive* if for all node pairs (s, t), the routing path $(s, \ldots, t)$ from s to t obtained by the strategy satisfies $||(s, \ldots, t)|| \leq c \cdot d(s, t)$, where $||(s, \ldots, t)||$ denotes the Euclidean length of $(s, \ldots, t)$ and $d(s, t)$ denotes the shortest Euclidean length of a path in UDG (V) from s to t.

We will focus on computing suitable abstractions of radio holes in the ad hoc network. The intuition behind that is simple: if there are no radio holes, then simple greedy routing (i.e., always take the neighbor that is closest to the destination) would already give us short routing paths to arbitrary destinations. Radio holes can be specified by the nodes along its boundary, but there can be many such nodes. Therefore, we look at more compact representations of radio holes like the (nodes forming the) convex hull of its boundary. Considering convex hulls as radio hole abstractions makes sense because in huge cities like New York City the shape of radio holes (caused by obstacles like buildings) is in many cases convex or close to a convex shape, and these shapes do not overlap. In order to obtain the desired abstraction, we will make use of ID-introductions in order to form an overlay network that allows us to compute these abstractions in a distributed manner using the long-range links. Since sending messages via long-range links is costly (in terms of money), our goal is to keep the long-range communication work of the nodes as low as possible.

1.3 Our Contributions

We consider any hybrid graph $G = (V, E, E_{AH})$ where the Unit Disk Graph of V is connected. Let H be the set of radio holes in G and C be the set of convex hulls of radio holes in H. $P(h)$ denotes the length of the perimeter of a radio hole $h \in H$. Further, $L(ch)$ denotes the circumference of a minimum bounding box of a convex hull $ch \in C$. Our main contribution is:

Theorem 1. *For any distribution of the nodes in V that ensures that UDG(V) is connected and of bounded degree and that the convex hulls of the radio holes do not overlap, our algorithm computes an abstraction of UDG(V) in $\mathcal{O}(\log^2 n)$ communication rounds using only polylogarithmic communication work at each node so that c-competitive paths between all source-destination pairs can be found in an online fashion.*

The space needed by the convex hull nodes of the radio holes is $\mathcal{O}\left(\sum_{c \in C} L(c)\right)$. Nodes lying on the boundary of radio holes need storage of size $\mathcal{O}\left(\max_{h \in H} P(h)\right)$. For every other node, the space requirement is constant.

The rest of this paper is dedicated to the proof of Theorem 1. For that, we use the following approach:

1. Given the Unit Disk Graph, we compute the 2-localized Delaunay Graph. This only needs $\mathcal{O}(1)$ communication rounds. The 2-localized Delaunay Graph allows the nodes to detect whether they are at the boundary of a radio hole. Nodes at the boundary can then form a ring.
2. We then develop a distributed algorithm that computes a convex hull of a ring of n nodes in expected $\mathcal{O}(\log n)$ communication rounds.
3. Afterwards, we introduce the nodes of the convex hulls to each other so that they form a clique. This will allow them to compute c-competitive paths for all source-destination pairs that are outside of a convex hull. The introduction requires $\mathcal{O}(\log^2 n)$ communication rounds.

Finally, we also consider the dynamic scenario (i.e., UDG(V) changes over time) in Sect. 6. All proofs can be found in Appendix B. The full version of this paper is available online on arXiv.

1.4 Related Work

Routing work in theory has mostly focused on approaches where routing paths do not have to be set up before sending out a message. Instead the focus has been on simple online routing strategies that are potentially based on a suitable overlay network consisting of a subset of the wireless connections available to the nodes. The simplest online strategy is to use a greedy strategy to route a message to a destination t: always forward the message to the neighbor closest to t (with respect to some metric). Unfortunately, greedy strategies like Compass routing [4] fail for graphs with radio holes, i.e., they might get stuck at a dead end. This can be avoided with the help of suitable virtual coordinates for the nodes (e.g., [10]), but computing these is quite expensive. Instead, Kuhn et al. [9] proposed GOAFR, a routing strategy that uses a combination of greedy and face routing, which can find paths with quadratic competitiveness [9]. They also proved that this is worst-case optimal, i.e., it is not possible to design routing strategies which use only local knowledge and achieve a better competitiveness than quadratic. Their lower bound is based on the fact that a radio hole might have a complex structure, like a maze, making it hard to find a short path to a destination in an online fashion. Some other examples of the many routing strategies that have been proposed are [12,14,16]. For example, Rührup and Schindelhauer considered routing strategies for grids that contain failed nodes [16]. This is similar to our scenario as failed nodes behave like radio holes in an ad hoc network. Their procedure uses a strategic search which distributes a message over multiple paths. They proved that their procedure is asymptotically optimal for their setting. However, it is not clear how the strategy can be generalized to node distributions in the Euclidean plane.

So the question arises: How to make use of long-range links to find a suitable abstraction of the radio holes with a local strategy such that we obtain c-competitive paths for any source-destination pair in the underlying ad hoc network?

A Hybrid Communication Network has been introduced in different contexts [5,13]. To the best of our knowledge, we are the first ones that consider these types of networks for the purpose of finding paths in ad hoc networks.

At the core of our algorithm is a 2-localized Delaunay Graph of the ad hoc network. A 2-localized Delaunay Graph is related to the Delaunay triangulation, which was introduced in [7]. The advantage of using Delaunay graphs is that they are Euclidean c-spanners, which means that they contain a path for any pair of nodes of length at most c times their Euclidean distance. The currently best known bound for c is 1.998 and was proven by Xia [17]. Because wireless communication is only possible for limited distances, Delaunay graphs are not directly applicable to ad hoc networks, but Delaunay graphs restricted to UDG edges, which are known as Restricted Delaunay Graphs [11]. Restricted Delaunay graphs are still hard to compute, so we will focus on the related 2-localized Delaunay Graph, which can be built in a constant number of rounds [11]. Based on that graph, one can use Chew's Algorithm [3] to efficiently route messages if there are no radio holes.

2   Preliminaries

Throughout this paper, we assume the set of nodes V to be in general position, i.e., there are no three nodes on a line and no four nodes on a cycle. Moreover, we assume that the coordinates of each node are unique and thus there are no two nodes on the same position. We consider a 2-localized Delaunay Graph $LDel^2(V)$ as topology for the ad hoc network which is related to the Delaunay Graph. The Delaunay Graph is a graph, where $\bigcirc(u,v,w)$ is the unique circle through the nodes u, v and w and $\triangle(u,v,w)$ be the triangle formed by the nodes u, v and w. For any $V \subseteq \mathbb{R}^2$, the *Delaunay Graph* Del (V) of V contains all triangles $\triangle(u, v, w)$ for which $\bigcirc(u, v, w)$ does not contain any further node besides u, v and w. The 2-localized Delaunay Graph is a structure that only allows edges which do not exceed the transmission range of a node. It can be constructed efficiently in a distributed manner. In k-localized Delaunay Graphs, a triangle $\triangle(u, v, w)$ for nodes u, v, w of V satisfies that all edges of $\triangle(u, v, w)$ have length at most 1 and the interior of the disk $\bigcirc(u, v, w)$ does not contain any node which can be reached within k hops from u, v or w in UDG(V). The *k-localized Delaunay Graph* $LDel^k(V)$ is defined to consist of all edges of k-localized triangles and all edges (u, v) for which the circle with diameter $\overline{uv}$ does not contain any further node $w \in V$. For $k = 2$, we obtain the 2-localized Delaunay Graph which is also a planar graph [11]. Since 2-localized Delaunay Graphs do not contain all edges of a corresponding Delaunay Graph, one cannot simply use routing strategies for Delaunay Graphs in our scenario. We denote faces of the 2-localized Delaunay Graph which are not triangles as *holes*. For the formal definition of holes, we distinguish between *inner* and *outer* holes.

Definition 1 (Hole). *Let $V \in \mathbb{R}^2$. An* inner hole *is a face of $LDel^2(V)$ with at least 4 nodes.*

Furthermore, let $CH(V)$ be the set of all edges of the convex hull of V. Define $\overline{LDel^2}(V)$ to be the graph that contains all edges of $LDel^2(V)$ and $CH(V)$. An outer hole *is a face in $\overline{LDel^2}(V)$ with at least 3 nodes, that contains an edge $e \in CH(V)$ with $\|e\| > 1$.*

Nodes lying on the perimeter of a hole are called *hole nodes*. Note that the hole nodes of the same hole form a ring, i.e., each hole node is adjacent to exactly two other hole nodes for each hole it is part of. The choice of the 2-localized Delaunay Graph as network topology is motivated by its *spanner*-property. The Delaunay Graph Del (V) contains paths between every pair of nodes v and w of V which are not longer than c times their Euclidean distance, i.e., a *path* $(v, \ldots, w)$ between two nodes v and w in a geometric graph G is a *geometric c-spanning path* between v and w, if its length is at most c times the Euclidean distance between w and v. Delaunay Graphs are proven to be geometric 1.998-spanners [17]. Xia argues that the bound of 1.998 also relates to 2-localized Delaunay Graphs [17]. However, these graphs are not spanners of the Euclidean metric but of the Unit Disk Graph. Bose et al. introduced the 5.9-competitive online routing strategy *Chew's Algorithm* for Delaunay Graphs which only considers edges of triangles that are intersected by the direct line segment between source and destination [2]. In case the source and the target node of the 2-localized Delaunay Graph are *visible* from each other, i.e., their direct line segment does not intersect any hole, Chew's Algorithm also finds 5.9-competitive paths in the 2-localized Delaunay Graph. To be able to find constant-competitive paths between any pair of nodes in the 2-localized Delaunay Graph, we take a look at results from computational geometry. If we abstract from the underlying 2-localized Delaunay Graph, our scenario is comparable to routing in polygonal domains. These kinds of routing problems usually consider a starting point s and a target point t in the Euclidean plane. The goal is to find a path in the plane from s to t. The challenging aspect of these problems is the presence of polygonal obstacles which avoid walking directly along the line segment $\overline{st}$. In our scenario, these polygonal obstacles are radio holes. De Berg et al. showed that it is enough to consider nodes of obstacle polygons for finding shortest paths in polygonal domains [1]: In the Visibility Graph $Vis\,(V)$ of a set of polygons, V represents the set of corners of the polygon, and there is an edge $\{v, w\}$ in $Vis\,(V)$ if and only if a line can be drawn from v to w without crossing any polygon, i.e., v is visible from w. The combination of the mentioned results implies that a shortest path in the Visibility Graph of hole nodes of the 2-localized Delaunay Graph yields to a 5.9-competitive path in the 2-localized Delaunay Graph by applying Chew's Algorithm between every pair of consecutive nodes on the path.

3   General Routing

We assume for now that every node which is located on the perimeter of a hole stores a Visibility Graph of all hole nodes. Two hole nodes are visible from each other if their direct line segment does not intersect any hole of the 2-localized Delaunay Graph. The routing protocol works as follows: A source node s that wants to send data to a target node t initially contacts t via a long-range link to ask for $t's$ geographical position, i.e., a tuple of coordinates (t_x, t_y). t responds with its position and s afterwards sends its message via Chew's Algorithm towards (t_x, t_y). We distinguish two cases: (1) The message reaches t via Chew's Algorithm and (2) the message reaches a hole node h_0, i.e., the direct line segment $\overline{st}$ intersects a hole. In case (1), we immediately obtain a 5.9-competitive path from s to t. Otherwise, h_0 inserts t into its Visibility Graph and applies a shortest path algorithm from itself to t. The resulting shortest path $(h_0, h_1, h_2, \ldots, h_k = t)$ is then used to transmit the message via ad hoc links. By applying Chew's Algorithm, a path of length $5.9 \cdot \|h_0 h_1\|$ is obtained. After reaching h_1, the procedure is repeated until the message finally reaches t. Let p_{st} be the shortest path between s and t in the Visibility Graph. In case h_0 lies on the shortest path between s and t in the Visibility Graph, the resulting path in the 2-localized Delaunay Graph has length at most $5.9 \cdot \|p_{st}\|$. Otherwise, the initial path to h_0 is a detour. Obviously, the detour increases the competitiveness only by a constant factor. As Chew's Algorithm did not reach t but a node h_0, the path taken from s to h_0 has length less or equal to $5.9 \cdot \|st\|$ which is less or equal to 5.9 times the shortest possible path between s and t in the 2-localized Delaunay Graph. Hence, the detour increases the competitive constant only by an additional factor of 3. We obtain an 17.7-competitive path between s and t.

An idea to reduce the number of edges to $\mathcal{O}\,(h)$ is to not compute the entire Visibility Graph but only a Delaunay Graph of all nodes lying on different holes. As Delaunay Graphs are planar graphs, this reduces the number of edges to $\mathcal{O}\,(h)$ and it also affects the obtained length of the paths. Delaunay Graphs do not contain the shortest geometric connection between two nodes in general but a path which is 1.998-competitive to such a path [17]. By using a Delaunay Graph instead of a Visibility Graph, we obtain a path length of $1.998 \cdot 17.7 \cdot \|p_{st}\| \leq 35.37 \cdot \|p_{st}\|$.

4 Routing for Convex Hulls as Hole Abstractions

In this section, we focus on the reduction of number of nodes in the Visibility Graph even further while still being able to compute competitive paths. Therefore, we consider *locally convex hulls*, where $(v_1, v_2, \ldots, v_k, v_1)$ is a cycle of nodes in $LDel^2(V)$ at the perimeter of some hole. We call $(v_{i_1}, v_{i_2}, \ldots, v_{i_\ell}, v_{i_1})$ for some $1 \leq i_1 < i_2 < \ldots, i_\ell \leq k$ a locally convex hull of that hole if (1) $\|v_{i_j} v_{i_{j+1}}\| \leq 1$ for all $j \in \{1, \ldots, \ell\}$ (where $v_{i_{\ell+1}} = v_{i_1}$), and (2) there are no 3 consecutive nodes u, v, w in that sequence where $\angle(u, v, w) \geq 180°$ and $\|uw\| \leq 1$.

For the locally convex hulls, we prove:

Lemma 1. *For any cycle $(v_1, v_2, \ldots, v_k, v_1)$ of hole nodes in $LDel^2(V)$ that covers an area of size A, any locally convex hull of that cycle contains $\mathcal{O}(A)$ nodes.*

Hence, locally convex hulls contain a number of nodes that is independent of the total number of nodes in the system and only depends on the area covered by the hole. A further reduction in the number of nodes can be achieved by the convex hull of a hole.

Lemma 2. *For any cycle $(v_1, v_2, \ldots, v_k, v_1)$ of hole nodes in $LDel^2(V)$ with a bounding box (i.e., the box of minimum size containing $v_1, \ldots, v_k$) of circumference L, the convex hull $(v_{i_1}, v_{i_2}, \ldots, v_{i_\ell}, v_{i_1})$ of the cycle contains $\mathcal{O}(L)$ nodes.*

4.1 c-Competitive Paths via Convex Hulls

In this section, we assume that the source and the target of a routing request lie outside of any convex hull and that the source and the target are not visible from each other as finding c-competitive paths for visible nodes can be found via Chew's Algorithm.

Lemma 3. *The shortest path between any pair of non-visible nodes of the 2-localized Delaunay Graph contains convex hull nodes.*

We define the *Overlay Delaunay Graph* to be a Delaunay Graph that contains all convex hulls of holes and connects the nodes of different convex hulls in a Delaunay Graph.

The following theorem is a conclusion of the so far mentioned properties:

Theorem 2. *Let s and t be two nodes of a 2-localized Delaunay Graph that do not lie inside of any convex hull. Further, let $(s = c_0, c_1, \ldots, c_{\ell-1}, c_\ell = t)$ be the shortest path in the Overlay Delaunay Graph via long-range links. Then we have*

1. *There is a $\left(1.998 \cdot \sum_{m=0}^{\ell-1} d_m\right)$-path in the 2-localized Delaunay Graph from s to t, where $d_m := \|c_m c_{m+1}\|$.*

2. *By applying Chew's Algorithm, we obtain a $\left(5.9 \cdot \sum_{m=0}^{\ell-1} d_m\right)$-path in the 2-localized Delaunay Graph from s to t, where $d_m := \|c_m c_{m+1}\|$.*

4.2   Routing Protocol

To investigate all cases of s and t's different geographical positions, we introduce *bay areas*. A bay area H_A of a hole consists of the nodes and edges of the 2-localized Delaunay Graph that are inside the convex hull and between two adjacent convex hull nodes. Bay areas allow us to describe all cases for s and t: (1) s and t are outside of convex hulls, (2) s or t is inside of a convex hull (3) s and t are inside different convex hulls (4) s and t are inside the same convex hull but in different bay areas and (5) s and t are inside the same convex hull and in the same bay area. Case 1 is solvable with few additional requirements to the routing protocol described in Sect. 3. Cases 2–5, need a more sophisticated solution and are postponed to Sect. 4.3.

For Case 1, we assume for now that each node located on the perimeter of a hole stores references to its two neighboring convex hull nodes and all nodes lying on convex hulls of holes store an Overlay Delaunay Graph of all convex hull nodes. The routing protocol for Case 1 works exactly as described in Sect. 3. A node s sends its message via Chew's Algorithm into the direction of t. In case the message arrives at a hole node, it is directed to a convex hull node. The convex hull node inserts t into its Visibility Graph and applies a shortest path algorithm. The resulting path is added to the message and used for forwarding the message in the ad hoc network. Between any pair of nodes on the received path, Chew's Algorithm is applied. Based on the results from Sect. 4.1, we obtain a c-competitive path in $LDel^2(V)$.

4.3   Limitations of Convex Hulls

In this section, we focus on routing from s to t, when their geographical coordinates fulfill the properties of Cases 2–5. Here, we only provide the routing algorithm, where both s and t are in the same bay area, i.e., Case 5. An analogous routing can be executed for Cases 2–4.

For computing c-competitive paths, we assume that each hole node knows the entire hole ring it belongs to. $\overline{st}$ denotes the direct line segment between s and t. We define S to be the first intersection point between $\overline{st}$ and the hole boundary, from the direction of s. Let T be the analogous intersection point from the direction of t. Let P_1 be the hole node with the shortest hop distance on the hole boundary to S and P_t the analogous hole node to T. We denote $H_{s,t}$ to be the set of all hole nodes that are located in this bay area between P_1 and P_t. We call the nodes of the convex hull of this set the *extreme points* $\{E_1, ..., E_k\}$. We define E_t to be the extreme point with the smallest index, where $\overline{E_t t}$ is visible to t.

The routing strategy works as follows: s executes Chew's Algorithm to send the message m in the direction of t until m either arrives at t (i.e., s and t are visible to each other) or at P_1. If it reaches P_1, then m is routed from P_1 to E_1, from E_1 to E_2,..., from E_i to E_t, for $i = 1, \ldots, t$. Finally m is routed from E_t to t. All these routing steps are done with Chew's Algorithm. Because Chew's Algorithm is 5.9-competitive and the provided path by the algorithm contains in total $2 + |E_{route}|$ direct lines, where $|E_{route}|$ denotes the number of extreme points that we route to, it is easy to see:

Lemma 4. *Let s and t be nodes with geographic coordinates in the same bay area, then the routing algorithm above provides a c-competitive routing path between s and t with $c = (2 + |E_{route}|) \cdot 5.9$.*

The convex hull node can locally decide to which convex hull the destination of the routing request belongs, but it cannot decide which convex hull node is responsible for the bay area the destination is located in. The convex hull node initiates two messages towards the destination. These messages are routed in opposite directions around the convex hull in which the target node of the routing request is located. Once the message arrives at the convex hull of the destination, each convex hull node there can locally decide whether the destination is located inside the bay area of the convex hull because we assume that each hole node is aware of all other hole nodes inside of its bay area (convex hull nodes are also hole nodes).

5   Computation and Information Dissemination

To compute the 2-localized Delaunay Graph, we use the distributed protocol described in [11]. The authors assume that an initial connected Unit Disk Graph of all ad hoc links is given. To obtain the initial connected Unit Disk Graph of all ad hoc links, every node executes a WiFi broadcast within its transmission range. Thus, each node is aware of all nodes in its transmission range and we obtain a Unit Disk Graph. The nodes execute the protocol of Li et al. which requires communication costs of $\mathcal{O}(n \log n)$ bits and only $\mathcal{O}(1)$ communication rounds [11]. The result is not a 2-localized Delaunay Graph but a supergraph of it called *Planar Localized Delaunay Graph*. As each edge has a length of at most 1 and the Planar Localized Delaunay Graph is a planar graph, our ideas of hole detection also work for these type of graphs. For convenience, we restrict ourselves to 2-localized Delaunay Graphs in the rest of this section.

5.1   Hypercube and Convex Hull Computation

In this section, we give a brief overview on a protocol that transforms a ring of nodes into a hypercube structure. This protocol is a prerequisite for a convex hull protocol and allows each node to detect whether it lies on the boundary of a hole. The full description can be found in Appendix C. In the beginning, each node of a ring chooses a predecessor and a successor. Then, pointer jumping is applied. In the first round, each node introduces its predecessor and successor to each other. Afterwards, the new neighbors are introduced to each other and so on. In addition to references, the minimal id of all so far seen neighbors is exchanged in each introduction step such that eventually every node is aware of the minimal id of all ring nodes. The node with minimal id becomes the leader and distributed hypercube ids afterwards. We summarize the results of this section in the following lemma:

Lemma 5. *A ring of k nodes can be transformed into a hypercube in $\mathcal{O}(\log k)$ communication rounds. The number of required messages is in $\mathcal{O}(\log k)$ per node.*

In the hypercube, we compute the convex hull of each hole. Initially, the coordinates of points have to be sorted. Sorting n points in a hypercube can be done in $\mathcal{O}(\log n)$ communication rounds on expectation with the algorithm of Reif and Valiant [15]. Upon termination, Miller's algorithm is applied which ensures that each node of the ring knows every convex hull node and especially each convex hull node identifies itself as a convex hull node. The following theorem follows:

Theorem 3. *Given a hole ring with k nodes, the convex hull of this hole ring can be calculated in $\mathcal{O}(\log k)$ communication rounds on expectation.*

5.2   Hole Detection

Boundary nodes locally cannot detect whether cycles are oriented clockwise or counterclockwise and hence cannot decide whether they are located on the outer boundary or on a hole. To let nodes distinguish these cases they sum up angles along each boundary into the direction of the orientation. Let v_1, v_2 be a predecessor and a successor along a boundary. In case walking from v_1 to v_2 requires a left turn, the angle between v_1 and v_2 is subtracted from the current sum. Angles of right turns are added. The result would be 360° for the outer boundary and $-360°$ for each hole [6]. The summation along a boundary could be done by a token passing technique initiated by a leader. The technique requires a linear number of communication rounds for each cycle. To improve the runtime, we sum angles in parallel to the hypercube protocol in the following way: In addition to the minimal ID, we also exchange the sum of angles with each edge of the pointer jumping procedure. At the end, every node of the ring knows the sum of all angles along the boundary. Hence, each node can decide whether it is a hole node in $\mathcal{O}(\log n)$ communication rounds. For determining outer holes, we need a second run of convex hull computations along the outer boundary. Outer holes are defined by an edge of the outer convex hull of the point set. After the convex hull of the outer boundary has been computed, a second run is started between every pair of consecutive convex hull nodes whose distance exceeds the transmission range of a node. All in all, we compute the convex hull of each hole and of the outer boundary to be able to distinguish the outer boundary and holes. Afterwards we start a second run of convex hull computations for each outer hole determined by the convex hull of the outer boundary from the first run.

5.3   Information Dissemination of Convex Hulls and Hole Rings

The main observation of this section is that nodes of a convex hull locally cannot decide in which directions other holes are located. Hence, we need to spread the information about convex hulls in the entire network. We use an Overlay

Network via the long-range links which only has a logarithmic diameter [8]. The protocol ensures that all nodes of the network are connected in a rooted tree via long-range links after $\mathcal{O}(\log^2 n)$ communication rounds. The tree has a height of $\mathcal{O}(\log n)$ and a constant degree. Consequently, the diameter of the tree is $\mathcal{O}(\log n)$. The tree allows us to distribute references of convex hull nodes in $\mathcal{O}(\log n)$ communication rounds, i.e., each convex hull node can direct its own reference both towards the root and into the subtree below itself. The root redirects the reference into every other subtree. This procedure avoids that nodes receive the same broadcast message multiple times. The total runtime of this step is $\mathcal{O}(\log^2 n)$ as the tree has to be established initially.

The protocol in which the source and the destination of a routing request can be located inside of a convex hull requires that each node of a hole ring stores references to every other node of the hole ring (see Sect. 4.3). We set this on top of the pointer jumping protocol described in Sect. 5.1. In addition to the leading coordinate, we also exchange references to all hole nodes which are bridged via an edge in the pointer jumping protocol. Hence, each node of a hole ring is aware of every other node of that hole ring after $\mathcal{O}(\log n)$ communication rounds.

6   Node Movement

We say that in a certain time interval, nodes cannot move more than $\frac{1}{2}$. We assume that only edges of length less or equal than $\frac{1}{2}$ are valid ad hoc links. This ensures that ad hoc links chosen by the routing protocol remain valid for the rest of the time interval as all nodes which have been in the communication range of each other at the beginning of the time interval stay inside of these communication ranges. Therefore, in every time interval, a re-execution of all protocols except the protocol for the distributed Overlay tree allows us to find competitive paths in a scenario where nodes are allowed to change their positions.

7   Future Work

In this paper, we considered non-intersecting convex hulls. Future work could be the design of routing strategies that can deal with finding competitive paths in areas of intersecting convex hulls. Besides, we concluded our paper with a dynamic scenario in which nodes are allowed to move. Our solution is to periodically recompute the entire Overlay Network. A model with bounded movement speed could be investigated in which only parts of the Overlay Network have to be recomputed. A further dynamic which could be considered, is joining and leaving nodes. Lastly, our model does not tackle physical aspects of wireless communication. Interesting aspects are for example wireless interference in crowded areas.

A   Visualization

See Fig. 1.

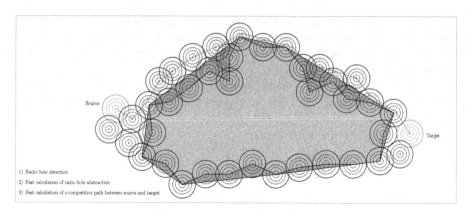

Source

Target

1) Radio hole detection
2) Fast calculation of radio hole abstraction
3) Fast calculation of c-competitive path between source and target

Fig. 1. An overview of our approach. It contains the detection of radio holes (1), the computation of a hole abstraction (2) and a routing algorithm that finds c-competitive paths (3). The blue regions are called *bay areas*. (Color figure online)

B Omitted Proofs and Lemmas

Proof (Proof of Lemma 1). Consider any locally convex hull $(v_{i_1}, v_{i_2}, \ldots, v_{i_\ell}, v_{i_1})$, and let u, v, w be 3 consecutive nodes in that sequence. I f $\angle(u, v, w) \geq 180°$, then we know from the definition of the locally convex hull that $\|uw\| > 1$. If $\angle(u, v, w) < 180°$, then $\|uw\| > 1$ as well since otherwise v would not be on the perimeter of the hole. This implies for the predecessor p of u and the successor s of w that $\|pv\| > 1$ and $\|vs\| > 1$. Also, there cannot exist any other node $x \in \{v_{i_1}, \ldots, v_{i_\ell}\}$ with $\|vx\| \leq 1$ as otherwise we had a shortcut in the perimeter, meaning that $(v_1, v_2, \ldots, v_k, v_1)$ cannot be the perimeter of a hole. Hence, the unit cycle around each v_{i_j} can contain at most 2 other nodes of the locally convex hull, which implies that $\ell = \mathcal{O}(A)$.

Proof (Proof of Lemma 2). Let B be the bounding box of the cycle and x be its center point. Let the points $w_{i_1}, \ldots, w_{e_\ell}$ be the projections of $v_{i_1}, v_{i_2}, \ldots, v_{i_\ell}$ from x onto the boundary of B, i.e., the points where the ray from x in the direction of v_{i_j} intersects the boundary of B. As is easy to check, the ℓ_1-distance of w_{i_j} and $w_{i_{j+1}}$ on B is at least as large as $\|v_{i_j} v_{i_{j+1}}\|$ for all j. Moreover, for any 3 consecutive points u, v, w on the convex hull it must hold that $\|uw\| > 1$. Hence, for any 3 consecutive points u', v', w' on the projection of the convex hull onto B it must hold that the ℓ_1-distance of u' and w' is more than 1, which implies that the convex hull contains only $\mathcal{O}(L)$ nodes.

Proof (Proof of Lemma 3). Let s, t be two nodes of the 2-localized Delaunay Graph, whose direct line segment intersects a hole. Starting from s, let ℓ be the first intersected line segment of the boundary of the intersected convex hull with endpoints v, w. We assume that the shortest path contains points of $(v, \ldots, w)$. Else, the argumentation must be repeated with the neighboring edges of the convex hull.

By contradiction, we assume that the shortest path from s to t contains a point $p \in (v, ..., w)$ from the interior of the convex hull, i.e., excluding v, w. Without loss of generality, we assume that the shortest path furthermore contains the point w (The same holds for v.).

Because of the triangle inequality, the following holds: $\|sw\| \leq \|sp\| + \|pw\|$ And we know that $\|(x, y)\| \leq 1.998 \cdot \|xy\|$ holds for any two nodes of a Delaunay triangulation.

Then:

$$\frac{\|(s, w)\|}{1.998} \leq \|sw\|$$
$$\leq \|sp\| + \|pw\|$$
$$\leq \|(s, p)\| + \|(p, w)\|$$
$$= \|(s, ..., p, ..., w)\|$$

Hence, points of the interior of convex hulls cannot be chosen as a path along a convex hull node would be shorter.

Lemma 6. *Let a and b be visible nodes of different convex hulls. Then there is a $1.998 \cdot \|ab\|$-spanning path between them in the 2-localized Delaunay Graph.*

Proof (Proof of Lemma 6). Since the Delaunay Graph is a 1.998-spanner of the complete Euclidean graph [17] and the 2-localized Delaunay Graph contains all edges of the original Delaunay Graph between a pair of visible nodes, there always exists a $1.998 \cdot \|ab\|$ path between two visible nodes a and b of two different convex hulls. This proves Lemma 6.

Lemma 7. *Let a and b be adjacent nodes on a convex hull, where $a \neq b$. Then there is a $1.998 \cdot \|ab\|$-spanning path in the 2-localized Delaunay Graph between a and b.*

Proof (Proof of Lemma 7). We use the observation of Xia that a 2-localized Delaunay Graph is a 1.998-spanner of the Unit Disk Graph. Thus there is a 1.998-competitive path between the to convex hull nodes. This proves Lemma 7.

Lemma 8. *Let a and b be visible nodes of different convex hulls. Then there is a $5.9 \cdot \|ab\|$-routing path between them in the 2-localized Delaunay Graph.*

Proof (Proof of Lemma 8). This fact follows immediately from [3] and the fact that for two visible nodes s and t, their direct line segment $\overline{st}$ intersects only triangles which are also part of the (standard) Delaunay Graph.

Lemma 9. *Let a and b be adjacent node on a convex hull, where $a \neq b$. Then there is a $5.9 \cdot \|ab\|$-routing path in the 2-localized Delaunay Graph between a and b.*

Proof (Proof of Lemma 9). The proof is the same as for Lemma 8 because two adjacent convex hull nodes are due to the assumption of non-intersecting convex hulls per definition visible from each other.

Proof (Proof of Theorem 2(1)). Theorem 2(1) follows immediately from Lemmas 6 and 7.

Proof (Proof of Theorem 2(2)). Recall that there exists an online routing strategy for Delaunay Graphs which finds a path between any source s and target t with length at most $5.9 \cdot \|st\|$. Furthermore, recall that the 2-localized Delaunay Graph contains all edges of the original Delaunay Graph between any pair of visible nodes. Thus, there is a routing strategy from any convex hull node a to any other convex hull node b in cases a and b are nodes of different convex hulls with length at most $5.9 \cdot \|ab\|$. This routing strategy can be applied to route in the 2-localized Delaunay Graph between to adjacent convex hull nodes a and b as well. This is due to the fact, that the routing strategy chooses the path along triangles in the 2-localized Delaunay Graph that are intersected by the line from a to b. As a and b are visible from each other, these edges would also be part of the original Delaunay Graph. Thus, the routing strategy applied on the hybrid communication model gives a path of length at most $\left(5.9 \cdot \sum_{m=0}^{\ell-1} d_m\right)$. All in all, we obtain Theorem 2.

C Hypercube Protocol

In this section, we describe a procedure that establishes a hypercube topology out of a ring with k nodes. On the one hand, this protocol is a prerequisite for the convex hull protocol. On the other hand, it allows a fast hole detection, i.e., enables nodes to quickly distinguish the outer boundary from a hole. More precisely, we execute the protocol both for holes and the outer boundary of the entire node set which are both connected in a ring topology. For the ease of notation, we summarize nodes of the outer boundary and hole nodes as *boundary nodes*. Note that each node can locally detect whether it part of an inner or outer hole by checking whether it is part of a triangle with a missing edge due to the restriction of the edge length. Each node v which is part of the convex hull of the entire node set detects that there are two consecutive neighbors v and w in the clockwise ordering of $v's$ neighbors such that $\angle(u, v, w) \geq 180°$.

Initially, each boundary node chooses a successor and a predecessor in each ring. This can be achieved as follows: Each boundary node sorts its boundary neighbors clockwise. Afterwards, for every pair of consecutive nodes in the sorting (also for the last and the first node) the first node is chosen as predecessor and the second node is chosen as successor. Now, every boundary is either oriented clockwise or counterclockwise. More precisely, the outer boundary is oriented clockwise and each hole is oriented counterclockwise. The orientation, however, is not important for the hypercube protocol but for the hole detection in Sect. 5.2.

We proceed with the hypercube protocol by giving a definition of a hypercube, where a d-dimensional *hypercube* consists of n nodes, where $n = 2^d$, such that each node has a unique bitstring $(x_1, \ldots, x_d) \in \{0, 1\}^d$ and there is an edge between two nodes if and only if their bitstring differs in only one bit. The decimal representation of a bitstring of a node h is denoted as $id(v)$. We assume the number of nodes in the ring to be a power of two. The techniques can be applied for an arbitrary number of nodes with a slight modification of the given protocol. For the construction of the hypercube we use pointer jumping. On the one hand, this technique enables us to build overlay edges for the hypercube fast and additionally it allows us to elect a leader in $\mathcal{O}(\log k)$ communication rounds which is responsible for setting up the hypercube IDs. The leader of the ring is the node with minimal ID. The ID of a node v is denoted as id_v. In addition, we assign two values to each edge $e = \{u, v\}$, which is created by the pointer jumping protocol. The first one, $\ell(e)$ defines the minimal ID of all ring nodes which are bridged by e, except id_u. The second value, $\text{level}(e) = \log(b)$, where b denotes the number of ring nodes between u and v.

The pointer jumping is used as follows: Let v be a node of the hole ring and let $pred_0$ be its predecessor and $succ_0$ its successor on the ring. In round 1 of the protocol, v introduces $succ_0$ to $pred_0$ to each other. Thus $succ_0$ and $pred_0$ become adjacent nodes and an overlay edge $e = \{pred_0, succ_0\}$ is established. Further, v assigns $\ell(e) = \min\{id_v, id_{succ_0}\}$ and $\text{level}(e) = 0$. As each node executes the protocol, v also gets introduced two nodes in round 1 which are denoted as $pred_1$ and $succ_1$. In particular, in round i, each node v of the hole ring introduces its predecessor $pred_{i-1}$ to its successor $succ_{i-1}$ and gets introduced $pred_i$ and $succ_i$. The node v that introduces $pred_{i-1}$ and $succ_{i-1}$ to each other also assigns $\ell(\{pred_{i-1}, succ_{i-1}\}) = \min\{\ell(\{pred_{i-1}, v\}), \ell(\{v, succ_{i-1}\})\}$ and $\text{level}(e) = \text{level}(\{pred_{i-1}, v\}) + 1$.

With pointer jumping, the hop distance between any pair of nodes halves from round to round. The protocol stops in a round i in which v gets introduced $succ_i$ and $pred_i$ and $\ell(\{pred_i, v\}) = \ell(\{v, succ_i\})$. At that point, each node (especially the leader itself) is locally aware of the minimal ID and hence knows the ID of the leader. As the distance between any pair of nodes halves from round to round, this protocol requires $\mathcal{O}(\log k)$ communication rounds. For the purpose of being able to emulate a hypercube, we do not only need the additional overlay edges, but also hypercube IDs. Recall that the node IDs of the hypercube are bitstrings of length $\log k$. To distribute the hypercube IDs to the corresponding boundary nodes, the leader v assigns for each hypercube edge $\{v, succ_i\}$ the binary representation of $\text{level}(\{v, succ_i\}) + 1$ as ID to $succ_i$. Each node that receives an ID from the leader repeats the ID distribution recursively, relative to its own ID. As the diameter of a hypercube of k nodes is $\mathcal{O}(\log k)$, the distribution of IDs requires $\mathcal{O}(\log k)$ communication rounds. Eventually, the nodes of the ring form a hypercube and we are able to apply every protocol designed for hypercubes.

References

1. de Berg, M., Cheong, O., van Kreveld, M., Overmars, M.: Computational Geometry: Algorithms and Applications, 3rd edn. Springer, Santa Clara (2008). https://doi.org/10.1007/978-3-540-77974-2
2. Bez, H.E., Edwards, J.: Distributed algorithm for the planar convex hull problem. Comput. Aided Des. **22**(2), 81–86 (1990)

3. Bonichon, N., Bose, P., Carufel, J.D., Perkovic, L., van Renssen, A.: Upper and lower bounds for online routing on delaunay triangulations. Discrete Comput. Geom. **58**(2), 482–504 (2017)
4. Bose, P., et al.: Online routing in convex subdivisions. In: Goos, G., Hartmanis, J., van Leeuwen, J., Lee, D.T., Teng, S.-H. (eds.) ISAAC 2000. LNCS, vol. 1969, pp. 47–59. Springer, Heidelberg (2000). https://doi.org/10.1007/3-540-40996-3_5
5. Cena, G., Valenzano, A., Vitturi, S.: Hybrid wired/wireless networks for real-time communications. IEEE Ind. Electron. Mag. **2**(1), 8–20 (2008)
6. Daymude, J.J., Gmyr, R., Richa, A.W., Scheideler, C., Strothmann, T.: Improved leader election for self-organizing programmable matter. In: Fernández Anta, A., Jurdzinski, T., Mosteiro, M.A., Zhang, Y. (eds.) ALGOSENSORS 2017. LNCS, vol. 10718, pp. 127–140. Springer, Cham (2017). https://doi.org/10.1007/978-3-319-72751-6_10
7. Delaunay, B.: Sur la sphère vide. A la Mémoire de Georges Voronoï. Bulletin de l'Académie des Sciences de l'URSS **6**, 793–800 (1934)
8. Gmyr, R., Hinnenthal, K., Scheideler, C., Sohler, C.: Distributed monitoring of network properties: the power of hybrid networks. In: Chatzigiannakis, I., Indyk, P., Kuhn, F., Muscholl, A. (eds.) 44th International Colloquium on Automata, Languages, and Programming (ICALP 2017). Leibniz International Proceedings in Informatics (LIPIcs), vol. 80, pp. 137:1–137:15. Schloss Dagstuhl-Leibniz-Zentrum fuer Informatik, Dagstuhl, Germany (2017)
9. Kuhn, F., Wattenhofer, R., Zollinger, A.: Worst-case optimal and average-case efficient geometric ad-hoc routing. In: Proceedings of the 4th ACM International Symposium on Mobile Ad Hoc Networking & Computing. MobiHoc 2003, pp. 267–278. ACM, New York (2003)
10. Li, S., Zeng, W., Zhou, D., Gu, X., Gao, J.: Compact conformal map for greedy routing in wireless mobile sensor networks. IEEE Trans. Mob. Comput. **15**(7), 1632–1646 (2016)
11. Li, X.Y., Calinescu, G., Wan, P.J.: Distributed Construction of a planar spanner and routing for ad hoc wireless networks. In: Proceedings of the 21st Annual Joint Conference of the IEEE Computer and Communications Societies, vol. 3, pp. 1268–1277. IEEE Press, New York (2002)
12. Lumelsky, V.J.: Algorithmic and complexity issues of robot motion in an uncertain environment. J. Complex. **3**(2), 146–182 (1987)
13. Murty, Y.S.N.: Hybrid communication networks for power utilities. In: Power Quality 1998, pp. 239–242. IEEE Press, New York, June 1998
14. Rao, N.S., Kareti, S., Shi, W., Iyengar, S.S.: Robot navigation in unknown terrains: introductory survey of non-heuristic algorithms. Technical report, Oak Ridge National Lab., TN (United States) (1993)
15. Reif, J.H., Valiant, L.G.: A logarithmic time sort for linear size networks. J. ACM (JACM) **34**(1), 60–76 (1987)
16. Rührup, S., Schindelhauer, C.: Online multi-path routing in a maze. In: Asano, T. (ed.) ISAAC 2006. LNCS, vol. 4288, pp. 650–659. Springer, Heidelberg (2006). https://doi.org/10.1007/11940128_65
17. Xia, G.: The stretch factor of the Delaunay triangulation is less than 1.998. SIAM J. Comput. **42**(4), 1620–1659 (2013)

On the Approximability and Hardness of the Minimum Connected Dominating Set with Routing Cost Constraint

Tung-Wei Kuo[✉]

Department of Computer Science, National Chengchi University,
No. 64, Sec. 2, ZhiNan Road, Taipei 11605, Taiwan (R.O.C.)
twkuo@cs.nccu.edu.tw

Abstract. In the problem of minimum connected dominating set with routing cost constraint, we are given a graph $G = (V, E)$ and a positive integer α, and the goal is to find the smallest connected dominating set D of G such that, for any two non-adjacent vertices u and v in G, the number of internal nodes on the shortest path between u and v in the subgraph of G induced by $D \cup \{u, v\}$ is at most α times that in G. For general graphs, the only known previous approximability result is an $O(\log n)$-approximation algorithm ($n = |V|$) for $\alpha = 1$ by Ding *et al.* For any constant $\alpha > 1$, we give an $O(n^{1 - \frac{1}{\alpha}} (\log n)^{\frac{1}{\alpha}})$-approximation algorithm. When $\alpha \geq 5$, we give an $O(\sqrt{n} \log n)$-approximation algorithm. Finally, we prove that, when $\alpha = 2$, unless $NP \subseteq DTIME(n^{poly \log n})$, for any constant $\epsilon > 0$, the problem admits no polynomial-time $2^{\log^{1-\epsilon} n}$-approximation algorithm, improving upon the $\Omega(\log \delta)$ bound by Du *et al.*, where δ is the maximum degree of G (albeit under a stronger hardness assumption).

Keywords: Connected dominating set · Spanner ·
Set cover with pairs · MIN-REP problem

1 Introduction

1.1 Motivation

In wireless network routing, a common approach is to select a set of nodes as the *virtual backbone*. The virtual backbone is responsible for relaying packets. Specifically, when a node s generates a packet destined to t, the packet is routed through path $(s, v_1, v_2, \cdots, v_k, t)$, where every internal node $v_i, 1 \leq i \leq k$, belongs to the virtual backbone. To realize this idea, we can model the wireless network as a graph $G = (V, E)$, where V is the set of nodes in the wireless network, and $(u, v) \in E$ if and only if u and v can communicate with each

This work is partially supported by the Ministry of Science and Technology of R.O.C. under contract No. MOST 106-2221-E-004-005-MY3.

© Springer Nature Switzerland AG 2019

S. Gilbert et al. (Eds.): ALGOSENSORS 2018, LNCS 11410, pp. 32–46, 2019.
https://doi.org/10.1007/978-3-030-14094-6_3

other directly. Thus, a connected dominating set of G is a virtual backbone for the wireless network.[1] One of the concerns in constructing the virtual backbone is the routing cost. Specifically, the routing cost of sending a packet from the source s to the destination t is the number of internal nodes (relays) in the routing path from s to t. For example, the routing cost is k if the routing path is $(s, v_1, v_2, \cdots, v_k, t)$. The routing cost should not be too high even if packets are only allowed to be routed through the virtual backbone. Next, we give the formal definition of the problem.

1.2   Problem Definition

Let $G[S]$ be the subgraph of $G = (V, E)$ induced by $S \subseteq V$. Let $m_G(u, v)$ be the number of internal vertices on the shortest path between u and v in G. For example, if u and v are adjacent, then $m_G(u, v) = 0$. If u and v are not adjacent and have a common neighbor, then $m_G(u, v) = 1$. Furthermore, given a vertex subset D of G, $m_G^D(u, v)$ is defined as $m_{G[D \cup \{u, v\}]}(u, v)$, i.e., the number of internal vertices on the shortest path between u and v through D. We use $n(G)$ to denote the number of vertices in graph G. When the graph we are referring to is clear from the context, we simply write n, $m(u, v)$, and $m^D(u, v)$ instead of $n(G)$, $m_G(u, v)$, and $m_G^D(u, v)$, respectively.

Definition 1. *Given a connected graph G and a positive integer α, the **Connected Dominating set problem with Routing cost constraint (CDR-α)** asks for the smallest connected dominating set D of G, such that, for every two vertices u and v, if u and v are not adjacent in G, then $m^D(u, v) \leq \alpha \cdot m(u, v)$.*

1.3   Preliminary

An Equivalent Problem: In the CDR-α problem, we need to consider all the pairs of non-adjacent nodes. Ding *et al.* discovered that to solve the CDR-α problem, it suffices to consider only vertex pairs (u, v) such that $m(u, v) = 1$, i.e., u and v are not adjacent but have a common neighbor [5]. We call the corresponding problem the 1-DR-α problem.

Definition 2. *Given a connected graph $G = (V, E)$ and a positive integer α, the 1-DR-α problem asks for the smallest dominating set D of G, such that, for every two vertices u and v, if $m(u, v) = 1$, then $m^D(u, v) \leq \alpha$.*

We say that u and v form a **target couple**, denoted by $[u, v]$, if $m(u, v) = 1$. We say that a set S **covers** a target couple $[u, v]$ if $m^S(u, v) \leq \alpha$. Hence, the 1-DR-α problem asks for the smallest dominating set that covers all the target couples. Notice that any feasible solution of the 1-DR-α problem must induce a connected subgraph of G. The equivalence between the CDR-α problem and the 1-DR-α problem is stated in the following theorem.

---

[1] A set $D \subseteq V$ is a *dominating set* of $G = (V, E)$ if every vertex in $V \setminus D$ is adjacent to D. Furthermore, if D induces a connected subgraph of G, then D is called a *connected dominating set* of G.

Theorem 1 (Ding et al. [5]). *D is a feasible solution of the CDR-α problem with input graph G if and only if D is a feasible solution of the 1-DR-α problem with input graph G.*

Corollary 1. *Any r-approximation algorithm of the 1-DR-α problem is an r-approximation algorithm of the CDR-α problem.*

In this paper, we thus focus on the 1-DR-α problem.

Feasibility of the 1-DR-α Problem for α ≥ 5: Next, we give the basic idea of finding a feasible solution of the 1-DR-α problem for $\alpha \geq 5$ used in previous researches, e.g., in [11]. One of our algorithms still uses this idea. First, find a dominating set D. Thus, for any target couple $[u, v]$, there exist u^d and v^d in D, such that u^d and v^d dominate u and v, respectively.[2] Let $D' = D$. We then add more vertices to D' so that D' becomes a feasible solution. For any two vertices u' and v' in D, if $m(u', v') \leq 3$, then we add the $m(u', v')$ internal vertices of the shortest path between u' and v' on G to D'. Observe that $m(u^d, v^d) \leq 3$. Hence, $m^{D'}(u, v) \leq 5$ and D' is a feasible solution of the 1-DR-α problem for $\alpha \geq 5$.

Lemma 1. *Let D be a dominating set of G. Let $D' \supseteq D$ be a vertex subset of G such that, for any two vertices u' and v' in D, if $m(u', v') \leq 3$, then $m^{D'}(u', v') \leq 3$. Then, D' is a feasible solution of the 1-DR-α problem with input G and $\alpha \geq 5$.*

1.4   Previous Result

Previous Result on General Graphs: When $\alpha = 1$, the 1-DR-α problem can be transformed into the set cover problem, i.e., cover all the vertices (to form a dominating set) and cover all the target couples. Observe that each target couple can be covered by a single vertex. The resulting approximation ratio is $O(\log n)$ [5]. When α is sufficiently large, e.g., $\alpha \geq n$, any connected dominating set is feasible for the CDR-α problem. Note that, for any α, the size of the minimum connected dominating set is a lower bound of the CDR-α problem. Since the connected dominating set can be approximated within a factor of $O(\log n)$ [12,21], the CDR-n problem can be approximated within a factor of $O(\log n)$. If α falls between these two extremes, e.g., $\alpha = 2$, the only known previous result is the trivial $O(n)$-approximation algorithm. On the hardness side, it has been proved that, unless $NP \subseteq DTIME(n^{O(\log \log n)})$, there is no polynomial-time algorithm that can approximate the CDR-α problem within a factor of $\rho \ln \delta$ ($\forall \rho < 1$) for $\alpha = 1$ [5] and $\alpha \geq 2$ [8,9], where δ is the maximum degree of G.

Open Question 1 (Du and Wan [8]). *Is there a polynomial-time $O(\log n)$-approximation algorithm for the CDR-α problem for $\alpha \geq 2$?*

---

[2] u^d dominates u if $u^d = u$ or u^d and u are adjacent.

Previous Result on Unit Disk Graph (UDG): Most of the studies on the CDR-α problem focused on UDG [5,8,9,11,19]. UDG exhibits many nice properties that enable constant factor approximation algorithms (or PTAS) in many problems where only $O(\log n)$-approximation algorithms (or worse) are known in general graphs, e.g., the minimum (connected) dominating set problem and the maximum independent set problem [3,4,20]. All the previous research on the CDR-α problem on UDG leveraged constant bounds of the maximum independent set or the minimum dominating set. However, all the previous research only solved the case where $\alpha \geq 5$ (by Lemma 1), and the best result so far is a PTAS by Du *et al.* [11]. When $1 < \alpha < 5$, the only known previous result is the trivial $O(n)$-approximation algorithm.

1.5   Our Result and Basic Ideas

In this paper, we first give an approximation algorithm of the 1-DR-α problem on general graphs for any constant $\alpha > 1$. A critical observation is that the 1-DR-2 problem is a special case of the Set Cover with Pairs (SCP) problem [13]. Hassin and Segev proposed an $O(\sqrt{t \log t})$-approximation algorithm for the SCP problem, where t is the number of targets to be covered. However, since there are $O(n^2)$ target couples to be covered, directly applying the $O(\sqrt{t \log t})$-approximation bound yields a trivial upper bound for the 1-DR-2 problem. We re-examine the analysis in [13] and find that, when applying the algorithm to the 1-DR-2 problem, the approximation ratio can also be expressed as $O(\sqrt{n \log n})$. Nevertheless, in this paper, we give a slightly simplified algorithm with an easier analysis for the SCP problem. The algorithm and analysis also make it easy to solve the generalized SCP problem. We obtain the following result, which is the first non-trivial result of the CDR-α problem for $\alpha > 1$ on general graphs and for $1 < \alpha < 5$ on UDG.

Theorem 2. *For any constant $\alpha > 1$, there is an $O(n^{1-\frac{1}{\alpha}}(\log n)^{\frac{1}{\alpha}})$-approximation algorithm for the 1-DR-α problem.*

Apparently, the above performance guarantee deteriorates quickly as α increases. In our second algorithm, we apply the aforementioned idea of finding a feasible solution when $\alpha \geq 5$, i.e., Lemma 1. We have the following result.

Theorem 3. *When $\alpha \geq 5$, there is an $O(\sqrt{n} \log n)$-approximation algorithm for the 1-DR-α problem.*

Finally, we answer Open Question 1 negatively. We improve upon the $\Omega(\log \delta)$ hardness result for the 1-DR-2 problem (albeit under a stronger hardness assumption) [8,9]. In this paper, we give a reduction from the MIN-REP problem [15].

Theorem 4. *Unless $NP \subseteq DTIME(n^{poly \log n})$, for any constant $\epsilon > 0$, the 1-DR-2 problem admits no polynomial-time $2^{\log^{1-\epsilon} n}$-approximation algorithm,* *even if the graph is triangle-free[3] or the constraint that the feasible solution must be a dominating set is ignored[4].*

1.6 Relation with the Basic k-Spanner Problem

When we ignore the constraint that any feasible solution must be a connected dominating set, the CDR-α problem is similar to the basic k-spanner problem. For completeness, we give the formal definition of the basic k-spanner problem. Given a graph $G = (V, E)$, a k-spanner of G is a subgraph H of G such that $d_H(u, v) \leq k d_G(u, v)$ for all u and v in V, where $d_G(u, v)$ is the number of edges in the shortest path between u and v in G. The basic k-spanner problem asks for the k-spanner that has the fewest edges. The CDR-α problem differs with the basic k-spanner problem in the following three aspects: First, in the CDR-α problem, we find a set of vertices D, and all the edges in the subgraph induced by D can be used for routing; while in the basic k-spanner problem, only edges in H can be used. Second, in the CDR-α problem, the objective is to minimize the number of chosen vertices; while in the basic k-spanner problem, the objective is to minimize the number of chosen edges. Finally, in the basic k-spanner problem, the distance is measured by the number of edges; while in the CDR-α problem, the distance is measured by the number of internal nodes. Despite the above differences, these two problems share similar approximability and hardness results. Althöfer *et al.* proved that every graph has a k-spanner of at most $n^{1+\frac{1}{\lfloor(k+1)/2\rfloor}}$ edges, and such a k-spanner can be constructed in polynomial time [1,7]. Since the number of edges in any k-spanner is at least $n - 1$, this yields an $O(n^{\frac{1}{\lfloor(k+1)/2\rfloor}})$-approximation algorithm for the basic k-spanner problem. For $k = 2$, there is an $O(\log n)$-approximation algorithm due to Kortsarz and Peleg [16], and this is the best possible [15]. For $k = 3$, Berman *et al.* proposed an $\tilde{O}(n^{1/3})$-approximation algorithm [2]. For $k = 4$, Dinitz and Zhang proposed an $\tilde{O}(n^{1/3})$-approximation algorithm [7]. On the hardness side, it has been proved that for any constant $\epsilon > 0$ and for $3 \leq k \leq \log^{1-2\epsilon} n$, unless $NP \subseteq BPTIME(2^{poly \log n})$, there is no polynomial-time algorithm that approximates the basic k-spanner problem to a factor better than $2^{(\log^{1-\epsilon} n)/k}$ [6].

2 Two Algorithms for the 1-DR-α Problem

2.1 The First Algorithm

We first give the formal definition of the Set Cover with Pairs (SCP) problem.

---

[3] If the graph is triangle-free, then any two vertices with a common neighbor form a target couple.

[4] One may drop the constraint that the solution must be a dominating set, and focuses on minimizing the number of vertices to cover all the target couples. This theorem also applies to such a problem.

Definition 3. *Let T be a set of t targets. Let V be a set of n elements. For every pair of elements $P = \{v_1, v_2\} \subseteq V$, $C(P)$ denotes the set of targets covered by P. The Set Cover with Pairs (SCP) problem asks for the smallest subset S of V such that $\bigcup_{\{v_1,v_2\}\subseteq S} C(\{v_1, v_2\}) = T$.*

Let OPT be the number of elements in the optimal solution. We only need to consider the case where $t > 1$ and $OPT > 1$.

Approximating the SCP Problem: Our algorithm is a simple greedy algorithm: in each round, we choose at most two elements u and v that maximize the number of covered targets. Specifically, S is an empty set initially. In each round, we select a set $P \subseteq V \setminus S$ such that $|P| \leq 2$ and P increases the number of covered targets the most, i.e., $P = \underset{P':|P'|\leq 2, P'\subseteq V\setminus S}{\operatorname{argmax}} g(P')$, where

$$g(P') = \left| \bigcup_{\{v_1,v_2\}\subseteq S\cup P'} C(\{v_1, v_2\}) \right| - \left| \bigcup_{\{v_1,v_2\}\subseteq S} C(\{v_1, v_2\}) \right|.$$

We then add P to S and repeat the above process until all the targets are covered. The algorithm terminates once all targets are covered.[5]

Theorem 5. *The above algorithm is an $O(\sqrt{n \log t})$-approximation algorithm for the SCP problem.*

Proof. Let R_i be the number of uncovered targets after round i. In the first round, some pair of elements in the optimal solution can cover at least $t/\binom{OPT}{2}$ targets. Since we choose a pair of elements greedily in each round, $R_1 \leq t(1 - 1/\binom{OPT}{2})$. In the second round, there exists a pair of elements in the optimal solution that can cover at least $R_1/\binom{OPT}{2}$ targets among the R_1 uncovered targets. Again, we choose the pair of elements that increases the number of covered targets the most. Hence, $R_2 \leq R_1 - R_1/\binom{OPT}{2} \leq t(1 - 1/\binom{OPT}{2})^2$. In general, $R_i \leq t(1 - 1/\binom{OPT}{2})^i$. After $r = \binom{OPT}{2} \ln t$ rounds, the number of uncovered targets is at most $t(1 - 1/\binom{OPT}{2})^r \leq t(e^{-1/\binom{OPT}{2}})^r \leq te^{-\ln t} = 1$. Hence, after $O(OPT^2 \ln t)$ rounds, all targets are covered. Let ALG be the number of elements chosen by the algorithm. Since we choose at most two elements in each round, $ALG = O(OPT^2 \ln t)$. Finally, since $ALG \leq n$, $ALG = O(\sqrt{n \cdot OPT^2 \ln t}) = O(\sqrt{n \ln t})OPT$. □

Note that, in Theorem 5, we can replace n with any upper bound of the size of solutions obtained by any polynomial-time algorithm $\mathcal{A}$ for the SCP problem. This is achieved by executing both $\mathcal{A}$ and our algorithm. Choosing the best between the two outputs yields the desired approximation ratio. For example, if we replace n with $2t$, we then get the result in [13].

---

[5] In [13], in each round, a set $P = \underset{P':|P'|\leq 2, P'\subseteq V\setminus S}{\operatorname{argmax}} g'(P')$ is added to S, where $g'(P') = \frac{g(P')}{|P'|}$.

Approximating the 1-DR-2 Problem: To transform the 1-DR-2 problem to the SCP problem, we treat each target couple as a target. Moreover, we treat each vertex as a target so that the output is a dominating set. The set of elements V in the SCP problem is the vertex set of G. $C(P)$ consists of all the vertices that are dominated by P in G and all the target couples that are covered by P in G. In this SCP instance, $n = n(G)$ and $t = O(n(G)^2)$. It is easy to verify the following result.

Theorem 6. *There is an $O(\sqrt{n \log n})$-approximation algorithm for the 1-DR-2 problem.*

The Set Cover with α-Tuples (SCT-α) Problem: In the 1-DR-2 problem, every target couple can be covered by no more than two vertices. In the 1-DR-α problem, every target couple can be covered by no more than α vertices. Hence, we consider the following generalization of the SCP problem.

Definition 4. *Let T be a set of t targets. Let V be a set of n elements. Let α be a positive integer constant greater than one. For every α-tuple $P = \{v_1, v_2, \cdots, v_\alpha\} \subseteq V$, $C(P)$ denotes the set of targets covered by P. The Set Cover with α-Tuples (SCT-α) problem asks for the smallest subset S of V such that* $\bigcup\limits_{\{v_1, v_2, \cdots, v_\alpha\} \subseteq S} C(\{v_1, v_2, \cdots, v_\alpha\}) = T$.

We only need to consider the case where $t > 1$ and $OPT \geq \alpha$ (α is a constant).

Approximating the SCT-α Problem and the 1-DR-α Problem: The algorithm for the SCT-α problem is a straightforward generalization of the algorithm for the SCP problem. The difference is that, in each round, we choose a set P of at most α elements that increases the number of covered targets the most. The transformation from the 1-DR-α problem into the SCT-α problem is also similar to the previous transformation. The value of α in the constructed SCT-α instance is equal to that in the 1-DR-α instance. Again, $n = n(G)$ and $t = O(n(G)^2)$ in the constructed SCT-α instance. Theorem 2 is a direct result of the following theorem.

Theorem 7. *There is an $O(n^{1-\frac{1}{\alpha}} \cdot (\ln t)^{\frac{1}{\alpha}})$-approximation algorithm for the SCT-α problem.*

Claim 1. *When $c = \frac{1}{\alpha} - \frac{\ln \ln(t^\alpha)}{\alpha \ln n}$, $n^{1-c} = \sqrt{n \cdot \alpha(n^c)^{\alpha-2} \ln t} = n^{1-\frac{1}{\alpha}} \cdot (\alpha \ln t)^{\frac{1}{\alpha}}$.*

Proof of Claim 1:

$$n^{1-c} = \sqrt{n \cdot \alpha(n^c)^{\alpha-2} \ln t}$$
$$\Leftrightarrow n^{2-2c} = n \cdot \alpha(n^c)^{\alpha-2} \ln t$$
$$\Leftrightarrow n^{2-2c-(1+c(\alpha-2))} = \alpha \ln t$$
$$\Leftrightarrow n^{1-c\alpha} = \alpha \ln t.$$

When $c = \frac{1}{\alpha} - \frac{\ln\ln(t^\alpha)}{\alpha\ln n}$,

$$n^{1-c\alpha} = n^{1-(1-\frac{\ln\ln(t^\alpha)}{\ln n})}$$

$$= n^{\frac{\ln\ln(t^\alpha)}{\ln n}} \tag{1}$$

$$= (n^{\ln(\ln(t^\alpha))})^{\frac{1}{\ln n}} \tag{2}$$

$$= ((\ln(t^\alpha))^{\ln n})^{\frac{1}{\ln n}} \tag{3}$$

$$= ((\alpha\ln t)^{\ln n})^{\frac{1}{\ln n}} \tag{4}$$

$$= \alpha\ln t. \tag{5}$$

Hence, when $c = \frac{1}{\alpha} - \frac{\ln\ln(t^\alpha)}{\alpha\ln n}$, $n^{1-c} = \sqrt{n \cdot \alpha(n^c)^{\alpha-2}\ln t}$.
Finally, when $c = \frac{1}{\alpha} - \frac{\ln\ln(t^\alpha)}{\alpha\ln n}$,

$$n^{1-c} = n^{1-\frac{1}{\alpha}+\frac{\ln\ln(t^\alpha)}{\alpha\ln n}}$$

$$= n^{1-\frac{1}{\alpha}} \cdot n^{\frac{\ln\ln(t^\alpha)}{\alpha\ln n}}$$

$$= n^{1-\frac{1}{\alpha}} \cdot (n^{\frac{\ln\ln(t^\alpha)}{\ln n}})^{\frac{1}{\alpha}}$$

$$= n^{1-\frac{1}{\alpha}} \cdot (\alpha\ln t)^{\frac{1}{\alpha}}.$$

In the last equality, we reuse Eqs.(1)–(5).   □

Proof of Theorem 7: Let R_i be the number of uncovered targets after round i. By a similar argument in the proof of Theorem 5, we get that $R_i \le t(1-1/(\frac{OPT}{\alpha}))^i$. After $r = (\frac{OPT}{\alpha})\ln t$ rounds, the number of uncovered targets is at most one. Hence, after $O(OPT^\alpha\ln t)$ rounds, all targets are covered. Let ALG be the number of elements chosen by the algorithm. Since we choose at most α elements in each round, $ALG = O(\alpha OPT^\alpha\ln t)$. Since $ALG \le n$, $ALG = O(\sqrt{n \cdot \alpha OPT^\alpha\ln t})$.

Let $c = \frac{1}{\alpha} - \frac{\ln\ln(t^\alpha)}{\alpha\ln n}$. When $OPT \ge n^c$, the approximation ratio is n^{1-c}. When $OPT \le n^c$, $ALG = O(\sqrt{n \cdot \alpha OPT^{\alpha-2}\ln t})OPT = O(\sqrt{n \cdot \alpha(n^c)^{\alpha-2}\ln t})OPT$. The proof then follows from Claim 1 and $\alpha^{\frac{1}{\alpha}} = O(1)$.   □

2.2 The Second Algorithm

The second algorithm is designed for the 1-DR-α problem when $\alpha \ge 5$. It has a better approximation ratio than that of the previous algorithm when $\alpha \ge 5$. The algorithm is suggested in Lemma 1: We first find a dominating set D by any $O(\log n)$-approximation algorithm. Let $D' = D$. For any two vertices u and v in D, if $m(u,v) \le 3$, we then add at most three vertices to D' so that $m^{D'}(u,v) \le 3$.

Proof of Theorem 3: By Lemma 1, D' is a feasible solution for the 1-DR-α problem when $\alpha \ge 5$. Let OPT_{DS} be the size of the minimum dominating set in G. Let OPT be the size of the optimum of the 1-DR-α problem. Since any feasible solution of the 1-DR-α problem must be a dominating set, $OPT_{DS} \le OPT$. Moreover, $|D'| \le |D| + 3\binom{|D|}{2} = O((\log n \cdot OPT_{DS})^2) = O((\log n \cdot OPT)^2)$. Since $|D'| \le n$, we have $|D'| = O(\sqrt{n \cdot (\log n \cdot OPT)^2}) = O(\sqrt{n}\log n)OPT$.   □

3   Inapproximability Result

3.1   The MIN-REP Problem

We prove Theorem 4 by a reduction from the MIN-REP problem [15]. The input of the MIN-REP problem consists of a bipartite graph $G = (X, Y, E)$, a partition of X, $\mathcal{P}_X = \{X_1, X_2, \cdots, X_{k_X}\}$, and a partition of Y, $\mathcal{P}_Y = \{Y_1, Y_2, \cdots, Y_{k_Y}\}$, such that $\bigcup_{i=1}^{k_X} X_i = X$ and $\bigcup_{i=1}^{k_Y} Y_i = Y$. Every $X_i \in \mathcal{P}_X$ (respectively, $Y_i \in \mathcal{P}_Y$) has size $|X|/k_X$ (respectively, $|Y|/k_Y$). $X_1, X_2, \cdots, X_{k_X}$ and $Y_1, Y_2, \cdots, Y_{k_Y}$ are called *super nodes*, and two super nodes X_i and Y_j are *adjacent* if some vertex in X_i and some vertex in Y_j are adjacent in G. If X_i and Y_j are adjacent, then X_i and Y_j form a *super edge*. In the MIN-REP problem, our task is to choose representatives for super nodes so that if X_i and Y_j form a super edge, then some representative for X_i and some representative for Y_j are adjacent in G. Note that a super node may have multiple representatives. Specifically, the goal of the MIN-REP problem is to find the smallest subset $R \subseteq X \cup Y$ such that if X_i and Y_j form a super edge, then R must contain two vertices x and y such that $x \in X_i$, $y \in Y_j$ and $(x, y) \in E$. In this case, we say that $\{x, y\}$ *covers* the super edge (X_i, Y_j). The inapproximability result of the MIN-REP problem is stated as the following theorem.

Theorem 8 (Kortsarz et al. [17]). *For any constant $\epsilon > 0$, unless $NP \subseteq DTIME(n^{poly \log n})$, there is no polynomial-time algorithm that can distinguish between instances of the MIN-REP problem with a solution of size $k_X + k_Y$ and instances where every solution is of size at least $(k_X + k_Y) \cdot 2^{\log^{1-\epsilon} n(G)}$, where $n(G)$ is the number of vertices in the input graph of the MIN-REP problem.*

3.2   The Reduction

Given inputs $G = (X, Y, E)$, $\mathcal{P}_X$, and $\mathcal{P}_Y$ of the MIN-REP problem, we construct a corresponding graph $G'(G, \mathcal{P}_X, \mathcal{P}_Y)$ of the 1-DR-2 problem. When G, $\mathcal{P}_X$, and $\mathcal{P}_Y$ are clear from the context, we simply write G' instead of $G'(G, \mathcal{P}_X, \mathcal{P}_Y)$. Initially, $G' = G$. Hence, G' contains X, Y, and E. For each super node X_i (respectively, Y_i), we create two corresponding vertices px_i^1 and px_i^2 (respectively, py_i^1 and py_i^2) in G'. If x is in super node X_i (respectively, y is in super node Y_i), then we add two edges (x, px_i^1) and (x, px_i^2) (respectively, (y, py_i^1) and (y, py_i^2)) in G'. If X_i and Y_j form a super edge, then we add two vertices $r_{i,j}^1$ and $r_{i,j}^2$ to G', and we add four edges $(px_i^1, r_{i,j}^1)$, $(r_{i,j}^1, py_j^1)$, $(px_i^2, r_{i,j}^2)$, $(r_{i,j}^2, py_j^2)$ to G'. $r_{i,j}^1$ (respectively, $r_{i,j}^2$) is called the *relay* of px_i^1 and py_j^1 (respectively, px_i^2 and py_j^2).

Before we complete the construction of G', we briefly explain the idea behind the construction so far. If two super nodes X_i and Y_j form a super edge, then px_i^I and py_j^I ($I \in \{1, 2\}$) have a common neighbor in G', i.e., the relay $r_{i,j}^I$. Because px_i^I and py_j^I are not adjacent, px_i^I and py_j^I form a target couple. To transform a solution D of the 1-DR-2 problem to a solution of the MIN-REP problem, we need to transform D to another feasible solution D' for the 1-DR-2 problem so that none of the relays is chosen, and only vertices in $X \cup Y$ are used to connect

px_i^I and py_j^I. This is the reason that we have two corresponding vertices for each super node (and thus two relays for each super edge). Under this setting, to connect px_i^1 to py_j^1 and px_i^2 to py_j^2, choosing two vertices in $X \cup Y$ is no worse than choosing the relays.

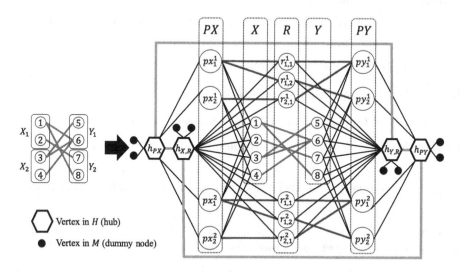

Fig. 1. An example of the reduction.

Let $PX = \{px_1^1, px_2^1, \cdots, px_{k_X}^1\} \cup \{px_1^2, px_2^2, \cdots, px_{k_X}^2\}$ be the set of vertices in G' corresponding to the super nodes in $\mathcal{P}_X$. Similarly, let $PY = \{py_1^1, py_2^1, \cdots, py_{k_Y}^1\} \cup \{py_1^2, py_2^2, \cdots, py_{k_Y}^2\}$. Let R be the set of all relays. To complete the construction, we add four vertices (hubs) $h_{X,R}$, $h_{Y,R}$, h_{PX}, and h_{PY} to G'. In G', all the vertices in X, Y, PX, and PY are adjacent to $h_{X,R}$, $h_{Y,R}$, h_{PX}, and h_{PY}, respectively. Moreover, every relay is adjacent to $h_{X,R}$ and $h_{Y,R}$. These four hubs induce a 4-cycle $(h_{PX}, h_{Y,R}, h_{PY}, h_{X,R}, h_{PX})$ in G'. Finally, for each hub h, we create two dummy nodes d_1 and d_2, and add two edges (h, d_1) and (h, d_2) to G'. This completes the construction of G'. Figure 1 shows an example of the reduction. Let H and M be the set of hubs and the set of dummy nodes, respectively. Hence, the vertex set of G' is $X \cup Y \cup PX \cup PY \cup R \cup H \cup M$. Let $N(u)$ be the set of neighbors of u in G'. We then have

$N(px) \subseteq X \cup R \cup \{h_{PX}\}$ if $px \in PX$. $N(py) \subseteq Y \cup R \cup \{h_{PY}\}$ if $py \in PY$.

$N(x) \subseteq PX \cup Y \cup \{h_{X,R}\}$ if $x \in X$. $N(y) \subseteq PY \cup X \cup \{h_{Y,R}\}$ if $y \in Y$.

$N(h_{X,R}) \setminus M = X \cup R \cup \{h_{PX}, h_{PY}\}$. $N(h_{Y,R}) \setminus M = Y \cup R \cup \{h_{PX}, h_{PY}\}$.

$N(h_{PX}) \setminus M = PX \cup \{h_{X,R}, h_{Y,R}\}$. $N(h_{PY}) \setminus M = PY \cup \{h_{X,R}, h_{Y,R}\}$.

$N(m) \subseteq H$ if $m \in M$. $N(r) \subseteq PX \cup PY \cup \{h_{X,R}, h_{Y,R}\}$ if $r \in R$.

Observe that $|R| = O(n(G)^2)$. We have the following lemma.

Lemma 2. $n(G') = O(n(G)^2)$.

It is easy to check that, for any two adjacent vertices u and v in G', u and v have no common neighbor. Hence, we have the following lemma.

Lemma 3. G' is triangle-free.

We say that a target couple $[a, b]$ is in $[A, B]$ if $a \in A$ and $b \in B$. It is easy to verify the following two lemmas.

Lemma 4. Only H can cover the target couples in $[M, M]$.

Lemma 5. H is a dominating set of G'.

The proof of the following lemma can be found in the appendix.

Lemma 6. H covers all the target couples except those in $[PX, PY]$.

Let px and py be vertices in PX and PY, respectively. Observe that, if (px, x, y, py) is a path in G', then $x \in X$ and $y \in Y$. We then have the following lemma.

Lemma 7. D covers target couples $[px_i^1, py_j^1]$ and $[px_i^2, py_j^2]$ if and only if at least one of the following conditions is satisfied.

1. There exist $x \in X$ and $y \in Y$ such that (px_i^1, x, y, py_j^1) and (px_i^2, x, y, py_j^2) are paths in G' and $\{x, y\} \subseteq D$.
2. $\{r_{i,j}^1, r_{i,j}^2\} \subseteq D$.

3.3   The Analysis

Let I_{MR} be an instance of the MIN-REP problem with inputs G, $\mathcal{P}_X$, and $\mathcal{P}_Y$. Let I_D be the instance of the 1-DR-2 problem with input $G'(G, \mathcal{P}_X, \mathcal{P}_Y)$. To prove the inapproximability result, we use the following two lemmas.

Lemma 8. If I_{MR} has a solution of size s, then I_D has a solution of size $s + 4$.

Lemma 9. If every solution of I_{MR} has size at least $s \cdot 2^{\log^{1-\epsilon} n(G)}$, then every solution of I_D has size at least $s \cdot 2^{\log^{1-\epsilon} n(G)} + 4$.

Proof of Theorem 4: By Theorem 8, for any constant $\epsilon > 0$, unless $NP \subseteq DTIME(n^{poly \log n})$, there is no polynomial-time algorithm that can distinguish between instances of the MIN-REP problem with a solution of size $k_X + k_Y$ and instances where every solution is of size at least $(k_X + k_Y) \cdot 2^{\log^{1-\epsilon} n(G)}$. By the above two lemmas, it is hard to distinguish between instances of the 1-DR-2 problem with a solution of size $k_X + k_Y + 4$ and instances in which every solution is of size at least $(k_X + k_Y) \cdot 2^{\log^{1-\epsilon} n(G)} + 4$. Therefore, for any constant $\epsilon > 0$, unless $NP \subseteq DTIME(n^{poly \log n})$, there is no polynomial-time algorithm that can approximate the 1-DR-2 problem by a factor better than $\frac{(k_X + k_Y) \cdot 2^{\log^{1-\epsilon} n(G)} + 4}{k_X + k_Y + 4}$. Lemma 2 implies that, for any constant $\epsilon' > 0$, unless

$NP \subseteq DTIME(n^{poly \log n})$, there is no $O(2^{\log^{1-\epsilon'} n(G')^{0.5}})$-approximation algorithm for the 1-DR-2 problem. By considering sufficiently large instances and a small enough ϵ', we have the hardness result claimed in Theorem 4. On the other hand, let 1-DR-2′ be the problem obtained by removing the constraint that any feasible solution must be a dominating set from the 1-DR-2 problem. Thus, in the 1-DR-2′ problem, we only focus on covering target couples. By Lemmas 4 and 5, a solution D is feasible for the 1-DR-2′ problem with input G' if and only if D is a feasible solution of I_D. Thus, the inapproximability result also applies to the 1-DR-2′ problem. Finally, the proof follows from Lemma 3.   □

Lemma 8 is a direct result of the following claim.

Claim 2. *If S is a feasible solution of I_{MR}, then $S \cup H$ is a feasible solution of I_D.*

Proof. Since H is a dominating set, by Lemma 6, it suffices to prove that every target couple $[u, v] = [px_i^{I_1}, py_j^{I_2}]$ in $[PX, PY]$ is covered by S. Note that $[px_i^{I_1}, py_j^{I_2}]$ cannot be a target couple if $I_1 \neq I_2$. This is because $px_i^{I_1}$ and $py_j^{I_2}$ do not have a common neighbor if $I_1 \neq I_2$. If $I_1 = I_2$, then the common neighbor must be $r_{i,j}^I$. By the construction of G', this implies that X_i and Y_j form a super edge. Since S is a feasible solution of I_{MR}, there exists $x \in X_i$ and $y \in Y_j$ such that x and y are adjacent in G and $\{x, y\} \subseteq S$. Again, by the construction of G', (u, x, y, v) is a path in G'. Hence, $S \supseteq \{x, y\}$ covers $[u, v]$.   □

To prove Lemma 9, we use the following claim.

Claim 3. *I_D has an optimal solution D^*, such that $D^* \setminus H$ is a feasible solution of I_{MR}.*

Proof of Claim 3: Let D_{OPT} be any optimal solution of I_D. By Lemmas 4, 6, and 7, $D_{OPT} \subseteq H \cup X \cup Y \cup R$. If $D_{OPT} \cap R = \emptyset$, by Lemma 7, each target couple $[px_i^I, py_j^I]$ is covered by some $x \in X$ and some $y \in Y$. By the construction of G', such x and y also cover the super edge (X_i, Y_j) in I_{MR}. Because each super edge in I_{MR} has a corresponding target couple in I_D, $D_{OPT} \setminus H$ is a feasible solution of I_{MR}.

If $D_{OPT} \cap R \neq \emptyset$, then some $r_{i,j}^I \in D_{OPT}$. We can further assume that both $r_{i,j}^1$ and $r_{i,j}^2$ are in D_{OPT}; otherwise, by Lemma 7, we can remove $r_{i,j}^I$ from D_{OPT}, the resulting solution is smaller and is still feasible. We then replace $r_{i,j}^1$ and $r_{i,j}^2$ with some $x \in X$ and some $y \in Y$ satisfying the first condition in Lemma 7. By Lemma 7, the resulting solution is still feasible, and the size remains the same. Repeat the above replacing process until the resulting solution does not contain any relay. The proof then follows from the argument of the case where $D_{OPT} \cap R = \emptyset$.   □

Proof of Lemma 9: Let S^* be the optimal solution of I_{MR}. By the assumption, we have $|S^*| \geq s \cdot 2^{\log^{1-\epsilon} n(G)}$. It suffices to prove that $S^* \cup H$ is an optimal solution for I_D, which implies that every feasible solution of I_D has size at least $|S^* \cup H| = |S^*| + 4 \geq s \cdot 2^{\log^{1-\epsilon} n(G)} + 4$. The feasibility of $S^* \cup H$ follows from Claim 2. For the sake of contradiction, assume that the optimal solution of I_D has size smaller than $|S^* \cup H| = |S^*| + 4$. Claim 3 and Lemma 4 then imply that S^* is not an optimal solution of I_{MR}, which is a contradiction.   □

A   Proof of Lemma 6

If $[u, v]$ is in $[PX, PX \cup \{h_{X,R}, h_{Y,R}\}]$, $[PY, PY \cup \{h_{X,R}, h_{Y,R}\}]$, $[X, X \cup R \cup \{h_{PX}, h_{PY}\}]$, $[Y, Y \cup R \cup \{h_{PX}, h_{PY}\}]$, or $[R, R \cup \{h_{PX}, h_{PY}\}]$, then $[u, v]$ can be covered by one vertex in H. If $[u, v]$ is in $[PX, Y], [PY, X], [X, \{h_{Y,R}\}]$, or $[Y, \{h_{X,R}\}]$, then $[u, v]$ can be covered by an edge in H. If $[u, v]$ is in $[PX, X \cup R \cup \{h_{PX}, h_{PY}\}]$, $[PY, Y \cup R \cup \{h_{PX}, h_{PY}\}]$, or $[X, Y]$, then $[u, v]$ cannot be a target couple (since u and v do not have a common neighbor). If $[u, v]$ is in $[X, \{h_{X,R}\}], [Y, \{h_{Y,R}\}]$, or $[R, \{h_{X,R}, h_{Y,R}\}]$, then $[u, v]$ cannot be a target couple (since u and v are adjacent[6]). Moreover, it is easy to see that H covers all the target couples in $[H, H]$ or $[V(G'), M]$, where $V(G')$ is the vertex set of G'. Finally, observe that if $[u, v]$ is in $[PX, PY]$, then H cannot cover $[u, v]$.   □

B   Reduction from the 1-DR-α Problem to Other Related Problems

In this section, we show that the 1-DR-α problem can be transformed into the submodular cost set cover problem and the minimum rainbow subgraph problem on multigraphs. We also summarize the approximability results of these two problems in the literature. Future progress in the approximability results of these two problems may lead to better approximation algorithms for the 1-DR-α problem.

Submodular Cost Set Cover Problem: The 1-DR-α problem can be considered as a special case of the submodular cost set cover problem [10,14,23]. In the set cover problem, we are given a set of targets $\mathcal{T}$ and a set of objects $\mathcal{S}$. Each object in $\mathcal{S}$ can cover a subset of $\mathcal{T}$ (specified in the input). The goal is to choose the smallest subset of $\mathcal{S}$ that covers $\mathcal{T}$. In the submodular cost set cover problem, there is a non-negative submodular function c that maps each subset of $\mathcal{S}$ to a cost, and the goal is to find the set cover with the minimum cost. To transform the 1-DR-α problem with input $G = (V, E)$ to the submodular cost set cover problem, let $\mathcal{T}$ be the union of V and the set of all target couples, and let $\mathcal{S}$ be the set of all subsets of V with size at most α. Hence, each object in $\mathcal{S}$ is a subset of V. An object $S \in \mathcal{S}$ can cover a vertex v if v is adjacent to some vertex in S or $v \in S$. An object $S \in \mathcal{S}$ can cover a target couple $[u, v]$ if $m^S(u, v) \leq \alpha$. The cost of a subset $\mathcal{S}'$ of $\mathcal{S}$ is simply the size of the union of objects in $\mathcal{S}'$, i.e., the number of distinct vertices specified in $\mathcal{S}'$.

Iwata and Nagano proposed a $|\mathcal{T}|$-approximation algorithm and an f-approximation algorithm, where f is the maximum frequency, $\max_{T \in \mathcal{T}} |\{S \in \mathcal{S} | S \text{ covers } T\}|$ [14]. Koufogiannakis and Young also proposed an f-approximation algorithm when the cost function c is non-decreasing [18]. It is easy to see that these algorithms give trivial bounds for the 1-DR-α problem. When the cost function c is integer-valued, non-decreasing, and satisfies $c(\emptyset) = 0$, Wan et al. proposed a $\rho H(\gamma)$-approximation algorithm, where $\rho = \min_{S^* : S^* \text{ is an optimal solution}} \frac{\sum_{S \in S^*} c(\{S\})}{c(S^*)}$, γ is the largest number of targets that can

---

[6] In addition, by Lemma 3, u and v do not have a common neighbor.

be covered by an object in $\mathcal{S}$, and $H(k)$ is the k-th Harmonic number [23]. Du *et al.* applied this algorithm to the 1-DR-α problem on UDG for $\alpha \geq 5$ and obtained a constant factor approximation algorithm [10]. It is unclear whether or not ρ can be upper bounded by $O(n^{1-\epsilon})$ for some $\epsilon > 0$ when applied to the 1-DR-α problem on general graphs.

Minimum Rainbow Subgraph Problem on Multigraphs: Given a set of p colors and a multigraph H, where each edge is colored with one of the p colors, the Minimum Rainbow Subgraph (MRS) problem asks for the smallest vertex subset D of H, such that each of the p colors appears in some edge induced by D. The 1-DR-2 problem can be transformed into the MRS problem as follows. Let $G = (V, E)$ be the input graph of the 1-DR-2 problem. Let T be the union of V and the set of all target couples. The set of colors for the MRS problem is $\{c_i | i \in T\}$. The input multigraph H of the MRS problem has the same vertex set as G. To form a dominating set, for each $v \in V$, v is incident to $d(v) + 1$ loops (v, v) in H, where $d(v)$ is the degree of v in G. Each of these loops receives a different color in $\{c_v\} \cup \{c_u | (u, v) \in E\}$. For each target couple $[u, v]$ in G, if w is a common neighbor of u and v in G, we add a loop (w, w) with color $c_{[u,v]}$ to H. Finally, for each target couple $[u, v]$ in G, if (u, w_1, w_2, v) is a path in G, we add an edge (w_1, w_2) with color $c_{[u,v]}$ to H. The MRS problem can be transformed into the SCP problem. When the input graph is simple, Tirodkar and Vishwanathan proposed an $O(n^{1/3} \log n)$-approximation algorithm [22].

References

1. Althöfer, I., Das, G., Dobkin, D., Joseph, D., Soares, J.: On sparse spanners of weighted graphs. Discrete Comput. Geom. **9**(1), 81–100 (1993). https://doi.org/10.1007/BF02189308
2. Berman, P., Bhattacharyya, A., Makarychev, K., Raskhodnikova, S., Yaroslavtsev, G.: Approximation algorithms for spanner problems and directed Steiner forest. Inf. Comput. **222**(Supplement C), 93–107 (2013). https://doi.org/10.1016/j.ic.2012.10.007
3. Cheng, X., Huang, X., Li, D., Wu, W., Du, D.Z.: A polynomial-time approximation scheme for the minimum-connected dominating set in ad hoc wireless networks. Networks **42**(4), 202–208 (2003). https://doi.org/10.1002/net.10097
4. Das, G.K., De, M., Kolay, S., Nandy, S.C., Sur-Kolay, S.: Approximation algorithms for maximum independent set of a unit disk graph. Inf. Process. Lett. **115**(3), 439–446 (2015). https://doi.org/10.1016/j.ipl.2014.11.002
5. Ding, L., Wu, W., Willson, J., Du, H., Lee, W., Du, D.Z.: Efficient algorithms for topology control problem with routing cost constraints in wireless networks. IEEE Trans. Parallel Distrib. Syst. **22**(10), 1601–1609 (2011). https://doi.org/10.1109/TPDS.2011.30
6. Dinitz, M., Kortsarz, G., Raz, R.: Label cover instances with large girth and the hardness of approximating basic k-spanner. ACM Trans. Algorithms **12**(2), 25:1–25:16 (2015). https://doi.org/10.1145/2818375
7. Dinitz, M., Zhang, Z.: Approximating low-stretch spanners. In: Proceedings of the Twenty-seventh Annual ACM-SIAM Symposium on Discrete Algorithms, SODA 2016, pp. 821–840. Society for Industrial and Applied Mathematics, Philadelphia (2016)

8. Du, D.Z., Wan, P.J.: Routing-cost constrained CDS. In: Du, D.Z., Wan, P.J. (eds.) Connected Dominating Set: Theory and Applications, vol. 77, pp. 119–131. Springer, New York (2013). https://doi.org/10.1007/978-1-4614-5242-3_7

9. Du, H., Wu, W., Ye, Q., Li, D., Lee, W., Xu, X.: CDS-based virtual backbone construction with guaranteed routing cost in wireless sensor networks. IEEE Trans. Parallel Distrib. Syst. **24**(4), 652–661 (2013). https://doi.org/10.1109/TPDS.2012.177

10. Du, H., Wu, W., Lee, W., Liu, Q., Zhang, Z., Du, D.Z.: On minimum submodular cover with submodular cost. J. Global Optim. **50**(2), 229–234 (2011). https://doi.org/10.1007/s10898-010-9563-3

11. Du, H., Ye, Q., Zhong, J., Wang, Y., Lee, W., Park, H.: Polynomial-time approximation scheme for minimum connected dominating set under routing cost constraint in wireless sensor networks. Theor. Comput. Sci. **447**(Supplement C), 38–43 (2012). https://doi.org/10.1016/j.tcs.2011.10.010. Combinatorial Algorithms and Applications (COCOA 2010)

12. Guha, S., Khuller, S.: Approximation algorithms for connected dominating sets. Algorithmica **20**(4), 374–387 (1998). https://doi.org/10.1007/PL00009201

13. Hassin, R., Segev, D.: The set cover with pairs problem. In: Sarukkai, S., Sen, S. (eds.) FSTTCS 2005. LNCS, vol. 3821, pp. 164–176. Springer, Heidelberg (2005). https://doi.org/10.1007/11590156_13

14. Iwata, S., Nagano, K.: Submodular function minimization under covering constraints. In: IEEE FOCS, pp. 671–680, October 2009. https://doi.org/10.1109/FOCS.2009.31

15. Kortsarz, G.: On the hardness of approximating spanners. Algorithmica **30**(3), 432–450 (2001). https://doi.org/10.1007/s00453-001-0021-y

16. Kortsarz, G., Peleg, D.: Generating sparse 2-spanners. J. Algorithms **17**(2), 222–236 (1994). https://doi.org/10.1006/jagm.1994.1032

17. Kortsarz, G., Krauthgamer, R., Lee, J.R.: Hardness of approximation for vertex-connectivity network design problems. SIAM J. Comput. **33**(3), 704–720 (2004). https://doi.org/10.1137/S0097539702416736

18. Koufogiannakis, C., Young, N.E.: Greedy δ-approximation algorithm for covering with arbitrary constraints and submodular cost. Algorithmica **66**(1), 113–152 (2013). https://doi.org/10.1007/s00453-012-9629-3

19. Liu, C., Huang, H., Du, H., Jia, X.: Performance-guaranteed strongly connected dominating sets in heterogeneous wireless sensor networks. In: IEEE INFOCOM 2016 - The 35th Annual IEEE International Conference on Computer Communications, pp. 1–9, April 2016. https://doi.org/10.1109/INFOCOM.2016.7524455

20. Nieberg, T., Hurink, J.: A PTAS for the minimum dominating set problem in unit disk graphs. In: Erlebach, T., Persinao, G. (eds.) WAOA 2005. LNCS, vol. 3879, pp. 296–306. Springer, Heidelberg (2006). https://doi.org/10.1007/11671411_23

21. Ruan, L., Du, H., Jia, X., Wu, W., Li, Y., Ko, K.I.: A greedy approximation for minimum connected dominating sets. Theor. Comput. Sci. **329**(1), 325–330 (2004). https://doi.org/10.1016/j.tcs.2004.08.013

22. Tirodkar, S., Vishwanathan, S.: On the approximability of the minimum rainbow subgraph problem and other related problems. Algorithmica **79**(3), 909–924 (2017). https://doi.org/10.1007/s00453-017-0278-4

23. Wan, P.J., Du, D.Z., Pardalos, P., Wu, W.: Greedy approximations for minimum submodular cover with submodular cost. Comput. Optim. Appl. **45**(2), 463–474 (2010). https://doi.org/10.1007/s10589-009-9269-y

On the Maximum Connectivity
Improvement Problem

Federico Corò[1(✉)], Gianlorenzo D'Angelo[1], and Cristina M. Pinotti[2]

[1] Gran Sasso Science Institute (GSSI), L'Aquila, Italy
{federico.coro,gianlorenzo.dangelo}@gssi.it
[2] Department of Computer Science and Mathematics, University of Perugia,
Perugia, Italy
cristina.pinotti@unipg.it

Abstract. In this paper, we define a new problem called the Maximum Connectivity Improvement (MCI) problem: given a directed graph $G = (V, E)$, a weight function $w : V \to \mathbb{N}_{\geq 0}$, a profit function $p : V \to \mathbb{N}_{\geq 0}$, and an integer B, find a set S of at most B edges not in E that maximises $f(S) = \sum_{v \in V} w_v \cdot p(R(v, S))$, where $p(R(v, S))$ is the sum of the profits of the nodes reachable from node v when the edges in S are added to G. We first show that we can focus on Directed Acyclic Graphs (DAG) without loss of generality. We prove that the MCI problem on DAG is *NP*-Hard to approximate to within a factor greater than $1 - 1/e$ even if we restrict to graphs with a single source or a single sink, and MCI remains *NP*-Complete if we further restrict to unitary weights. We devise a polynomial time algorithm based on dynamic programming to solve the MCI problem on trees with a single source. We propose a polynomial time greedy algorithm that guarantees $(1-1/e)$-approximation ratio on DAGs with a single source or a single sink.

Keywords: Graph augmentation · Approximation algorithms · Greedy algorithms · Submodularity · DAG · Trees · Dynamic programming

1 Introduction

In this paper, we consider the problem of improving the reachability of a graph. We approach the problem from a graph augmentation perspective, in which a set of non-existing edges are added to the graph to increase the overall number of reachable nodes. There are several recent possible application scenarios for this problem. For example, suggesting friends in a social network with the objective of increasing the spreading of information [2,5] or performing faster network simulations by reducing the convergence time of random walk processes [13,14]. Graph augmentation problems are also well known in traditional graph theory. In [7], Tarjan et al. consider the problems of adding a minimum (or minimum-weight) set of edges to a graph so as to satisfy a given connectivity condition,

© Springer Nature Switzerland AG 2019
S. Gilbert et al. (Eds.): ALGOSENSORS 2018, LNCS 11410, pp. 47–61, 2019.
https://doi.org/10.1007/978-3-030-14094-6_4

such as to make a directed graph strongly connected or to make an undirected graph bridge-connected or biconnected. They have already proved that some variants of augmentation problems are *NP*-Complete.

More recently, several optimization problems related to graph augmentation have been addressed. Demaine and Zadimoghaddam [6] study the problem of minimising the eccentricity of a graph by adding a limited number of new edges. A 4-approximation algorithm is introduced and it is proven that the problem is *NP*-hard to be approximated within a factor smaller than 3/2. The problem of minimising the average all-pairs shortest path distance – characteristic path length – of the whole graph has been studied by Papagelis in [14]. The author considers the problem of adding a small set of edges to minimise the character-istic path length, and proves that the problem is *NP*-Hard. He proposes a path screening technique to select the edges to be added. The problem of adding a small set of links in order to maximise the centrality of a given node in a network has been addressed for different centrality measures: page-rank [1,13], eccentric-ity [6], average distance [11], harmonic and betweenness centrality [3,4], some measures related to the number of paths passing through a given node [10].

In this paper, we study the problem of adding at most B edges to a directed graph in order to maximise the overall weighted number of reachable nodes, which we call the Maximum Connectivity Improvement (MCI) problem. We first show that we can focus on Directed Acyclic Graphs (DAG) without loss of generality (Sect. 2). Then, we focus on the complexity of the problem (Sect. 3) and we prove that the MCI problem is *NP*-Hard to approximate to within a factor greater than $1 - \frac{1}{e}$. This result holds even if the DAG has a single source or a single sink. Moreover, the problem remains *NP*-complete if we further restrict to the unweighted case. In Sect. 4, we give a dynamic programming algorithm for the case in which the graph is a rooted tree, where the root is the only source node. In Sect. 5, we present a greedy algorithm which guarantees a $(1 - 1/e)$-approximation factor for the case in which the DAG has a single source or a single sink. We end with some concluding remarks in Sect. 6.

2   Preliminaries

Let $G = (V, E)$ be a directed graph. Each node $v \in V$ is associated with a weight $w_v \in \mathbb{N}_{\geq 0}$ and a profit $p_v \in \mathbb{N}_{\geq 0}$. Given a node $v \in V$, we denote by $R(v, G)$ the set of nodes that are reachable from v in G, that is $R(v, G) = \{u \in V : \exists \text{ path from } v \text{ to } u \text{ in } G\}$. Moreover, we denote by $p(R(v, G)) = \sum_{u \in R(v,G)} p_u$ the sum of the profits of the nodes reachable from v in G. In the rest of the paper, we also use the form $p(R(v, G) \setminus R(u, G)) = \sum_{u \in R(v,G) \setminus R(u,G)} p_u$ to denote the sum of the profits of the nodes in G that are reachable from v, but not from u. Note that, in the case $R(u, G) \subseteq R(v, G)$, it holds $p(R(v, G) \setminus R(u, G)) = p(R(v, G)) - p(R(u, G))$. Given a set S of edges not in E, we denote by $G(S)$ the graph augmented by adding the edges in S to G, i.e., $G(S) = (V, E \cup S)$. Let $R(v, G(S))$ and $p(R(v, G(S)))$ be, respectively, the set of nodes that are reachable from v in $G(S)$ and the sum of the profits of the nodes in $R(v, G(S))$. Note that, augmenting G the connectivity cannot be worse, and thus: $R(u, G) \subseteq R(u, G(S))$. Let $f(G) = \sum_{v \in V} w_v p(R(v, G))$ be a weighted measure of the connectivity of G. When weights and profits are unitary, $f(G)$ represents the overall number of connected pairs in G.

In this paper, we aim to augment G by adding a set S of edges of at most size B, i.e., $|S| \leq B$ and $B \in \mathbb{N}_{\geq 0}$, that maximises the weighted connectivity of $f(G(S))$. We call this problem the *Maximum Connectivity Improvement* (MCI) problem because maximising $f(G(S))$ is the same as to maximise $f(G(S)) - f(G)$.

From now on, for simplicity, we omit from the notations the original graph G. So, we simply use $R(v)$ and $R(v, S)$ to denote $R(v, G)$ and $R(v, G(S))$, respectively. Similarly, we simply denote with f and $f(S)$ the value of the weighted connectivity in G and in $G(S)$, respectively.

At first, we will show how to transform any directed graph G with cycles into a Directed Acyclic Graph (DAG) $G' = (V', E')$ and how to transform any solution for G' into a feasible solution for G.

Graph $G' = (V', E')$ has as many nodes as the number of strongly connected components of G. Specifically, G' selects one representative node for each strongly connected component of G and G' adds one directed edge between two nodes u' and v' of G' if there is a directed edge in G connecting any vertex of the strongly connected component represented by u' with any vertex of the strongly connected component represented by v'. Graph G' is called *condensation* of G and can be computed in $O(|V| + |E|)$ time by using Tarjan's algorithm which consists in performing a DFS visit [16].

The weight and the profit of a node v' in G' is given by the sum of the weight and profit of the nodes of G that belong to the strongly connected component $C_{v'}$ that is represented by v', i.e., $w_{v'} = \sum_{v \in C_{v'}} w_v$ and $p_{v'} = \sum_{v \in C_{v'}} p_v$.

Since the condensation preserves the connectivity of G, the following lemma can be proved:

Lemma 1. *Given a graph G and its condensation G', it yields: $f(G') = f(G)$.*

Proof. See Appendix.

$\square$

Given a solution S' for the MCI problem in G', we can build a solution S with the same value for the MCI problem in G as follows: for each edge (u', v') in S', we add an edge (u, v) in S, where u and v are two arbitrary nodes in the connected component corresponding to u' and v', respectively.

This derives from the fact that applying the condensation algorithm to $G' \cup S'$ or to $G \cup S$ we obtain the same condensed graph, say G''. From Lemma 1, we can conclude that $f(G' \cup S') = f(G'') = f(G \cup S)$.

Observe that if we add an edge e within the same strongly connected component in G, we do not add any edge to G'. Since the condensation G'' of $(G \cup \{e\})$ is the same as G', we have $f(G \cup \{e\}) = f(G') = f(G)$. As a consequence, in the remainder of the paper, we will assume that the graph is a DAG.

Given a DAG, a node with no incoming edges is called a *source*, while a node with no outgoing edges is called a *sink*. The next lemma allows us to focus on solutions that contain only edges connecting sink nodes to source nodes.

Lemma 2. *Let S be a solution to the MCI problem, then there exists a solution S' such that $|S| = |S'|$, $f(S) \leq f(S')$, and all edges in S' connect sink nodes to source nodes.*

Proof. We show how to modify any solution S in order to find a solution S' with properties of the statement. To obtain S', we start from S and we repeatedly apply the following modifications to each edge (u, v) of S such that u is not a sink or v is not a source: (1) If u is not a sink then there exists a path from u to some sink u' and we swap edge (u, v) with edge (u', v). The objective function does not decrease and increases at least by the sum of the weights on a path from u to u'. Namely, after adding the edge (u', v), any node z on the path from u to u' will now reach v passing through u'. Note that the objective function will not decrease and, instead, may increase due to the fact that the nodes z now are able to reach the node v. (2) If v is not a source then there exists a path from a source v' to v and we swap edge (u, v) with edge (u, v'). The objective function does not decrease and increases at least by the number of nodes in a path from v' to v multiplied by w_u. Note that in both cases the gain of a node on the path we are extending can be zero if it was already able to reach the source/sink from another edge in the solution.   □

3 NP-Hardness and Hardness of Approximation

In this section, we first show that the MCI problem is *NP*-Complete, even in the case in which all the weights and profits are unitary and the graph contains a single sink node or a single source node. Then, we show that it is *NP*-hard to approximate MCI to within a factor greater than $1 - \frac{1}{e}$. This last result holds also in the case of graphs with a single sink node or a single source node, but not in the case of unitary weights.

Theorem 1. *MCI is NP-Complete, even in the case in which all the weights and profits are unitary and the graph contains a single sink node or a single source node.*

Proof. We consider the decision version of MCI in which all the weights and profits are unitary (i.e., $w_v = p_v = 1$): Given a directed graph $G = (V, E)$ and two integers $M, B \in \mathbb{N}_{\geq 0}$, the goal is to find a set of additional edges $S \subseteq (V \times V) \setminus E$ such that $f(S) \geq M$ and $|S| = B$. The problem is in *NP* since it can be checked in polynomial time if a set of nodes S is such that $f(S) \geq M$ and $|S| = B$. We reduce from the Set Cover (SC) problem which is known to be *NP*-Complete [9]. Consider an instance of the SC problem $I_{SC} = (X, F, k)$ defined by a collection of subsets $F = \{S_1, \ldots, S_m\}$ for a ground set of items $X = \{x_1, \ldots, x_n\}$. The problem is to decide whether there exist k subsets whose union is equal to X. We define a corresponding instance $I_{MCI} = (G, M, B)$ of MCI as follows: (1) $B = k$; (2) $G = (V, E)$, where $V = \{v_{x_j} \mid x_j \in X\} \cup \{v_{S_i} \mid S_i \in F\} \cup \{v\}$ and $E = \{(v_{S_i}, v_{x_j}) \mid x_j \in S_i\} \cup \{(v_{x_j}, v) \mid x_j \in X\}$; (3) $M = (n + 1 + B)^2 + (m - B)(n + B + 2)$.

See Fig. 1 (left, top) for an example. Note that G is a DAG. By Lemma 2, we can assume that any solution S of MCI contains only edges (v, v_{S_i}) for some $S_i \in F$. In fact, v is the only sink node and v_{S_i} are the only source nodes. Assume that there exists a set cover F', then we define a solution S to the MCI instance as $S = \{(v, v_{S_i}) \mid S_i \in F'\}$. It is easy to show that $f(S) = M$ and $|S| = k = B$. Indeed, all the nodes in G can reach: node v, all the nodes v_{x_j} (since F' is a set cover), and all the nodes v_{S_i} such that $S_i \in F'$. Moreover, each node v_{S_i} such that $S_i \notin F'$ can reach itself. Therefore there are $n + B + 1$ nodes that reach $n + B + 1$ nodes and $m - B$ that reach $n + B + 2$ nodes, that is $f(S) = M$. On the other hand, assume that there exists a solution for MCI then S is in the form $\{(v, v_{S_i}) \mid S_i \in F\}$ and we define a solution for the set cover as $F' = \{S_i \mid (v, v_{S_i}) \in S\}$. We show that F' is a set cover. By contradiction, if we assume that F' is not a set cover and it cover only $n' < n$ elements of X then $f(S) = (n' + B + 1)^2 + (n - n' + m - B)(n' + B + 2) < M$. Note that in the above reduction, the graph G has a single sink node. We can prove the NP-hardness of the case of graphs with a single source node by using the same arguments on an instance of MCI made of the inverse graph of G, $M = (B + n + 1)(n + m + 1) + m - B$, and $B = k$ (see Fig. 1 (left, bottom) for an example). $\square$

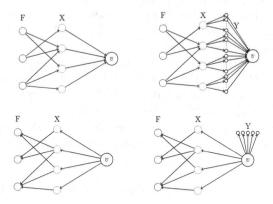

Fig. 1. (left) Example of reduction from SC to MCI used in Theorem 1. (right) Example of reduction from MC to MCI used in Theorem 2. (top) Single sink. (Bottom) single source.

Theorem 2. *MCI is NP-hard to approximate to within a factor $1 - \frac{1}{e} + \epsilon$, for any $\epsilon > 0$, even if graph contains a single sink node or a single source node.*

Proof. We give two approximation factor preserving reductions from the Maximum coverage problem (MC), which is known to be *NP*-hard to approximate to within a factor greater than $1 - \frac{1}{e}$ [8].

The MC problem is defined as follows: given a ground set of items $X = \{x_1, \ldots, x_n\}$, a collection of subsets $F = \{S_1, \ldots, S_m\}$ of subsets of X, and an integer k, find k sets in $\mathcal{F}$ that maximise the cardinality of their union.

We first focus on the single sink problem. Given an instance of the MC problem $I_{MC} = (X, F, k)$ we define an instance of the (maximisation) MCI problem $I_{MCI} = (G, k)$ similar to the one used in Theorem 1, but where we modify the weights and add Y paths of one node between each v_{x_j} and v, where Y is an arbitrarily high number (polynomial in $n + m$).

In detail I_{MCI} is defined as follows: (1) $B = k$; (2) $G = (V, E)$, where $V = \{v_{x_j} \mid x_j \in X\} \cup \{v_{S_i} \mid S_i \in F\} \cup \{v_{x_j}^l \mid x_j \in X$ and $l = 1, \ldots, Y\} \cup \{v\}$ and $E = \{(v_{S_i}, v_{x_j}) \mid x_j \in S_i\} \cup \{(v_{x_j}, v_{x_j}^l) \mid x_j \in X, l = 1, \ldots, Y\} \cup \{(v_{x_j}^l, v) \mid x_j \in X, l = 1, \ldots, Y\}$; (3) $w(v) = 1$ and $w(u) = 0$, for each $u \in V \setminus \{v\}$; (4) $p_v = 1$ for any node $v \in V$.

See Fig. 1 (right, top) for an example. We first show that there exists a solution $F' \subseteq F$ to I_{MC} that covers n' elements of X if and only if there exists a solution S to I_{MCI} such that $f(S) = n'(Y + 1) + B + 1$. Moreover, we can compute F' from S and vice versa in polynomial time. Indeed, given F', we define S as $S = \{(v, v_{S_i}) \mid S_i \in F'\}$. We can verify that $f(S) = n'(Y+1) + B + 1$ and $|S| = k = B$. Indeed, only node v as a weight different from 0, and then $f(S) = R(v, S) = n'(Y + 1) + B + 1$, since v can reach the $n'(Y + 1)$ nodes v_{x_j} corresponding to the items x_j covered by F', the B nodes v_{S_i} it is connected to, and itself. On the other hand, given a solution S to I_{MCI}, by Lemma 2 we can assume that it has only edges from v to nodes v_{S_i}. Let n' be the number of nodes v_{x_j} in $R(v, S)$, then $f(S) = n'(Y + 1) + B + 1$ and $F' = \{S_i \mid (v, v_{S_i}) \in S\}$ covers n' elements in X.

If $OPT(I_{MC})$ and $OPT(I_{MCI})$ denote the optimum value for I_{MC} and I_{MCI}, respectively, then $OPT(I_{MCI}) \geq OPT(I_{MC})(Y + 1) + B + 1 \geq Y \cdot OPT(I_{MC})$. Moreover, given the above definition of S and F', then for any $\epsilon' > 0$ there exists a value of $Y = O(poly(n + m))$ such that $f(S) \leq (n' + \epsilon')Y$.

Let us assume that there exists a polynomial-time algorithm that guarantees an α approximation for I_{MCI}, then we can compute a solution S such that $f(S) \geq \alpha OPT(I_{MCI})$. It follows that:

$$\alpha Y \cdot OPT(I_{MC}) \leq \alpha OPT(I_{MCI}) \leq f(S) \leq (n' + \epsilon')Y,$$

Where n' is the number of nodes covered by the solution F' to MC obtained from S. Therefore we obtained an algorithm that approximates the MC problem with a factor α (up to lower order terms). Since it is *NP*-hard to approximate to within a factor greater than $1 - \frac{1}{e}$ [8], then the statement follows.

Let us now focus on the single source case. Given I_{MC}, we define $I_{MCI} = (G, B)$ as follows: (1) $B = k$; (2) $G = (V, E)$, where $V = \{v_{x_j} \mid x_j \in X\} \cup \{v_{S_i} \mid S_i \in F\} \cup \{v\} \cup \{v^l \mid l = 1, \ldots Y\}$ and $E = \{(v_{x_j}, v_{S_i}) \mid x_j \in S_i\} \cup \{(v, v_{x_j}) \mid x_j \in X\} \cup \{(v, v^l) \mid l = 1, \ldots Y\}$; (3) $w(v_{x_j}) = 1$, for each $x_j \in X$ and

$w(u) = 0$, for each $u \in V \setminus \{v_{x_j} \mid x_j \in X\}$; (4) $p_v = 1$ for any node $v \in V$. Where Y is an arbitrarily high polynomial value in $m + n$. See Fig. 1 (right, bottom) for an example. We use similar arguments as above. In detail, there exists a solution $F' \subseteq F$ to I_{MC} that covers n' elements of X if and only if there exists a solution S to I_{MCI} such that $f(S) = n'(n + m - m' + Y) + n(m' + 1)$, where m' is the number of sets in F that do not cover any of the n' elements covered by F'. Moreover, we can compute F' from S and vice versa in polynomial time. Given F', we define S as $S = \{(v, v_{S_i}) \mid S_i \in F'\}$ and we can verify that $f(S) = \sum_{x_j \in X} p(R(v_{x_j}, S)) = \sum_{x_j \in X} |R(v_{x_j}, S)| = n'(n + m + Y + 1) + (n - n')(m' + 1) = n'(n + m - m' + Y) + n(m' + 1)$ and $|S| = B$. Given S, if n' is the number of nodes v_{x_j} such that $v \in R(v_{x_j}, S)$, then $f(S) = n'(n + m - m' + Y) + n(m' + 1)$ and $F' = \{S_i \mid (v_{S_i}, v) \in S\}$ covers n' elements in X.

As above, we can show that $OPT(I_{MCI}) \geq Y \cdot OPT(I_{MC})$ and that there exists a value of $Y = O(poly(n + m))$ such that $f(S) \leq (n' + \epsilon')Y$, for any $\epsilon' > 0$. Then, the statement follows by using the same arguments as above. □

4 Polynomial-Time Algorithm for Trees

In this section, we focus on the case of directed weighted rooted trees in which the root of the tree is the only source node and all the edges are directed towards the leaves. We give a polynomial time algorithm based on dynamic programming that requires $O(|V|B^2)$ time and $O(|V|B)$ space. By Lemma 2, we can focus on edges that connect leaves to the root. We first assume that the tree is binary and give an algorithm to solve this special case, then we show how to transform any tree into a binary tree in such a way that each solution for the transformed instance has the same value as the corresponding solution in the original instance. The algorithm for binary trees requires $O(|V|B^2)$ time and $O(|V|B)$ space while the transformation requires $O(|V|)$ time and $O(|V|)$ space.

4.1 Binary Trees

We are given a directed weighted binary tree $T = (V, E)$, where all the edges are directed towards the leaves, the root $r \in V$ is the only source node, and $w : V \to \mathbb{N}_{\geq 0}$, $p : V \to \mathbb{N}_{\geq 0}$. Let us denote by $\psi(v)$ (left child) and $\delta(v)$ (right child) the children of node $v \in T$, moreover, we denote as $T(v)$ the sub-tree rooted at v. In the following, we introduce our dynamic-programming algorithm to solve the MCI problem starting from the leaves of T. Given a node v, let S_v be a solution that connects some leaves of $T(v)$ to r. The *gain* of solution S_v in $T(v)$ is the increase in weighted reachability of all the nodes in $T(v)$, that is the gain of S_v in $T(v)$ is equal to $\sum_{u \in T(v)} w_u(p(T(u, S_v)) - p(T(u)))$. For each node v and for each budget $b = 0, 1, \ldots, B$, the algorithm computes a solution that connects b leaves of $T(v)$ to r and maximises the gain in $T(v)$. We define $g(v, b)$ as the maximum gain in $T(v)$ achievable by adding at most b edges from b leaves of $T(v)$ to node r. Note that $g(v, 1) \leq g(v, 2) \leq \ldots \leq g(v, b)$. We now show how to compute $g(v, b)$ for each node v and for each budget $b = 0, 1, \ldots, B$ by

---

Algorithm 1. Dynamic programming algorithm for MCI

    Input: $T = (V, E)$, B, w_v, p_v $\forall v \in V$

    Output: Set S of edges

1: **for** each node v **do**

2:      $g(v, 0) := 0$;

3:      $S_{(v,b)} := \emptyset$;

4: **for** each leaf v and budget b, with $\{1, \ldots, B\}$ **do**

5:      $g(v, b) := w_v \cdot (p(T(r)) - p(T(v)))$;

6:      $S_{(v,b)} := \{(v, r)\}$;

7: **for** each node v in post-ordering **do**

8:      **for** $b \in 1, \ldots, B$ **do**

9:          $g(v, b) := \displaystyle\max_{\substack{b_l, b_r \in \{0, \ldots, b\}, \\ b_l + b_r = b}} \{g(\psi(v), b_l) + g(\delta(v), b_r)\} + w_v \cdot (p(T(r)) - p(T(v)))$;

     ▷ Let b_l and b_r the budgets that maximise Line 9,

10:      $S_{(v,b)} := S_{(\psi(v), b_l)} \cup S_{(\delta(v), b_r)}$;

11: $S := S_{(r, B)}$;

---

using dynamic programming. For each leaf $v \in T$ and for each $b = 1, 2, \ldots, B$, $g(v, b) = w_v \cdot (p(T(r)) - p(T(v)))$, that is, the sum of profits p of the new nodes that v can reach thanks to the new edge (v, r). Moreover $g(v, 0) = 0$ for each $v \in V$. Then, the algorithm visits T in post order. For each internal node v we compute $g(v, b)$ by using the solutions of its sub-trees, i.e., $T(\psi(v))$ and $T(\delta(v))$. Let us assume that we have computed $g(\psi(v), b)$ and $g(\delta(v), b)$, for each $b = 0, 1, \ldots, B$. Note that if a solution adds an edge between any leaf of $T(v)$ and r, then the gain of node v is $w_v(p(T(r)) - p(T(v)))$ since v will now reach all the nodes in T. This gain is independent of the number of edges that are added from the leaves of $T(v)$ to r. In fact, given $g(\psi(v), b_l)$ be the maximum gain for $T(\psi(v))$ and budget $b_l \in \{1, 2, \ldots, B\}$, then the gain in $T(v)$ of a solution that connects b_l leaves of $T(\psi(v))$ to r is equal to $g(\psi(v), b_l) + w_v \cdot (p(T(r)) - p(T(v)))$. Similarly the gain in $T(v)$ of the solution that connects, for some $b_r \in \{1, 2, \ldots, B\}$, b_r leaves of $T(\delta(v))$ to r is equal to $g(\delta(v), b_l) + w_v \cdot p(T(r)) - p(T(v))$.

Then, to compute $g(v, b)$ once we have decided how many edges to add in $\psi(v)$ and how many edges in $\delta(v)$, we increase the reachability function of the same quantity, i.e., $w_v \cdot (p(T(r)) - p(T(v)))$.

Hence, $g(v, b)$ is given by the combination of b_l and b_r such that $b_l + b_r = b$ that maximises the sum $g(\psi(v), b_l) + g(\delta(v), b_r) + w_v \cdot (p(T(r)) - p(T(v)))$.

Precisely:

$$g(v, b) = \max_{\substack{b_l, b_r \in \{0, \ldots, b\} \\ b_l + b_r = b}} \{g(\psi(v), b_l) + g(\delta(v), b_r)\} + w_v \cdot (p(T(r)) - p(T(v))). \quad (1)$$

The optimal value of the problem $f(S) = g(r, B) + f(T)$, where $f(T)$ is the value of the objective function on T (i.e., when no edges have been added). The pseudocode of the algorithm is reported in Algorithm 1.

Theorem 3. *Algorithm 1 finds an optimal solution for MCI if the graph is a binary tree.*

Proof. Let us assume by contradiction that v and b are, respectively, the first node and the first budget for which Algorithm 1 computes a non-maximum gain at line 9 of Algorithm 1, that is $g(v, b) < g^*(v, b)$, where $g^*(v, b)$ is the maximum gain for tree $T(v)$ and budget b. Let S^* be an optimal solution that achieves $g^*(v, b)$ and let S_l^*, S_r^* be the edges in S^* that starts from leaves in $T(\psi(v))$ and $T(\delta(v))$, respectively. Let $b_l^* = |S_l^*|$ and $b_r^* = |S_r^*|$. Then, the gain of the optimal solution S^* is: $g^*(v, b) = g(\psi(v), b_l^*) + g(\delta(v), b_r^*) + w_v \cdot (p(T(r)) - p(T(v)))$.

Since by hypothesis $g(v, b)$ is the first time for which Algorithm 1 does not find the maximum gain and since the cost $(p(T(r)) - p(T(v))) \cdot w_v$ does not depend on the edges selected in the left and right sub-trees of v, this implies that at line 9 Algorithm 1 must select $g(v, b) = g(\psi(v), b_l^*) + g(\delta(v), b_r^*) + (p(T(r)) - p(T(v))) \cdot w_v = g^*(v, b)$. Thus contradicting $g(v, b) < g^*(v, b)$.   □

For each node v, the algorithm computes $B + 1$ values. From Eq. 1, it follows that Algorithm 1 takes $O(|V|B^2)$ time. Note that $B \in O(|V|)$ because we limit the new edges to be of the form leaf-root.

4.2   General Trees

We can transform a generic rooted tree $T = (V, E)$ into an equivalent binary tree $T' = (V \cup U, E')$, following a tree transformation proposed in [15] by adding at most $|V| - 3$ *dummy* node.

The transformation requires $O(|V|)$ time and space (see Appendix 1.C for a detailed description of the algorithm). The nodes in T' will have $w_v' = w_v$ and $p_v' = p_v$ if $v \in V$ and $w_v' = p_v' = 0$ for any dummy node. Note that $p(R(v, T')) = p(R(v, T))$ for any node v in T due to the fact that the added dummy nodes have $p_v' = 0$, moreover, dummy nodes do not increase the objective function because they have the weight set to zero, i.e., any dummy node v will have $w_v' \cdot p(R(v, T')) = 0$.

For each node $v \in T'$ and solution S to MCI in T', let $f'(S) = \sum_{v \in V} w_v p(T(v, S))$. It is easy to see that by applying Algorithm 1 to T' we will obtain an optimal solution with respect to f'. Moreover, for each solution S to T', $f'(S) = f(S)$. Note that a solution S for T' that connects leaves of T' to its root are feasible solution also for T since T and T' have the same root and leaves.

5   Polynomial-Time Algorithm for DAG with a Single Source or a Single Sink

In this section, we focus on the case of weighted DAG in which we have a single source node or a single sink node. We first describe our greedy algorithm to approximate MCI on DAGs with a single source. Then, we will show how to modify the algorithm for the case of DAGs with a single sink.

In the case of single source, by Lemma 2, we restrict our choices to the edges that connect sinks nodes to the source. Recall S is the set of edges added to G. With a little abuse of notation, we also use S to denote the set of sinks from which the edges in S start. The Greedy algorithm for MCI on DAGs with a single source (see Algorithm 2), starts from the empty solution, i.e., $S = \emptyset$ and repeatedly adds to S the edge e' that maximises the function $f(S \cup \{e'\})$. The edge e' is chosen from the set E' of edges (u, s), where u is a sink in V, not already inserted in S, and s is the single source in V (see lines 2 and 4).

To implement the Greedy Algorithm with single source, some preprocessing is required. First, for each node $v \in V$, we perform a DFS on G to compute $R(v)$ and $p(R(v))$. We store $p(R(v, S))$ in a vector ρ of size $|V|$. Every time a new edge is added to the solution S, each entry of ρ is updated in constant time because for each node v, $p(R(v, S))$ is either equal to $p(R(v))$ or $p(R(s))$, as we explain below. To compute the gain of adding the edge $e = (u, s)$, we need to find all the nodes $R^T(u, S)$ that reach u in $G(S)$. $R^T(u, S)$ is computed by performing a DFS starting from u on the reverse graph $G^T(S)$ of $G(S)$. Note that the reverse graph G^T of G is initially computed in a preprocessing phase in $O(|V| + |E|)$ time, and after every new edge is added to S, $G^T(S)$ is updated in constant time. Finally, to compute $f(S \cup \{e\})$ for $e = (u, s)$, observe that $f(S \cup \{e\}) = f(S) + \sum_{z \in R^T(u,S)} w_z(p(R(s)) - p(R(z, S)))$. After selecting the edge $e' = (u, s)$ that maximise $f(S \cup \{e'\})$, we update S and we set $p(R(z, S)) = p(R(s))$ for each node $z \in R^T(u, S)$ in vector ρ because z reaches s traversing the edge $e' = (u, s)$ and inherits the reachability of $R(s)$.

The Greedy algorithm with a single source requires $O(B|V||E|)$ time. Namely, for each edge $e = (u, s) \in E'$, to compute $f(S \cup \{e\})$ it is required $O(|E|)$ time to compute $R^T(u, S)$ on G^T, and $O(|V|)$ time to compute $\sum_{z \in R^T(u,S)} w_z(p(R(s)) - p(R(z, S)))$. Since there are $O(|V|)$ sinks, the computation of the maximum value at line 4 costs $O(|V||E|)$ time. Selected e', $O(|V|)$ time is spent to update ρ. Since at most B edges are added, the Greedy algorithm requires $O(B|V||E|)$.

In the case of a single sink, we use the same Greedy Algorithm 2 only substituting E' in Line 2 with the set of edges (d, v), where v is a source in V and d is the only sink in V (see line 2) by Lemma 2. However, the implementation in the case of DAGs with a single sink is slightly different. In fact, $R^T(d, S) = V$ because (by definition) all the nodes reach the single sink, but at the same time, the reachability of a node v does not assume only the values V or $R(v)$. As a consequence to compute $f(S \cup \{(d, v)\})$ it is not required to perform any DFS visit of G^T, but the vicinity of any other node depends on the set of added edges and has to be recomputed by performing a DFS on the augmented graph. Hence, $f(S \cup \{(d, v)\}) = f(S) + \sum_{z \in V} w_z \cdot p(R(v, S) \setminus R(d, S))$. Since for computing $f(S \cup \{(d, v)\})$, we must compute $|V|$ DFSs, the overall cost of the Greedy algorithm with a single sink increases by a factor of $|V|$ with respect to the case of DAG with a single source, thus becoming $O(B|V|^2|E|)$.

Observe that Algorithm 2 with a single source can also be used on trees in place of Algorithm 1. However, the complexity of the Greedy algorithm will be $O(|V|^2B)$ that is greater than the complexity of Algorithm 1, which is $O(|V|B^2)$.

Algorithm 2. Greedy Algorithm for single source

    Input: DAG $G = (V, E)$, B, w_v, p_v $\forall v \in V$
    Output: set S of edges
1: $S := \emptyset$;
2: $E' := \{e = (u, s) \mid u$ is sink and s is the only source$\}$;
3: **while** $|S| \leq B$ **do**
4:     $e' := \underset{e \in E' \setminus S}{\arg\max}\ f(S \cup \{e\})$;
5:     $S := S \cup \{e'\}$;

To give a lower bound on the approximation ratio of Algorithm 2, we show that the objective function $f(S)$ is *monotone and submodular*[1]. This allows us to apply the result by Nemhauser et al. [12]:

Given a finite set N, an integer k', and a real-valued function z defined on the set of subsets of N, the problem of finding a set $S \subseteq N$ such that $|S| \leq k'$ and $z(S)$ is maximum can be $1 - \frac{1}{e}$ approximated by starting with the empty set, and repeatedly adding the element that gives the maximal marginal gain, if z is monotone and submodular. We recall that $f(S) = \sum_{v \in V} w_v p(R(v, S))$ and $w_v, p_v \in \mathbb{N}_{\geq 0}$, therefore in order to prove that $f(S)$ is monotone increasing and submodular, we show that $p(R(v, S))$ is monotone increasing and submodular, for each $v \in V$ and solution S, because a non-negative linear combination of a monotone submodular functions is also monotone and submodular.

Theorem 4. *Function $f(S)$ is monotone and submodular with respect to any feasible solution for MCI on DAGs with a single source.*

Proof. To prove that $f(S)$ is monotone, we prove that for each $v \in V$, $S \subseteq E'$, and $e = (t', s) \in E' \setminus S$, we have $p(R(v, S \cup \{e\})) \geq p(R(v, S))$. We first notice that for each $v \in V$ and solution S, if there exists an edge $(t, s) \in S$ such that $t \in R(v)$, then $p(R(v, S)) = p(R(s))$; otherwise, $p(R(v, S)) = p(R(v))$. The same holds for $p(R(v, S \cup \{e\}))$.

We analyse the following cases recalling that $e = (t', s)$:

- If there exists an edge $(t, s) \in S$ such that $t \in R(v)$, then, $p(R(v, S \cup \{e\})) = p(R(v, S)) = p(R(s))$;
- Otherwise,
  - If $t' \in R(v)$, then, $p(R(v, S \cup \{e\})) = p(R(v))$ and $p(R(v, S)) = p(R(v))$
  - If $t' \notin R(v)$, then, $p(R(v, S \cup \{e\})) = p(R(v, S)) = p(R(v))$

It follows that $p(R(v, S \cup \{e\})) \geq p(R(v, S))$.

To prove that $f(S)$ is submodular, we prove that for any node $v \in V$, any two solutions S, T of MCI such that $S \subseteq T$, and any edge $e = (t', s) \notin T$, where t' is a sink node, it holds:

$$p(R(v, S \cup \{e\})) - p(R(v, S)) \geq p(R(v, T \cup \{e\})) - p(R(v, T)). \qquad (2)$$

---

[1] For a ground set N, a function $z : 2^N \to \mathbb{R}$ is submodular if for any pair of sets $S \subseteq T \subseteq N$ and for any element $e \in N \setminus T$, $z(S \cup \{e\}) - z(S) \geq z(T \cup \{e\}) - z(T)$.

We analyse the following cases:

- If there exists an edge $(t, s) \in S$ such that $t \in R(v)$, then, $p(R(v, S \cup \{e\})) = p(R(v, S)) = p(R(v, T \cup \{e\})) = p(R(v, T)) = p(R(s))$;
- Otherwise,
  - If there exists $(t'', s) \in T$ such that $t'' \in R(v)$, then $p(R(v, S \cup \{e\})) - p(R(v, S)) \geq 0 = p(R(v, T \cup \{e\})) - p(R(v, T))$ because $p(R(v, T \cup \{e\})) = p(R(v, T)) = p(R(s))$;
  - If for each $(t'', s) \in T$, $t'' \notin R(v)$, then $p(R(v, S \cup \{e\})) = p(R(v, T \cup \{e\}))$ and $p(R(v, S)) = p(R(v, T))$.

In all the cases Inequality (2) holds.   □

Theorem 5. *Function $f(S)$ is monotone and submodular with respect to any feasible solution for MCI on DAGs with a single sink.*

Proof. To prove that $f(S)$ is monotone, we show that for each $v \in V$, $S \subseteq E'$, and $e = (d, s') \in E' \setminus S$, we have $p(R(v, S \cup \{e\})) \geq p(R(v, S))$. We observe that

$$R(v, S) = R(v) \cup \bigcup_{(d,s_i) \in S} R(s_i)$$

$$R(v, S \cup \{e\}) = R(v) \cup \bigcup_{(d,s_i) \in S} R(s_i) \cup R(s') = R(v, S) \cup R(s') \qquad (3)$$

Thus, $p(R(v, S)) \leq \sum_{u \in R(v,S) \cup R(s')} p_u$.

To prove that $f(S)$ is submodular, we prove that for any node $v \in V$, any two solutions S, T of MCI such that $S \subseteq T$, and any edge $e = (d, s') \notin T$, where s' is a source node:

$$p(R(v, S \cup \{e\})) - p(R(v, S)) \geq p(R(v, T \cup \{e\})) - p(R(v, T)). \qquad (4)$$

We first make the following observations based on Eq. 3:

- $R(v, S \cup \{e\}) = R(v, S) \cup R(s') = R(v, S) \cup (R(s') \setminus R(v, S))$ and
- $R(v, T \cup \{e\}) = R(v, T) \cup R(s') = R(v, T) \cup (R(s') \setminus R(v, T))$

Then, $p(R(v, S \cup \{e\})) - p(R(v, S)) = p(R(s') \setminus R(v, S))$ and $p(R(v, T \cup \{e\})) - p(R(v, T)) = p(R(s') \setminus R(v, T))$.

Inequality (4) follows by observing that $R(v, S) \subseteq R(v, T)$.   □

Corollary 1. *Algorithm 2 provides a $(1 - \frac{1}{e})$-approximation for the MCI problem either on DAG with a single source or with a single sink.*

6   Conclusion and Future Works

In this paper, we first defined the maximum connectivity improvement problem, that is the problem of adding k edges to a directed graph in order to maximise the overall weighted number of reachable nodes. We proved that the MCI problem on DAGs with a single source or a single sink is *NP-Complete* and *NP-Hard* to approximate to within a factor greater than $1 - 1/e$. We proposed a polynomial time greedy algorithm that guarantees a $1 - 1/e$ approximation ratio on DAG with a single source or a single sink. For rooted trees, to solve the MCI problem on single source, we devised a polynomial time algorithm based on dynamic programming faster than the greedy algorithm.

As future works, we plan to extend our approach to general DAG, i.e., with multiple sources and multiple sinks. Another possible extension is to solve the MCI problem by considering the budgeted version of the problem in which each edge can be added at a different budget cost.

Appendix 1.A Omitted Proofs

Lemma 3. *Given a graph G and its condensation G', it yields: $f(G') = f(G)$.*

Proof. First, consider two nodes u and v that belong to the same strongly connected component $C_{v'}$ in G'. Clearly, $R(u, G) = R(v, G)$.

Moreover, it holds $p(R(v, G)) = p(R(v', G'))$ because $R(v', G')$ contains one node for each different strongly connected component in $R(u, G)$ and thus:

$$p(R(v', G')) = \sum_{u' \in R(v',G')} p_{u'} = \sum_{u' \in R(v',G')} \sum_{u \in C_{u'}} p(u) = \sum_{u \in R(v,G)} p(u) = p(R(v, G))$$

Denoted $C_{v'}$ the strongly connected component represented by v', we have:

$$f(G') = \sum_{v' \in V'} w_{v'} p(R(v', G'))$$

$$= \sum_{v' \in V'} w_{v'} \left(\sum_{u' \in R(v',G')} \sum_{u \in C_{u'}} p(u) \right)$$

$$= \sum_{v':C_{v'} \in \mathcal{C}} w_{v'} \left(\sum_{u' \in R(v',G')} \sum_{u \in C_{u'}} p(u) \right)$$

$$= \sum_{v':C_{v'} \in \mathcal{C}} \left(\sum_{v \in C_{v'}} w_v \right) \left(\sum_{u' \in R(v',G')} \sum_{u \in C_{u'}} p(u) \right)$$

$$= \sum_{v':C_{v'} \in \mathcal{C}} \sum_{v \in C_{v'}} \left(w_v \left(\sum_{u' \in R(v',G')} \sum_{u \in C_{u'}} p(u) \right) \right)$$

$$= \sum_{v':C_{v'} \in \mathcal{C}} \sum_{v \in C_{v'}} \left(w_v \sum_{u \in R(v,G)} p(u) \right)$$

$$= \sum_{v':C_{v'} \in \mathcal{C}} \sum_{v \in C_{v'}} w_v p(R(v, G))$$

$$= \sum_{v \in V} w_v p(R(v, G)) = f(G)$$

$\square$

Appendix 1.B Omitted Images

$$T = (V, E).\qquad\qquad T = (V, E \cup S)$$

Fig. 2. Example of Algorithm 1. Consider the node c with $w_c = 2$, $p_v = 1\ \forall v \in V$ and $B = 2$. We have: $g(d, 2) = 19$, $g(d, 1) = 12$, $g(e, 1) = 7 + 2(6) = 19$. Therefore $g(c, 2) = g(e, 1) + g(d, 1) + w_c \cdot (p(T(r)) - p(T(v))) = 35$.

Appendix 1.C Generic Trees Algorithm

Given a generic rooted tree $T = (V, E)$, let us transform it into a rooted binary tree $T' = (V \cup U, E')$ with weights w', p' by adding *dummy* nodes U as follows:

1. Let the root r of T be the root of T'.
2. For each non-leaf node v, let $v_1, v_2, \ldots v_l$ be the children of v:
   (a) Add edge (v, v_1) to E';
   (b) If $l = 2$ add (v, v_2) to E';
   (c) If $l > 2$, add $l - 2$ dummy nodes $u_{v_2}, u_{v_3}, \ldots, u_{v_{l-2}}, u_{v_{l-1}}$
   (d) Add edge (v, u_{v_2}) and edges $(u_{v_i}, u_{v_{i+1}})$ to E', for each $2 \leq i \leq l - 2$;
   (e) Add edge (u_{v_i}, v_i) to E', for each $2 \leq i \leq l - 1$;
   (f) If $l > 2$, add edge $(u_{i_{l-1}}, v_l)$ to E'.
3. If $v \in V$, then $w'_v = w_v$, otherwise $w'_v = 0$ and $p'_v = p_v$, otherwise $p'_v = 0$.

See Fig. 3 for an example of the transformation.

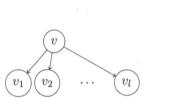

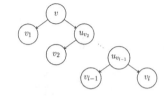

(a) General tree $T = (V, E)$

(b) Transformed binary tree $T' = (V \cup U, E')$

Fig. 3. Example of transformation from general tree to binary tree

References

1. Avrachenkov, K., Litvak, N.: The effect of new links on Google PageRank. Stoc. Models **22**(2), 319–331 (2006)
2. Cordasco, G., et al.: Whom to befriend to influence people. CoRR abs/1611.08687 (2016). http://arxiv.org/abs/1611.08687
3. Crescenzi, P., D'Angelo, G., Severini, L., Velaj, Y.: Greedily improving our own centrality in a network. In: Bampis, E. (ed.) SEA 2015. LNCS, vol. 9125, pp. 43–55. Springer, Cham (2015). https://doi.org/10.1007/978-3-319-20086-6_4
4. Crescenzi, P., D'Angelo, G., Severini, L., Velaj, Y.: Greedily improving our own closeness centrality in a network. TKDD **11**(1), 9:1–9:32 (2016)
5. D'Angelo, G., Severini, L., Velaj, Y.: Selecting nodes and buying links to maximize the information diffusion in a network. In: 42nd International Symposium on Mathematical Foundations of Computer Science, MFCS 2017, LIPIcs, vol. 83, pp. 75:1–75:14 (2017)
6. Demaine, E.D., Zadimoghaddam, M.: Minimizing the diameter of a network using shortcut edges. In: Kaplan, H. (ed.) SWAT 2010. LNCS, vol. 6139, pp. 420–431. Springer, Heidelberg (2010). https://doi.org/10.1007/978-3-642-13731-0_39
7. Eswaran, K.P., Tarjan, R.E.: Augmentation problems. SIAM J. Comput. **5**(4), 653–665 (1976)
8. Feige, U.: A threshold of ln n for approximating set cover. J. ACM **45**(4), 634–652 (1998)
9. Garey, M.R., Johnson, D.S.: Computers and Intractability: A Guide to the Theory of NP-Completeness. W.H. Freeman and Company, New York (1979)
10. Ishakian, V., Erdös, D., Terzi, E., Bestavros, A.: A framework for the evaluation and management of network centrality. In: Proceedings of the 12th SIAM International Conference on Data Mining (SDM), pp. 427–438. SIAM (2012)
11. Meyerson, A., Tagiku, B.: Minimizing average shortest path distances via shortcut edge addition. In: Dinur, I., Jansen, K., Naor, J., Rolim, J. (eds.) APPROX/RANDOM -2009. LNCS, vol. 5687, pp. 272–285. Springer, Heidelberg (2009). https://doi.org/10.1007/978-3-642-03685-9_21
12. Nemhauser, G.L., Wolsey, L.A., Fisher, M.L.: An analysis of approximations for maximizing submodular set functions-i. Math. Program. **14**(1), 265–294 (1978)
13. Olsen, M., Viglas, A.: On the approximability of the link building problem. Theor. Comput. Sci. **518**, 96–116 (2014)
14. Papagelis, M.: Refining social graph connectivity via shortcut edge addition. ACM Trans. Knowl. Discovery Data (TKDD) **10**(2), 12 (2015)
15. Tamir, A.: An o(pn^2). algorithm for the p-median and related problems on tree graphs. Oper. Res. Lett. **19**(2), 59–64 (1996)
16. Tarjan, R.: Depth-first search and linear graph algorithms. SIAM J. Comput. **1**(2), 146–160 (1972)

Average Case - Worst Case Tradeoffs for Evacuating 2 Robots from the Disk in the Face-to-Face Model

Huda Chuangpishit[(⊠)], Konstantinos Georgiou, and Preeti Sharma

Department of Mathematics, Ryerson University, 350 Victoria Street, Toronto, ON M5B 2K3, Canada
{h.chuang,konstantinos,preeti.sharma}@ryerson.ca

Abstract. The problem of evacuating two robots from the disk in the face-to-face model was first introduced in [16], and extensively studied (along with many variations) ever since with respect to worst case analysis. We initiate the study of the same problem with respect to average case analysis, which is also equivalent to designing randomized algorithms for the problem. First we observe that algorithm $\mathscr{B}_2$ of [16] with worst case cost $\mathrm{Wrs}\,(\mathscr{B}_2) := 5.73906$ has average case cost $\mathrm{Avg}\,(\mathscr{B}_2) := 5.1172$. Then we verify that none of the algorithms that induced worst case cost improvements in subsequent publications has better average case cost, hence concluding that our problem requires the invention of new algorithms. Then, we observe that a remarkable simple algorithm, $\mathscr{B}_1$, has very small average case cost $\mathrm{Avg}\,(\mathscr{B}_1) := 1 + \pi$, but very high worst case cost $\mathrm{Wrs}\,(\mathscr{B}_1) := 1 + 2\pi$. Motivated by the above, we introduce constrained optimization problem $_2\mathrm{EVAC}_{F2F}^w$, in which one is trying to minimize the average case cost of the evacuation algorithm given that the worst case cost does not exceed w. The problem is of special interest with respect to practical applications, since a common objective in search-and-rescue operations is to minimize the average completion time, given that a certain worst case threshold is not exceeded, e.g. for safety or limited energy reasons.

Our main contribution is the design and analysis of families of new evacuation parameterized algorithms $\mathscr{A}(p)$ which can solve $_2\mathrm{EVAC}_{F2F}^w$, for every $w \in [\mathrm{Wrs}\,(\mathscr{B}_1), \mathrm{Wrs}\,(\mathscr{B}_2)]$. In particular, by letting parameter(s) p vary, we obtain parametric curve $(\mathrm{Avg}\,(\mathscr{A}(p)), \mathrm{Wrs}\,(\mathscr{A}(p)))$ that induces a continuous and strictly decreasing function in the mean-worst case space, and whose endpoints are $(\mathrm{Avg}\,(\mathscr{B}_1), \mathrm{Wrs}\,(\mathscr{B}_1))$, $(\mathrm{Avg}\,(\mathscr{B}_2), \mathrm{Wrs}\,(\mathscr{B}_2))$. Notably, the worst case analysis of the problem, since it's introduction, has been relying on technical numerical, computer-assisted, calculations, following tedious robots trajectories' analysis. Part of our contribution is a novel systematic procedure, which, given *any evacuation algorithm*, can derive it's worst and average case performance in a clean and unified way.

K. Georgiou—Research supported in part by NSERC Discovery Grant.

© Springer Nature Switzerland AG 2019
S. Gilbert et al. (Eds.): ALGOSENSORS 2018, LNCS 11410, pp. 62–82, 2019.
https://doi.org/10.1007/978-3-030-14094-6_5

Keywords: Evacuation · Disk · Face-to-Face model ·
Average case analysis

1   Introduction

Search problems are concerned with the exploration of a domain, aiming to identify the location of a hidden object. More particularly, in evacuation-type problems where the domain is the unit disk, introduced recently by Czyzowicz et al. in [16], a group of mobile agents collectively search for a hidden item (the exit) placed on the perimeter of the disk, attempting to expedite the time it takes for the last agent to evacuate, i.e. reach the exit. As it was the case in [16], as well as in a series of follow-up improvements and problem variations, the main objective was the design of evacuation algorithms that minimize the *worst case performance*. In contrast, real-life search-and-rescue operations, in which current problems find applications, are mostly concerned with good *average performance*. Keeping also in mind that, in realistic search tasks, mobile agents do not have unbounded resources and at the same time it is imperative that the search terminates successfully with probability 1, one is motivated to study average case - worst case trade-offs for evacuation search problems.

In this direction, we initiate the study of the traditional evacuation problem first introduced in [16] from the perspective of average case analysis, which in our case is equivalent to designing efficient randomized algorithms. More specifically, we introduce problem $2\mathrm{EVAC}^w_{F2F}$ which, at a high level, asks for efficient evacuation algorithms that perform well on average, given that their worst case performance does not exceed w (which can be thought as the maximum time robots can operate, e.g. due to energy restrictions). The problem seems particularly challenging given that the worst case performance analysis of all known evacuation algorithms require tedious analysis, tailored to robots' trajectories, and followed by intense, computer-assisted calculations, which are always numerical. Our results pertain to new families of evacuation algorithms, whose worst case performance analysis can be done rigorously, and whose average case analysis requires again intense computer-assisted calculations, achieving average case - worst case trade-offs for a wide spectrum of values. Our computer-assisted calculations rely on a novel theoretical and unified approach to compute the cost of *any evacuation algorithm* and for *any placement of the hidden item* without relying on tedious analysis specific to robots' trajectories. Equipped with these techniques, we also verify, somehow surprisingly, that the best evacuation algorithms known prior to this work, designed to perform well in the worst case, *do not perform well* for $2\mathrm{EVAC}^w_{F2F}$, adding this way to the motivation of our problem.

1.1   Related Work

In search problems, mobile agents, commonly referred as robots, aim to locate efficiently a hidden item placed in some geometric domain. Numerous search-types problems have been introduced and studied since the 60's, when two seminal papers on probabilistic search, [8] and [9], were concerned with minimizing the *expected time* to locate the item. The number of search-type variants, along with the difficulty of the underlying mathematical problems and the elegance of many results soon gave rise to what is known nowadays as Search Theory. Many of the variants have been classified in surveys, e.g. [10,23], while a number of books provide a comprehensive study for similar problems, e.g. see [1,4,35] and the most recent [5].

Search-type problems have also been studied under the perspective of exploration in [2,3,22,29] by a single robot, and in [12,36,37] by multiple robots. Terrain mapping has been the main search task even in problems where exploration is not the primary objective, e.g. [30,32,34]. Numerous other search-type problems have been introduced and classified as hide-and-seek and pursuit-evasion games, e.g. see [14,25,31,33]. Overall the list of search-type problems is enormous, and having given a representative list above, in what follows we refer only to the most relevant ones.

The perception of a search-type problem as an evacuation problem, from a theoretical perspective, appeared almost a decade ago, e.g. in [7,24]. The problem we study here is a direct follow-up to the evacuation problem $_2\text{EVAC}_{F2F}$ (a search-type problem) first introduced in [16], which included many variants based on the number of robots and the communication model between them. In the variant $_2\text{EVAC}_{F2F}$ which is relevant to our work, two robots start from the center of the unit disk, while an exit is hidden somewhere on the perimeter. The robots move at speed 1, their perception of their environment is restricted to their location and they can exchange information only by meeting. The goal is to minimize the worst case evacuation time, i.e. the time it takes the last robot to reach the exit, over all exit placements. The upper bound of 5.73906 in [16] was later improved to 5.628 in [21], and further to the currently best known 5.625 in [11], while the best lower bound known for the problem is 5.255 due to [21].

Since the introduction of $_2\text{EVAC}_{F2F}$ in [16], a number of variants emerged, focusing on different geometric domains, different number of robots and robots' specifications, different communication models etc. Examples include evacuation from the disk with more than 1 exits in the wireless model [15], evacuation of a group of robots on a line [13] (generalization of the celebrated Cow-Path problem [6]), evacuation in the presence of faulty robots in a line [20] and in a disk [17], evacuation with advice [28] while more recently evacuation with combinatorial requirements on the robots that need to evacuate, e.g. [18,19,26,27].

1.2   Outline of Our Results and Paper Organization

We initiate the study of evacuating 2 robots from the disk in the face-to-face model from an average case complexity perspective. In particular we introduce problem $_2\text{EVAC}_{F2F}^w$ in which one tries to minimize the expected performance of randomized evacuation algorithms, subject to that the worst case performance does not exceed w. The problem is particularly challenging given that existing positive results, from a worst case complexity perspective, rely on tedious theoretical analysis tailored to algorithmic solutions, and supported by intense computer-assisted calculations. One of our main contributions is a unified and simple approach to quantify the performance of any evacuation algorithm and for any input. Equipped with this technique, we first verify that none of the previously known evacuation algorithms has good average case performance. Then, we introduce families of evacuation algorithms that have competitive average case performance, given that their worst case performance does not exceed w, for a wide range of w's. Our results rely on rigorous and technical worst case performance analysis for the newly proposed algorithms. Building upon our new technique for efficiently evaluating the cost of evacuation algorithms for any input, we are able to numerically compute the average case performance of our algorithms, as well as to quantify formally the induced average case -worst case trade-offs.

In Sect. 2.1 we formally define $_2\text{EVAC}_{F2F}^w$ and we give a high-level outline of the results we establish. Section 2.2 contains one of our main contributions, which is a systematic process to compute the performance of any evacuation algorithm, given that robots' trajectories have convenient representations, described in Sect. 2.3. In Sect. 3 we analyze two benchmark algorithms for $_2\text{EVAC}_{F2F}^w$, as well as we motivate further the problem for certain values of w, among others showing, somehow surprisingly, that none of the previously proposed evacuation algorithms is efficient for our problem. Section 4 describes our main contributions in the form of new families of evacuation algorithms. Then, in Sect. 5 we perform rigorous worst case analysis for all new algorithms and in Sect. 6 we perform average case analysis, using our results from Sect. 2.2 along with heavy computer-assisted calculations. In the same section, we also quantify formally all our results for $_2\text{EVAC}_{F2F}^w$. Finally, in Sect. 7 we conclude with some open problems. All omitted proofs of statements in the main body of the paper can be found in the Appendix A.

2 Preliminaries

2.1 Problem Definition and Main Results

In $_2\text{EVAC}_{F2F}$, two searchers (robots) start from the center of the unit disk. Moving at maximum speed 1, the two robots can move anywhere on the plane. Somewhere on the perimeter of the disk there is a hidden object (exit) that can be located by any of the robots only if the robot is co-located with the exit.

The two robots do not see each other from distance, neither can they exchange messages unless they meet (face-to-face model), but they can agree in advance on each other's trajectories. A *feasible evacuation* algorithm is determined by the trajectories of the robots, in which eventually both robots reach the exit. For simplicity, we also require, w.l.o.g. that eventually any robot stays idle. For convenience, we think that the center of the unit disk lies at the origin $(0,0)$ of a Cartesian system, and we denote by cycle(x) the point $(\cos(x), \sin(x))$, which will be referred to as an instance of $_2\text{EVAC}_{F2F}$ when the exit is placed at cycle(x). Given instance cycle(x), we define the *evacuation time* $C(x)$ of the feasible evacuation algorithm as the time it takes the last robot to reach the exit.

In this work we are concerned with determining tradeoffs between the worst case and the average case performance (of uniform placements of the exit) of evacuation algorithms for $_2\text{EVAC}_{F2F}$. More specifically, we say that an evacuation algorithm $\mathscr{A}$ with evacuation cost $C(x)$ on instance cycle(x) is (a, w)-*efficient* if

$$\text{Avg}\,(\mathscr{A}) := \mathbb{E}_{x \in [0, 2\pi)}[C(x)] \leq a,$$
$$\text{Wrs}\,(\mathscr{A}) := \sup_{x \in [0, 2\pi)}\,\{C(x)\} \leq w.$$

where the expectation is with respect to the uniform distribution over $[0, 2\pi)$. Special to our problem is that $\text{Avg}\,(\mathscr{A})$ can also be interpreted as the expected performance of a randomized algorithm based on $\mathscr{A}$. Indeed, consider an algorithm which first performs a random rotation of the disk around the origin of angle θ, where θ is chosen uniformly at random from $[0, 2\pi)$, and then simulates $\mathscr{A}$. This random step is equivalent to choosing a deployment point uniformly at random on the disk. Due to the symmetry of the domain, it is irrelevant where the adversary will place the unique exit, and hence the expected performance of this randomized algorithm equals $\text{Avg}\,(\mathscr{A})$.

For algorithms $\mathscr{A}(p)$ parameterized by parameter(s) p, the pair $(\text{Avg}\,(\mathscr{A}(p)), \text{Wrs}\,(\mathscr{A}(p)))$ will correspond to a subset of $\mathbb{R}^2$ (and a curve if p is only one parameter), that we will refer to as the *Efficient Frontier*. We also adopt an optimization perspective of the problem, and we introduce the following optimization problem $_2\text{EVAC}_{F2F}^w$ on parameter w:

$$\min \frac{1}{2\pi} \int_0^{2\pi} C(x)dx \qquad\qquad (_2\text{EVAC}_{F2F}^w)$$
$$\text{s.t.} \quad C(x) \leq w, \quad \forall x \in [0, 2\pi).$$

Due to an analysis we perform later, $_2\text{EVAC}_{F2F}^w$ is interesting as long as $w_1 \leq w \leq w_2$. At a high level, values w_1, w_2 above are obtained from two benchmark algorithms, $\mathscr{B}_1, \mathscr{B}_2$, where $\text{Wrs}\,(\mathscr{B}_1) = w_1 \approx 5.739, \text{Avg}\,(\mathscr{B}_1) = a_1 \approx 5.1172, \text{Wrs}\,(\mathscr{B}_2) = w_2 \approx 7.283, \text{Avg}\,(\mathscr{B}_1) = a_2 \approx 7.28319$, hence $\mathscr{B}_1$ being efficient in worst case and inefficient in average case, while $\mathscr{B}_2$ being efficient in average case and inefficient in worst case. As it is common for $_2\text{EVAC}_{F2F}$ (and many follow-up variation problems) closed forms for the cost of best-solutions known do not exist, and upper and lower bounds are given numerically. Our results involve upper bounds for a continuous spectrum of parameters w for problem $_2\text{EVAC}_{F2F}^w$. In particular we propose families of algorithms $\mathscr{A}$ (over some parameters) so that, as their parameters vary, we obtain $\text{Wrs}\,(\mathscr{A}) = w$ and $\text{Avg}\,(\mathscr{A}) = g(w)$, for each $w \in [w_1, w_2]$. The curve $(g(w), w)$ summarizing our results is depicted in Fig. 1, and it is later quantified in Theorem 7 (see Sect. 5).

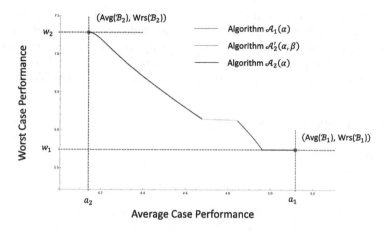

Fig. 1. Illustration of the performance of our solution to $_2\text{EVAC}^w_{F2F}$, for every $w \in [w_1, w_2]$. Depicted curve corresponds to parametric curve $(g(w), w)$, where $w, g(w)$ are the worst case performance and average case performance of three different families of evacuation algorithms $\mathscr{A}_1, \mathscr{A}'_2, \mathscr{A}_2$, discussed formally in Sect. 4. Note that the magenta curve is not a straight line and, as we show next, induces decreasing worst case performance (as the average case performance increases). (Color figure online)

Note that an (a, w)-efficient algorithm gives a solution of value a for $_2\text{EVAC}^w_{F2F}$. Our approach to prove Theorem 7 is to define families of evacuations algorithms $\mathscr{A}(p)$ parameterized by parameter(s) p. We will prove that these algorithms are $(u(p), v(p))$-efficient for some functions $u(p), v(p)$, and in particular the evaluation of the worst case performance will be exact and monotone in p, while the computation of $v(p)$ will be computer-assisted. Then we will set $p = v^{-1}(w)$, and will be able to describe the average case performance as a function of w as $g(w) := u(v^{-1}(w))$.

2.2   Computing Evacuation Times

For any feasible evacuation algorithm, we define by $\mathcal{S}(x)$, the first time that cycle(x) is visited by any robot. Clearly, when a robot, say $\mathscr{R}_1$ locates the exit at cycle(x), it may attempt to catch $\mathscr{R}_2$ while moving along $\mathscr{R}_2$'s trajectory along the shortest line segment, say of length $\mathcal{E}(x)$. Once robots meet, they return together to cycle(x), inducing total evacuation cost $\mathcal{C}(x) = 1 + \mathcal{S}(x) + 2\mathcal{E}(x)$.

All existing results for $_2\text{EVAC}_{F2F}$, from a worst case complexity perspective, rely on numerical computer-assisted estimation of $\sup_x \mathcal{C}(x)$, after identifying properties of the maximizer. In this section, we elevate existing arguments, and we propose a generalized and unified approach for computing $\mathcal{C}(x)$, for any x and for any robots' trajectories. For the sake of formality, as well as for practical purposes, robots' trajectories will be defined by parametric functions $\mathcal{F}(t) = (f(t), g(t))$, where $f, g : \mathbb{R} \mapsto \mathbb{R}$ are continuous and piecewise differentiable. In particular, search protocols for the two robots will be given by trajectories

$\mathscr{R}_1(t), \mathscr{R}_2(t)$, where $\mathscr{R}_i(t)$ will denote the position of robot $\mathscr{R}_i$ at time $t \geq 0$. Therefore, any evacuation algorithm will be identified by a tuple $(\mathscr{R}_1, \mathscr{R}_2)$. To simplify notation, we will only determine the trajectories from the moment the two robots reach the perimeter of the circle, and until the entire circle is searched, and we will silently assume that robots stay put after exploration is over.

Lemma 1. *Consider instance* cycle(x) *of* $_2$EVAC$_{F2F}$, *and suppose that for a feasible evacuation algorithm* $(\mathscr{R}_1, \mathscr{R}_2)$, *robot 1 is the first robot that finds the exit. Then* $\mathcal{E}(x) = \bar{t} - \mathcal{S}(x)$, *where* $\bar{t} = \bar{t}(x)$ *is the smallest root, no less than* $\mathcal{S}(x)$, *of function*

$$h_x(t) := \|\mathscr{R}_2(t) - \mathscr{R}_1(\mathcal{S}(x))\| - t + \mathcal{S}(x). \tag{1}$$

Proof. First observe that $h_x(t)$ is continuous, and assuming that the two robots are not co-located when the exit is found, we have $h_x(\mathcal{S}(x)) > 0$. At the same time, since the evacuation algorithm is feasible, $\mathscr{R}_2(t)$ is eventually a constant, and hence for big enough t we have that $h_x(t)$ becomes eventually negative. By the mean value theorem, there is $t_0 > 0$ for which $h_x(t_0) = 0$.

Now consider the smallest positive root $\bar{t}$ of h_x, no less than $\mathcal{S}(x)$. At time $\bar{t}$, $\mathscr{R}_2$ is located at point $\mathscr{R}_2(\bar{t})$, and it is $\|\mathscr{R}_2(\bar{t}) - \mathscr{R}_1(\mathcal{S}(x))\|$ away from the location cycle(x) of the discovered exit. At the same time, $\mathscr{R}_1$ moves with speed 1 along the shortest path to catch $\mathscr{R}_2$ in her trajectory. Hence it takes $\mathscr{R}_1$ some $\bar{t} - \mathcal{S}(x)$ extra time from the moment the exit is found till she reaches point $\mathscr{R}_2(\bar{t})$. By definition we have $\mathscr{R}_1(\bar{t}) = \mathscr{R}_2(\bar{t})$, and therefore $\mathcal{E}(x) = \bar{t} - \mathcal{S}(x)$ as claimed. □

For some special trajectories, $\mathcal{E}(x)$ admits a simpler description that we describe next. Before that, we introduce some notation pertaining to a function $\delta : [0, \pi] \mapsto \mathbb{R}_+$, which we widely use in the remaining of the paper:

$$\delta(x) := \text{ unique non-negative root (w.r.t. } d\text{) of} \quad \text{``}2\sin\left(x + \frac{d}{2}\right) = d\text{''.} \tag{2}$$

To simplify notation, we will also abbreviate $\delta(x)$ by δ_x. The fact that δ_x is well defined follows easily from the monotonicity of $\sin$ in $[0, \pi]$.

Lemma 2. *For some instance* cycle(x) *of* $_2$EVAC$_{F2F}$, *suppose that for a feasible evacuation algorithm* $(\mathscr{R}_1, \mathscr{R}_2)$, $\mathscr{R}_1$ *is the founder of the exit, say at time* $t_0 = \mathcal{S}(x)$. *Assume that both* $\mathscr{R}_1(t_0), \mathscr{R}_2(t_0)$ *lie on the circle at arc distance* 2α, *and suppose that* $\mathscr{R}_2$'s *movement is along the perimeter of the circle toward the complementary arc of length* $2\pi - \alpha$. *Then,* $\mathcal{E}(x) = \delta_\alpha$.

Proof. The lemma follows by applying transformation $t - \mathcal{S}(x) = d$ in the definition of $h_x(t)$ in Lemma 1, so that $\mathcal{E}(x) = t - \mathcal{S}(x) = d$. □

We are ready to conclude with a corollary that will be handy for computing evacuation times numerically, and without relying on excessive case analysis, as it was the case before.

Corollary 1. *Consider feasible evacuation algorithm* $(\mathscr{R}_1, \mathscr{R}_2)$ *for* $_2\text{EVAC}_{F2F}$. *For any instance* $\text{cycle}(x)$ *for which* $\mathscr{R}_1$ *is the exit founder, the evacuation cost can be computed as* $\mathcal{C}(x) = 1 + 2\bar{t} - \mathcal{S}(x)$, *where* $\bar{t} = \bar{t}(x)$ *is the smallest root, at least* $\mathcal{S}(x)$, *of* $h_x(t) := \|\mathscr{R}_2(t) - \mathscr{R}_1(\mathcal{S}(x))\| - t + \mathcal{S}(x)$.

2.3 Trajectories' Description

Robots' trajectories will be described in phases. We will always omit the "deployment phase", i.e. the movement from the circle center to its perimeter, and we will only describe the trajectories from the moment robots start searching the circle. In each phase, robot $\mathscr{R}$, will be moving between two explicit points, either along an arc, or along a line segment (chord of an arc), see Observations 1 and 2 below. We will summarize robot's trajectories in tables of the following format.

Robot	Phase #	Trajectory	Duration
$\mathscr{R}$	1	$\mathscr{R}(t)$	t_1
	2	$\mathscr{R}(t)$	t_2
	$\vdots$		$\vdots$

In order to ease notation, trajectory $\mathscr{R}(t)$ of phase i will be described with parametric equations as if the time is reset to 0 after time $t_0 + t_1 + t_2 + \ldots + t_{i-1}$, where $t_0 = 1$ (this is the time that robots reach the circle). The two fundamental trajectory components are movements along arcs and movements along line segments.

Observation 1. *Let* $b \in [0, 2\pi)$ *and* $\sigma \in \{-1, 1\}$. *The trajectory of an object moving at speed 1 on the perimeter of a unit circle with initial location* $\text{cycle}(b)$ *is given by the parametric equation* $\text{cycle}(\sigma t + b) = (\cos(\sigma t + b), \sin(\sigma t + b))$. *If* $\sigma = 1$ *the movement is counter-clockwise (ccw), and clockwise (cw) otherwise.*

Observation 2. *Consider distinct points* $A = (a_1, a_2), B = (b_1, b_2)$ *in* $\mathbb{R}^2$. *The trajectory of a speed 1 object moving along the line passing through* A, B *and with initial position* A *is given by the parametric equation* $\text{line}(A, B, t) := \left(\frac{b_1 - a_1}{\|A - B\|}t + a_1, \frac{b_2 - a_2}{\|A - B\|}t + a_2\right)$.

Finally, the analysis of our algorithms' trajectories will give rise to a number of constants. For the reader's convenience, we list here the numerical values of the most common constants that will be encountered later; $w_1 \approx 5.73906, w_0 \approx 6.11953, w' \approx 6.12851, w_2 \approx 7.28319, \alpha' \approx 1.15468, \bar{\alpha} \approx 1.54419, \beta' \approx 0.0241653, \beta_0 \approx 0.04388$. All constants are formally defined when they are first introduced.

3   Two Benchmark Algorithms and Motivation

In this section we describe two benchmark algorithms for $_2\text{EVAC}_{F2F}$, as well as perform average case analysis to algorithms previously proposed in the literature. The reader may consult Fig. 2 for the algorithms analyzed in this section.

Czyzowicz et al. [16] were the first to introduce an evacuation algorithm for $_2\text{EVAC}_{F2F}$, which we denote here by $\mathscr{B}_1$ (see Fig. 2 on the left).

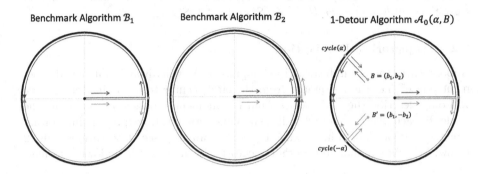

Fig. 2. Robots' Trajectories for algorithms $\mathscr{B}_1, \mathscr{B}_2, \mathscr{A}_0$. The depicted trajectories show the search of the circle, and not the evacuation step that is performed once the exit is found.

Definition 1 (Benchmark Algorithm $\mathscr{B}_1$). *For all* $t \in [0, \pi]$, $\mathscr{R}_1(t) = cycle(t)$ *and* $\mathscr{R}_2(t) = cycle(-t)$.

Observation 3. *Benchmark Algorithm $\mathscr{B}_1$ is* $(5.1172, 5.73906)$*-efficient.*

$\mathscr{B}_1$ should be understood as being efficient in the worst case, but inefficient on average. The claim becomes transparent by introducing the following *naive* algorithm for $_2\text{EVAC}_{F2F}$ that we depict in the middle of Fig. 2.

Definition 2 (Benchmark Algorithm $\mathscr{B}_2$). *For each* $t \in [0, 2\pi]$, $\mathscr{R}_1(t) = \mathscr{R}_2(t) = cycle(t)$.

Observation 4. *Benchmark Algorithm $\mathscr{B}_2$ is* $(1 + \pi, 1 + 2\pi)$*-efficient.*

$\mathscr{B}_2$ should be understood as highly efficient on average, but inefficient in the worst case. Moreover, it should be clear that $\mathscr{B}_1, \mathscr{B}_2$ are feasible solutions to $_2\text{EVAC}_{F2F}^w$, for $w = 5.1172$ and $w = 1 + 2\pi$, respectively. We conjecture that $\mathscr{B}_1$ is indeed the optimal evacuation algorithm among all algorithms with worst case performance no more than $1 + 2\pi$. At the same time, below we show that $\mathscr{B}_2$ is the best algorithm for $_2\text{EVAC}_{F2F}^w$, when $w = 5.1172$, among those previously used to improve upon the worst case performance. The importance of this observation is twofold; first we are motivated to study $_2\text{EVAC}_{F2F}^w$ for the entire spectrum of $w \in [\text{Wrs}(\mathscr{B}_1), \text{Wrs}(\mathscr{B}_2)]$, and second we deduce that in order to perform well on average, we need to devise and analyze new evacuation algorithms.

Upper bounds for the worst case performance of $\mathscr{B}_1$ were later improved in [11, 21], first to 5.628, and then to 5.625, using refined algorithms, respectively. The main idea behind the improvement is to understand the monoticity of $\mathcal{C}(x)$ for algorithm $\mathscr{B}_1$. Indeed, the following lemma was implicit in both [11, 21], and can be obtained numerically.

Lemma 3. *There is α_0, where $\alpha_0 \approx 0.96782$, so that evacuation cost $\mathcal{C}(x)$ of $\mathcal{B}_1$ for $_2\mathrm{EVAC}_{F2F}$ on instance cycle(x) is strictly increasing for $x \in [0, \alpha_0]$, and strictly decreasing in $x \in [\alpha_0, \pi]$. In particular, $\mathrm{Wrs}\,(\mathcal{B}_1) = \mathcal{C}(\alpha_0) \approx 5.73906$.*

Consider now an execution of $\mathcal{B}_1$ in which one of the robots, say $\mathcal{R}_2$ continues searching on the circle and is close to approach a location that would be the meeting point if the instance was cycle(α_0). In an attempt to help expedite a potential meeting (in case $\mathcal{R}_1$ is approaching) and effectively reducing the cost of the worst case, $\mathcal{R}_2$ would make a minor detour toward the interior of the disk, before returning back to the exploration of the circle. This simple idea was explored in [21] where the following family of algorithms were introduced, parameterized by $\alpha \in [0, \pi]$ and point B within the unit disk, see also right of Fig. 2.

Definition 3 (1-Detour Algorithm $\mathscr{A}_0(\alpha, B)$). *For all $t \in [0, \pi + 2\,\|$cycle $(\alpha) - B\|]$, the trajectory of $\mathcal{R}_1$ is defined as*

Robot	Phase #	Trajectory	Duration
$\mathcal{R}_1$	1	cycle(t)	α
	2	line(cycle$(\alpha), B, t$)	$\|$cycle$(\alpha) - B\|$
	3	line(B, cycle$(\alpha), t$)	$\|$cycle$(\alpha) - B\|$
	4	cycle$(t + \alpha)$	$\pi - \alpha$

The trajectory of $\mathcal{R}_2$ is symmetric with respect to the horizontal axis.

The crux of the contribution of [21] was to prove that there exists α, B for which the worst case performance is no more than 5.644 (and a delicate refinement is needed to achieve 5.628). Notably, their analysis is tedious and lengthy, whereas we can obtain the same result, relying again on numerical calculations, with minimal effort. Then, [11] introduced variations of $\mathscr{A}_0(\alpha, B)$ in which each robot performs more than 1 detours (see Phases 2,3 of $\mathscr{A}_0(\alpha, B)$). Hence, t-detour algorithms are parameterized by a sequence $\alpha_1, \ldots, \alpha_t$, where $\alpha_i \geq 0$ and $\sum_i \alpha_i \leq \pi$, and points B_i in the disk. Even 2-detour algorithms achieve worst case performance 5.625, while for each $t \geq 2$, t-detour algorithms do induce strictly improved performance (for appropriate choices of the parameters) but the improvement is negligible.

Motivated by the results in [11,21], one is tempted to ask whether any algorithm in the family $\mathscr{A}_0(\alpha, B)$ improves upon $\mathcal{B}_1$ with respect to the average case analysis.

Theorem 5. *For every $\alpha \in [0, \pi)$ and for every B in the unit disk $\mathrm{Avg}\,(\mathscr{A}_0(\alpha, B)) \geq \mathrm{Avg}\,(\mathcal{B}_1)$.*

Theorem 5 provides strong motivation for studying problem $_2\mathrm{EVAC}_{F2F}^w$, since it shows that in oder to establish good upper bounds, i.e. our main results depicted in Fig. 1 and quantified later in Theorem 7, one needs to employ new evacuation algorithms. Recall that even $\mathrm{Wrs}\,(\mathcal{B}_1)$ that was first calculated in [16], or $\mathrm{Wrs}\,(\mathscr{A}_0(\alpha, B))$ first calculated in [21] for various α, B, were all estimated with computer-assisted calculations. Due to the nature of the problem, we are bound to rely on computer-assisted calculations as well. Notably, our much more intense computational work is feasible only because we employ the brand new method for computing evacuation times due to Corollary 1 and Definition 3 of $\mathscr{A}_0(\alpha, B)$ trajectories. Overall, in order to verify Theorem 5 we compute pairs $(\mathrm{Avg}\,(\mathscr{A}_0(\alpha, B)), \mathrm{Wrs}\,(\mathscr{A}_0(\alpha, B)))$ for more than 500,000 different parameter values and we depict them in Fig. 3.

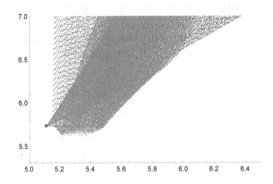

Fig. 3. Performance analysis of $\mathscr{A}_0(\alpha, B)$ for various values of parameters α, B. Blue points (a, w) correspond to (a, w)-efficient algorithms $\mathscr{A}_0(\alpha, B)$. The red point is $(\mathrm{Avg}\,(\mathscr{B}_1), \mathrm{Wrs}\,(\mathscr{B}_1))$, i.e. the performance of $\mathscr{B}_1$ in the average-worst case space. Note that no algorithm $\mathscr{A}_0$ performs better on average than $\mathscr{B}_1$, while all $\mathscr{A}_0(t, \mathrm{cycle}(t))$ is exactly $\mathscr{B}_1$ for every point $t \in [0, \pi]$. (Color figure online)

4   New Evacuation Algorithms

In this section we propose families of evacuation algorithms for problem $_2\mathrm{EVAC}^w_{F2F}$, for the entire spectrum of $w \in [\mathrm{Wrs}\,(\mathscr{B}_1), \mathrm{Wrs}\,(\mathscr{B}_2)]$. Our algorithms are summarized in Fig. 4.

First we define families of evacuation algorithms that, as we show next, perform well for $_2\mathrm{EVAC}^w_{F2F}$ in the "neighborhood of $\mathscr{B}_1$", i.e. for w close to $\mathrm{Wrs}\,(\mathscr{B}_1)$. Our algorithms are parameterized by α, and their circle exploration lasts $2\pi - \alpha$.

Definition 4 (Algorithm $\mathscr{A}_1(\alpha)$). *For all $t \in [0, 2\pi - \alpha]$, the trajectory of $\mathscr{R}_1$ is defined as*

Robot	Phase #	Trajectory	Duration
$\mathscr{R}_1$	1	$\mathrm{cycle}(t)$	α
	2	$\mathrm{line}(\mathrm{cycle}(\alpha), \mathrm{cycle}(-\alpha - \delta_\alpha), t)$	δ_α
	3	$\mathrm{cycle}(-\alpha - \delta_\alpha - t)$	$2\pi - 2\alpha - \delta_\alpha$

where δ_a is defined in (2). The trajectory of $\mathscr{R}_2$ is defined as $\mathscr{R}_2(t) = \mathrm{cycle}(-t)$, for all $t \in [0, 2\pi - \alpha]$.

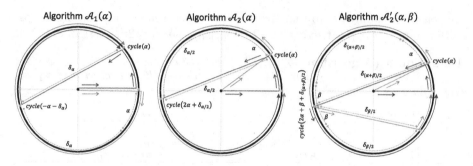

Fig. 4. Robots' Trajectories for algorithms $\mathscr{A}_1, \mathscr{A}_2, \mathscr{A}_2'$. The depicted trajectories show the search of the circle, and not the evacuation step that is performed once the exit is found. Arcs that are searched by both robots are also searched simultaneously, i.e. robots are co-located and search together.

$\mathscr{A}_1$ is depicted in Fig. 4 on the left. At a high level $\mathscr{A}_1(\alpha)$ is a modification of $\mathscr{B}_1$ that is based on the following idea. The execution of $\mathscr{A}_1(\alpha)$ is the same as in $\mathscr{B}_1$ till each robot searches an arc of length α (and hence $\mathscr{A}(\pi)$ coincides with $\mathscr{B}_1$). After time α, $\mathscr{R}_1$ abandons her trajectory and catches $\mathscr{R}_2$, on the perimeter of the circle resembling a trajectory as if the exit was located at $\mathscr{R}_1(\alpha)$. It is not difficult to see that the definition of δ_α above satisfies $\mathscr{R}_1(\alpha + \delta_\alpha) = \mathscr{R}_2(\alpha + \delta_\alpha) = \text{cycle}(-\alpha - \delta_\alpha)$.

Next we define a family of algorithms $\mathscr{A}_2$ which, as we show later, perform well in the "neighborhood of $\mathscr{B}_2$", i.e. for w close to $\text{Wrs}(\mathscr{B}_2)$. For this recall definition (2) of δ_a. We let $\gamma_0 \approx 2.2412$ be the root of $2\alpha + \delta_{\alpha/2} = 2\pi$. For every $\alpha \leq \gamma_0$ we define a family of algorithms on parameter α whose circle exploration lasts $2\pi - \alpha$.

Definition 5 (Algorithm $\mathscr{A}_2(\alpha)$). *For all $t \in [0, 2\pi - \alpha]$, the trajectory of $\mathscr{R}_1$ is defined as*

Robot	Phase #	Trajectory	Duration
$\mathscr{R}_1$	1	cycle(t)	α
	2	line(cycle(α), cycle($2\alpha + \delta_{\alpha/2}$), t)	$\delta_{\alpha/2}$
	3	cycle($2\alpha + \delta_{\alpha/2} + t$)	$2\pi - 2\alpha - \delta_{\alpha/2}$

The trajectory of $\mathscr{R}_2$ is defined as $\mathscr{R}_2(t) = \text{cycle}(\alpha + t)$, for all $t \in [0, 2\pi - \alpha]$.

$\mathscr{A}_2$ is depicted in the middle of Fig. 4. The condition that $\alpha \leq \gamma_0$ is added for simplicity to ensure that the latest catching point occurs while the other robot is still searching, and is not mandatory. At a high level $\mathscr{A}_2(\alpha)$ is a generalization of $\mathscr{B}_2$ (note that $\mathscr{A}_2(0) = \mathscr{B}_2$). For the first α time units, robots search in the same direction till $\mathscr{R}_1$ arrives at the deployment point of $\mathscr{R}_2$. Then, $\mathscr{R}_1$ catches $\mathscr{R}_2$ on the circle, as if the exit was located at $\mathscr{R}_1(\alpha)$ (which by Lemma 2 happens in $\delta_{\alpha/2}$ extra time).

Finally we introduce a family of evacuation algorithms which will perform well for $_2\text{EVAC}_{F2F}^w$ for intermediate values of $w \in [\text{Wrs}(\mathscr{B}_1), \text{Wrs}(\mathscr{B}_2)]$. For this we generalize family $\mathscr{A}_2$ so that the two robots perform two alternating jumps, with parameters α, β satisfying $2\alpha + 2\beta + \delta_{(\alpha+\beta)/2} + \delta_{\beta/2} \leq 2\pi$, see right of Fig. 4.

Definition 6 (Algorithm $\mathscr{A}_2'(\alpha, \beta)$). *For notational convenience, we set $\zeta_{\alpha,\beta} := 2\alpha + \beta + \delta_{(\alpha+\beta)/2}$. For all $t \in [0, 2\pi - \alpha - \beta]$, the trajectories of $\mathscr{R}_1, \mathscr{R}_2$ are defined as follows*

Robot	Phase #	Trajectory	Duration
$\mathscr{R}_1$	1	cycle(t)	α
	2	line(cycle(α), cycle $(\zeta_{\alpha,\beta})$, t)	$\delta_{(\alpha+\beta)/2}$
	3	cycle $(\zeta_{\alpha,\beta} + t)$	$2\pi - 2\alpha - \beta - \delta_{(\alpha+\beta)/2}$
$\mathscr{R}_2$	1	cycle($\alpha + t$)	$\alpha + \beta + \delta_{(\alpha+\beta)/2}$
	2	line(cycle $(\zeta_{\alpha,\beta})$, cycle $(\zeta_{\alpha,\beta} + \delta_{\beta/2})$, t)	$\delta_{\beta/2}$
	3	cycle $(\zeta_{\alpha,\beta} + \beta + \delta_{\beta/2} + t)$	$2\pi - 2\alpha - 2\beta - \delta_{(\alpha+\beta)/2} - \delta_{\beta/2}$

Robots' trajectories α, β have the following meaning. As in the family of algorithms $\mathscr{A}_2$, parameter α represents the arc distance the two robots have before the one preceding decides to jump ahead. In $\mathscr{A}_2$ the two robots meet again once the jumper reaches the perimeter of the circle. In $\mathscr{A}_2'$ the jumper deploys a little further away on the circle so that when the other robot reaches the deployment point of the jumper, the two robots are at arc distance β. As a result, the time it takes both robots to complete searching the entire circle is $2\pi - \alpha - \beta$, as well as $\mathscr{A}_2'(\alpha, 0)$ coincides with $\mathscr{A}_2(\alpha)$. Finally, note that even though $\mathscr{A}_2'$ will be eventually invoked for seemingly restricted values of β ($\beta \leq \beta_0 \approx 0.04388$), the deviation in the performance will be significant enough (e.g. $\delta_{\beta_0/2} \approx 0.977997$) to account for its utilization in our upper bounds.

5 Worst Case Performance Analysis

In this section we perform worst case analysis for all algorithmic families $\mathscr{A}_1, \mathscr{A}_2, \mathscr{A}_2'$ with respect to their parameters. Notably, results in this section are quantified formally and exactly by closed formulas. At a high level, each of $\mathscr{A}_1, \mathscr{A}_2, \mathscr{A}_2'$ will be invoked to solve $_2\text{EVAC}_{F2F}^w$ for different values of $w \in [\text{Wrs}(\mathscr{B}_1), \text{Wrs}(\mathscr{B}_2)]$, and each of them will have competitive average case performance for the corresponding worst case performance w. As an easy warm-up, we analyze $\mathscr{A}_1$.

Lemma 4 (Worst Case Analysis for $\mathscr{A}_1$). *Let $\bar{\alpha} = 1 + 2\pi - w_1$, where $w_1 = \text{Wrs}(\mathscr{B}_1)$. Then, for all $\alpha \in [0, \pi]$, we have that*

$$\text{Wrs}(\mathscr{A}_1(\alpha)) = \begin{cases} 1 + 2\pi - \alpha, & \forall \alpha \in [0, \bar{\alpha}) \\ \text{Wrs}(\mathscr{B}_1), & \forall \alpha \in [\bar{\alpha}, \pi] \end{cases}.$$

In a similar fashion, we can easily analyze $\mathscr{A}_2$.

Lemma 5 (Worst Case Analysis for $\mathscr{A}_2$). *For all $\alpha \leq \pi - 2$, we have* $\text{Wrs}(\mathscr{A}_2(\alpha)) = 1 + 2\pi - \alpha$.

Next, our goal is to analyze $\mathscr{A}_2'(\alpha, \beta)$, which is much more technical. For this we will invoke $\mathscr{A}_2'$ only for special parameters, whose choice is motivated by the following observation pertaining to the performance of $\mathscr{A}_2$ (whose generalization is $\mathscr{A}_2'$). From the proof of Lemma 5, it follows that among all algorithms $\mathscr{A}_2(\alpha)$, where $\alpha \leq \gamma_0$ (see discussion before Definition 5), the one with minimum worst case evacuation cost is $\mathscr{A}_2(\pi - 2)$, and the cost becomes $3 + \pi$. In fact, for all $w \in [3+\pi, 1+2\pi]$ there are two different values of α for which $\text{Wrs}\,(\mathscr{A}_2(\alpha)) = w$, and we restrict $\alpha \in [0, \pi - 2]$ so that we obtain evacuation algorithms with minimum average case cost. Moreover, $\alpha = \pi - 2$ is the only parameter for which $\text{Wrs}\,(\mathscr{A}_2(\alpha)) = 3 + \pi$ and as a byproduct, it is the algorithm in the family $\mathscr{A}_2$ that minimizes the worst case.

By Lemma 5 we know that as $\beta \to 0$, the value of α that minimizes $\text{Wrs}\,(\mathscr{A}_2'(\alpha, \beta))$ approaches $\pi - 2$. That value of α is what made the evacuation cost of $\mathscr{A}_2(\alpha)$ attain the same value in two different (worst case) exit placements. Motivated by this, and for values of $\beta > 0$ not too big, we still find the optimal choices of α that minimize the worst case performance.

Lemma 6 (Worst Case Analysis for $\mathscr{A}_2'$). *Let $\beta_0 = 0.0438855$, and set $\alpha_\beta := \pi - \beta/2 - 2\cos\,(\beta/4)$. Then for all $\beta \in [0, \beta_0]$ we have $\text{Wrs}\,(\mathscr{A}_2'(\alpha_\beta, \beta)) = 1 + \pi - \beta/2 + 2\cos\,(\beta/4)$.*

6   Average Case Performance Analysis and the Efficient Frontier

In this section we perform average case analysis for all algorithmic families $\mathscr{A}_1, \mathscr{A}_2, \mathscr{A}_2'$, with respect to their parameters. For the sake of exposition of our results, we set $w_1 = \text{Wrs}\,(\mathscr{B}_1) \approx 5.73906, w_2 = \text{Wrs}\,(\mathscr{B}_2) = 1 + 2\pi \approx 7.28319$ and for $\beta_0 \approx 0.04388$, as in Lemma 6, we set $w_0 := \text{Wrs}\,(\mathscr{A}_2'(\alpha_{\beta_0}, \beta_0)) \approx 6.11953$. We also recall $\bar{\alpha} \approx 1.54419$ of Lemma 4. Finally, we set

$$v(\alpha) := 1 + 2\pi - \alpha$$
$$v_2(\beta) := 1 + \pi - \beta/2 + 2\cos\,(\beta/4)$$
$$u_1(\alpha) := 0.00889\alpha^3 - 0.16944\alpha^2 + 0.71518\alpha + 4.23089$$
$$u_2'(\beta) := 530.673\beta^3 - 78.5498\beta^2 + 7.36219\beta + 4.70493$$
$$u_2(\alpha) := 0.093056\alpha^2 + 0.346659\alpha + 4.1719$$

Combined with our findings of Sect. 5, the main result of the current section is the following.

Theorem 6. *For every $w \in [w_1, w_2]$ there is algorithm $\mathscr{A} \in \{\mathscr{A}_1, \mathscr{A}_2', \mathscr{A}_2\}$ and unique parameter(s) p such that $\text{Wrs}\,(\mathscr{A}(p)) = w$. In particular,*

- *for all $\alpha \in [1, \bar{\alpha}]$, $\mathscr{A}_1(\alpha)$ is $(u_1(\alpha), v(\alpha))$-efficient, and $v([1, \bar{\alpha}]) = [w_1, 2\pi]$,*
- *for all $\beta \in [0, \beta_0]$, $\mathscr{A}_2'(\alpha_\beta, \beta)$ is $(u_2'(\beta), v_2(\beta))$-efficient, and $v_2([0, \beta_0]) = [w_0, 3 + \pi]$,*

– *for all* $\alpha \in [0, \pi - 2]$, $\mathscr{A}_2(\alpha)$ *is* $(u_2(\alpha), v(\alpha))$-*efficient, and* $v([0, \pi - 2]) = [3 + \pi, w_2]$.

Finally, we aim to formally quantify the efficient frontier of our algorithms as depicted in Fig. 1 (see Sect. 2.1). The parametric curves described in Theorem 6 provide, strictly speaking, an upper bound for the parametric curve of Fig. 1. Next, we compute $g : \mathbb{R} \mapsto \mathbb{R}$, so that the parametric curves of Theorem 6 are written in the form $\{(g(w), w)\}_{w \in [w_1, w_2]}$. That would also imply that there is a solution to $_2\mathrm{EVAC}^w_{F2F}$ of cost at most $g(w)$.

In that direction, we study each evacuation algorithm family $\mathscr{A}(p)$ with worst case performance, say, $v(p)$, and average case upper bound, say, $u(p)$. For each $w \in [w_1, w_2]$ in the range of $\mathscr{A}(p)$, we set $p = v^{-1}(w)$ so that the average case performance achieved becomes $u(v^{-1}(w))$.

Recall that $\mathrm{Wrs}\,(\mathscr{A}_i(\alpha)) = v(\alpha)$, so that $v^{-1}(w) = 1 + 2\pi - w$, and hence for algorithms $\mathscr{A}_i$ we can easily compute $u_i(v^{-1}(w))$, $i = 1, 2$. For $\mathscr{A}'_2$ we recall that $\mathrm{Avg}\,(\mathscr{A}'_2(\alpha_\beta, \beta))$ is decreasing in β. Since v_2^{-1} does not admit a closed form, we need to observe that $2.999 + \pi - \beta/2 \le v_2(\beta) \le 3 + \pi - \beta/2$ for all $\beta \in [0, \beta_0]$ so that an upper bound for $\mathrm{Avg}\,(\mathscr{A}'_2(\alpha_\beta, \beta))$ admitting worst case performance w can be computed by $u'_2(12.2812 - 2w)$.

Now for each $w \in [w_1, w_2]$ we need to specify which of the evacuation algorithms we will invoke. Note that in Theorem 6 we chose the range of α in $\mathscr{A}_1$ to start from 1 so that as to guarantee that $\mathrm{Wrs}\,(\mathscr{A}_1(1)) \ge w_0$. We note that $u'_2(12.2812 - 2w) = u_1(1 + 2\pi - w)$ for $w' \approx 6.12851$, so algorithm $\mathscr{A}_1$ should be invoked for $w \in [w_1, w']$ (and w' is obtained for $\alpha' := 1 + 2\pi - w' \approx 1.15468$), then $\mathscr{A}'_2$ for $w \in [w', 3 + \pi]$ (and w' is obtained for β' so that $v_2(\beta') = w'$, where $\beta' \approx 0.0241653$), and $\mathscr{A}_2$ for $w \in [3 + \pi, w_2]$. We conclude with the next Theorem (for convenience, the values of all constants are summarized at the end of Sect. 2.3).

Theorem 7. *For every* $w \in [w_1, w_2]$, *the optimal solution to* $_2\mathrm{EVAC}^w_{F2F}$ *is at most* $g(w)$, *where*

$$g(w) = \begin{cases} -0.00889w^3 + 0.0248026w^2 + 0.338241w + 3.88629, & w \in [w_1, w']\ (\mathscr{A}_1(\alpha),\ \alpha \in [\alpha', \bar{\alpha}]) \\ -4245.38w^3 + 77893.3w^2 - 476397.w + 971235, & w \in [w', 3 + \pi]\ (\mathscr{A}_2(\alpha_\beta, \beta),\ \beta \in [0, \beta']) \\ 0.093056w^2 - 1.70215w + 11.6328, & w \in [3 + \pi, w_2]\ (\mathscr{A}_2(\alpha),\ \alpha \in [0, \pi - 2]) \end{cases}$$

7　Conclusion and Open Problems

Our work suggests a number of open problems directly aiming to understand $_2\mathrm{EVAC}^w_{F2F}$ better. Apart from generally improving our upper bounds, we find the following list of questions particularly interesting and challenging:

(a) Note that when $w = \mathrm{Wrs}\,(\mathscr{B}_1)$, we presented algorithm $\mathscr{A}_1(\alpha)$ which, for certain value of α, has worst case performance equal to w and average case performance less that $\mathrm{Avg}\,(\mathscr{B}_1)$. Is there an algorithm whose average case performance is no more than $\mathrm{Avg}\,(\mathscr{B}_1)$, and worst case performance strictly less than w?

(b) Is it true that the best possible efficient frontier is given by a smooth transition between families of evacuation algorithms? Note that $\mathscr{A}_2$ naturally extends $\mathscr{B}_2$, $\mathscr{A}_2'$ naturally extends $\mathscr{A}_2$, and that $\mathscr{A}_1$ naturally extends $\mathscr{B}_1$. However, $\mathscr{A}_1$ and $\mathscr{A}_2'$ behave differently, even though their efficient frontier agrees for certain values of the parameters.

(c) $\mathrm{Avg}\,(\mathscr{B}_2) = 1 + \pi$, and none of our algorithms beat this performance. We conjecture that this is the best possible average evacuation time, even in the wireless model, and for any number of robots.

Apart from the list above, we believe that the direction of studying randomized algorithms for evacuation-type problems, especially with respect to average case/worst case trade-offs is of special interest, and should be considered for existing as well as for new search problems in the area.

A   Appendix

A.1   Observation 3

Proof (Observation 3). Note that it takes time π to search the entire circle, and that the two trajectories are symmetric with respect to horizontal axis. Therefore, we may assume that the instance cycle(x) satisfies $x \in [0, \pi]$.

Clearly, for any such x, we have that $\mathcal{S}(x) = x$. By Lemma 2, we have that $\mathcal{C}(x) = 1 + \mathcal{S}(x) + 2\mathcal{E}(x) = 1 + x + 2\delta_x$. Numerical calculations (software assisted) show that

$$\mathrm{Wrs}\,(\mathscr{B}_1) = \sup_{x \in [0,\pi]} \{\mathcal{C}(x)\} = \sup_{x \in [0,\pi]} \{1 + x + 2\delta_x\} \approx 5.73906,$$

$$\mathrm{Avg}\,(\mathscr{B}_1) = \mathrm{E}_{x \in [0,\pi]}[\mathcal{C}(x)] = \frac{1}{\pi} \int_{x=0}^{\pi} (1 + x + 2\delta_x)\,dx \approx 5.1172.$$

$\square$

A.2   Observation 4

Proof (Observation 4). It is easy to see that for all $x \in [0, 2\pi)$ we have $\bar{t}(x) = \mathcal{S}(x) = x$ and $\mathcal{E}(x) = 0$. Therefore $\mathcal{C}(x) = 1 + x$, and hence

$$\mathrm{Wrs}\,(\mathscr{B}_2) = \sup_{x \in [0,2\pi)} \{\mathcal{C}(x)\} = 1 + 2\pi,$$

$$\mathrm{Avg}\,(\mathscr{B}_2) = \mathrm{E}_{x \in [0,2\pi)}[\mathcal{C}(x)] = \int_{x=0}^{2\pi} (1 + x)\,dx = 1 + \pi.$$

$\square$

A.3 Lemma 4

Proof (Lemma 4). First it is easy to show that the worst case evacuation time is induced either when $\mathscr{R}_1$ finds the exit while moving from cycle(0) to cycle(α), or while $\mathscr{R}_1, \mathscr{R}_2$ are exploring the circle together (after having met). By Lemma 2, the cost in the first case would be

$$\max_{0 \le x \le \alpha} \{1 + x + 2\delta_x\} = \begin{cases} 1 + \alpha + 2\delta_\alpha, & \text{if } \alpha \le \alpha_0 \\ \text{Wrs}(\mathscr{B}_1), & \text{otherwise} \end{cases}$$

where the values of the piecewise function above follow from Lemma 3. In the other case, the worst placement of exit is obtained using instances cycle($\alpha + \epsilon$) for arbitrary small values of $\epsilon > 0$ in which case the evacuation cost becomes $1 + 2\pi - \alpha$.

Overall, is easy to see that $1 + \alpha_0 + 2\delta_{\alpha_0} \le 1 + 2\pi - \alpha_0$ showing that the dominant evacuation cost when $\alpha \le \bar{\alpha}$ is $1 + 2\pi - \alpha$. For $\alpha > \bar{\alpha}$ the evacuation cost becomes equal to w_1. □

A.4 Lemma 5

Proof (Lemma 5). We distinguish three cases as to where the exit is. If $x \in [0, \alpha)$, then the worst instance cycle(x) is when $x = \alpha - \epsilon$ for arbitrarily small $\epsilon > 0$, and the cost is $1 + \alpha + 2\delta_{\alpha/2}$. In the second case $x \in [\alpha, 2\alpha + \delta_{\alpha/2})$ and it is not difficult to see that the worst case induced cost in this case is not more than that of the first case. Finally, in the third case $x \in [2\alpha + \delta_{\alpha/2}, 2\pi)$, and the two robots move together, so the total cost, in the worst case, is $1 + 2\pi - \alpha$, when $x = 2\pi - \epsilon$ for arbitrarily small $\epsilon > 0$. It is not difficult to see that the dominant case is actually the third one, and in fact the two cases induce the same cost when $\pi = \alpha + \delta_{\alpha/2}$. By the definition of $\delta_{\alpha/2}$ we know that $\delta_{\alpha/2} = 2 \sin\left(\frac{\alpha + \delta_{\alpha/2}}{2}\right) = 2 \sin(\pi/2) =$ '2. Hence the costs become equal when $\alpha = \pi - 2$. □

A.5 Lemma 6

Proof (Lemma 6). Let $w(\beta) = 1 + \pi - \beta/2 + 2\cos(\beta/4)$. First we show that $w(\beta)$ is the worst case performance of $\mathscr{A}'_2(\alpha_\beta, \beta)$ for two specific placements of the exit.

We proceed by describing evacuation cost $\mathcal{C}(x)$ assuming two arbitrary α, β for two different instances cycle(x). Using Lemma 2, we see that

$$\lim_{\epsilon \to 0^+} \mathcal{C}(\alpha - \epsilon) = 1 + \lim_{\epsilon \to 0^+} \mathcal{S}(\alpha - \epsilon) + 2 \lim_{\epsilon \to 0^+} \mathcal{E}(\alpha - \epsilon) = 1 + \alpha + 2\delta_{\alpha/2}. \quad (3)$$

Since the total search time is $2\pi - \alpha - \beta$, we also see that

$$\lim_{\epsilon \to 0^+} \mathcal{C}(2\pi - \epsilon) = 1 + 2\pi - \alpha - \beta. \quad (4)$$

Now we claim that (3), (4) are equal when $\alpha = \alpha_\beta$. Indeed, equating (3), (4) gives

$$a + \delta_{\alpha/2} = \pi - \beta/2. \quad (5)$$

But then, using (2), we see that

$$\delta_{\alpha/2} = 2\sin\left(\frac{\alpha + \delta_{\alpha/2}}{2}\right) = 2\sin\left(\frac{\pi - \beta/2}{2}\right) = 2\cos\left(\beta/4\right). \tag{6}$$

Substituting (6) into (5), we see that the value of α for which (3), (4) are equal satisfies $\alpha = \pi - \beta/2 - 2\cos\left(\beta/4\right)$, as promised. Substituting this special value of $\alpha = \alpha_\beta$ either in (3) or in (4) induces evacuation cost $w(\beta) = 1 + \pi - \beta/2 + 2\cos\left(\beta/4\right)$.

Next we show that as long as β is not too big, $w(\beta)$ is indeed the worst case evacuation cost. We consider the following cases $x \in I_i$, $i = 1, \ldots, 4$ for possible instances cycle(x); $I_1 := [0, \alpha)$, $I_2 := [\alpha, 2\alpha + \beta + \delta_{(\alpha+\beta)/2})$, $I_3 := [2\alpha + \beta + \delta_{(\alpha+\beta)/2}, 2\alpha + 2\beta + \delta_{(\alpha+\beta)/2} + \delta_{\beta/2})$, $I_4 := [2\alpha + 2\beta + \delta_{(\alpha+\beta)/2} + \delta_{\beta/2}, 2\pi)$. Clearly, (3), (4) demonstrate the worst case evacuation costs for instances in I_1, I_4, respectively, and the cost in both cases, for $\alpha = \alpha_\beta$ is equal to $w(\beta)$.

If $x \in I_2$ then $\mathcal{C}(x) = 1 + \mathcal{S}(x) + 2\mathcal{E}(x)$. It is easy to see that both $\mathcal{S}(x), \mathcal{E}(x)$ are monotone in I_2, so the worst case evacuation in this case is

$$\lim_{\epsilon \to 0^+} \mathcal{C}(2\alpha_\beta + \beta + \delta_{(\alpha_\beta+\beta)/2} - \epsilon) = 1 + \alpha_\beta + \beta + \delta_{(\alpha_\beta+\beta)/2} + 2\delta_{\beta/2}. \tag{7}$$

Denote $\delta_{\beta/2}$ satisfying (2) by δ_β'. Using (2) and the definition of α_β, we see that

$$\delta_{(\alpha_\beta+\beta)/2} = 2\sin\left(\frac{\alpha_\beta + \beta + \delta_{(\alpha_\beta+\beta)/2}}{2}\right) = 2\cos\left(\cos\left(\beta/4\right) - \beta/4 - \delta_{(\alpha_\beta+\beta)/2}\right)$$

For simplicity, we denote $\delta_{(\alpha_\beta+\beta)/2}$ that satisfies the equation above by δ_β''. Then, continuing from (7), the worst case evacuation cost when $x \in I_2$ becomes $1 + \pi + \beta/2 - 2\cos\left(\beta/4\right) + \delta_\beta'' + 2\delta_\beta'$, an expression that depends exclusively on β. The latter cost is no more than $w(\beta)$ if and only if $4\cos\left(\beta/4\right) - \beta - \delta_\beta'' - 2\delta_\beta' \geq 0$, and numerically we verify that this is satisfied as long as $\beta \leq \beta_0$ (see also Fig. 5).

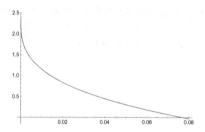

Fig. 5. The behavior of expression $4\cos\left(\beta/4\right) - \beta - \delta_\beta'' - 2\delta_\beta'$, for $\beta = 0, \ldots, 0.8$.

Finally, it is easy to verify that $\delta_{\beta/2}$ and $|I_4|$ are increasing and decreasing respectively for $\beta \leq \beta_0$ and that $\delta_{\beta_0/2} = 0.977997 \leq 1.01099 = |I_4|$ (for $\beta = \beta_0$). As a result, the worst case evacuation cost of case $x \in I_3$ cannot exceed that of case $x \in I_4$, and hence the lemma follows. □

A.6   Theorem 6

Proof (Theorem 6). The claims for the worst case performances of $\mathscr{A}_1, \mathscr{A}_2', \mathscr{A}_2$ follow directly from Lemmata 4, 6 and 5, respectively. Next we argue that as the parameters vary in their specified range, we obtain the entire spectrum of $w \in [w_1, w_2]$, and this for unique values of the parameters. For this, we will rely on that for all evacuation algorithm families, the worst case cost is monotone in the parameters.

First, we argue about $\mathscr{A}_1$. We observe that by the definition of $\bar{\alpha}$, $\mathrm{Wrs}\,(\mathscr{A}_1(\bar{\alpha})) = w_1$, and $\mathrm{Wrs}\,(\mathscr{A}_1(1)) = 1 + 2\pi - 1 = 2\pi$. Together with the fact that $v(\alpha)$ is strictly decreasing, we see that $\mathrm{Wrs}\,(\mathscr{A}_1(\alpha))$ is 1-1 and onto to $[w_1, 2\pi]$ as α ranges in $[1, \bar{\alpha}]$.

Second, we study $\mathscr{A}_2'$ whose worst case cost $v_2(\beta)$ is strictly decreasing in β. Moreover, by definition of β_0, we have $\mathrm{Wrs}\,(\mathscr{A}_2(\alpha_{\beta_0}, \beta_0)) = w_0$. Then we note that for $\beta = 0$, $\mathscr{A}_2(\alpha_\beta, \beta)$ coincides with $\mathscr{A}_2(\pi - 2)$, and in particular the induced worst case cost becomes $3 + \pi$. Therefore $\mathrm{Wrs}\,(\mathscr{A}_2'(\alpha_\beta, \beta))$ is 1-1 and onto to $[w_0, 3 + \pi]$ as β ranges in $[0, \beta_0]$.

Third, we study $\mathscr{A}_2$, for which we know that $\mathrm{Wrs}\,(\mathscr{A}_2(\pi - 2)) = 3 + \pi$. Again, the worst case cost is monotone in α and $\mathscr{A}_2(0)$ coincides with benchmark algorithm $\mathscr{B}_2$, that is $\mathrm{Wrs}\,(\mathscr{A}_2(0)) = w_2$. Hence, $\mathrm{Wrs}\,(\mathscr{A}_2(\alpha))$ is 1-1 and onto to $[3 + \pi, w_2]$ as α ranges in $[0, \pi - 2]$.

Finally, we argue that

$$\mathrm{Avg}\,(\mathscr{A}_1(\alpha)) \leq u_1(\alpha), \forall \alpha \in [1, \bar{\alpha}]$$
$$\mathrm{Avg}\,(\mathscr{A}_2'(\alpha_\beta, \beta)) \leq u_2'(\beta), \forall \beta \in [0, \beta_0]$$
$$\mathrm{Avg}\,(\mathscr{A}_2(\alpha)) \leq u_2(\alpha), \forall \alpha \in [0, \pi - 2]$$

For this, we numerically compute $\mathrm{Avg}\,(\mathscr{A}_1(\alpha)), \mathrm{Avg}\,(\mathscr{A}_2'(\alpha_\beta, \beta)), \mathrm{Avg}\,(\mathscr{A}_2(\alpha))$ for various values of parameters α, β, and we heuristically choose u_1, u_2', u_2 so as to upper bound the average case performance of $\mathscr{A}_1, \mathscr{A}_2', \mathscr{A}_2$, effectively verifying our claim numerically. For each evacuation algorithm, we utilize Corollary 1, which together with the analytic description of our evacuation algorithms (see Definitions 4, 6, and 5) allow us to compute their average case performance using computer-assisted calculations. Our numerical calculations are depicted in Fig. 6.

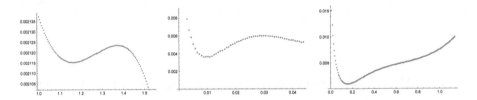

Fig. 6. On the right $u_1(\alpha) - \mathrm{Avg}\,(\mathscr{A}_1(\alpha))$, for $\alpha' \leq \alpha \leq \bar{\alpha}$. In the middle, $u_2'(\beta) - \mathrm{Avg}\,(\mathscr{A}_2'(\alpha_\beta, \beta))$, for $0 \leq \beta \leq \beta_0$. On the right $u_2(\alpha) - \mathrm{Avg}\,(\mathscr{A}_2(\alpha))$, for $0 \leq \alpha \leq \pi - 2$.

$\square$

References

1. Ahlswede, R., Wegener, I.: Search Problems. Wiley-Interscience, Hoboken (1987)
2. Albers, S., Henzinger, M.R.: Exploring unknown environments. SIAM J. Comput. **29**(4), 1164–1188 (2000)
3. Albers, S., Kursawe, K., Schuierer, S.: Exploring unknown environments with obstacles. Algorithmica **32**(1), 123–143 (2002)
4. Alpern, S., Gal, S.: The Theory of Search Games and Rendezvous, vol. 55. Kluwer Academic Publishers (2002)
5. Alpern, S.: Ten open problems in rendezvous search. In: Alpern, S., Fokkink, R., Gąsieniec, L., Lindelauf, R., Subrahmanian, V. (eds.) Search Theory, pp. 223–230. Springer, New York (2013). https://doi.org/10.1007/978-1-4614-6825-7_14
6. Baeza Yates, R., Culberson, J., Rawlins, G.: Searching in the plane. Inf. Comput. **106**(2), 234–252 (1993)
7. Baumann, N., Skutella, M.: Earliest arrival flows with multiple sources. Math. Oper. Res. **34**(2), 499–512 (2009)
8. Beck, A.: On the linear search problem. Israel J. Math. **2**(4), 221–228 (1964)
9. Bellman, R.: An optimal search. SIAM Rev. **5**(3), 274–274 (1963)
10. Benkoski, S., Monticino, M., Weisinger, J.: A survey of the search theory literature. Nav. Res. Logist. (NRL) **38**(4), 469–494 (1991)
11. Brandt, S., Laufenberg, F., Lv, Y., Stolz, D., Wattenhofer, R.: Collaboration without communication: evacuating two robots from a disk. In: Fotakis, D., Pagourtzis, A., Paschos, V.T. (eds.) CIAC 2017. LNCS, vol. 10236, pp. 104–115. Springer, Cham (2017). https://doi.org/10.1007/978-3-319-57586-5_10
12. Burgard, W., Moors, M., Stachniss, C., Schneider, F.E.: Coordinated multi-robot exploration. IEEE Trans. Rob. **21**(3), 376–386 (2005)
13. Chrobak, M., Gąsieniec, L., Gorry, T., Martin, R.: Group search on the line. In: Italiano, G.F., Margaria-Steffen, T., Pokorný, J., Quisquater, J.-J., Wattenhofer, R. (eds.) SOFSEM 2015. LNCS, vol. 8939, pp. 164–176. Springer, Heidelberg (2015). https://doi.org/10.1007/978-3-662-46078-8_14
14. Chung, T.H., Hollinger, G.A., Isler, V.: Search and pursuit-evasion in mobile robotics. Auton. Rob. **31**(4), 299–316 (2011)
15. Czyzowicz, J., Dobrev, S., Georgiou, K., Kranakis, E., MacQuarrie, F.: Evacuating two robots from multiple unknown exits in a circle. Theoret. Comput. Sci. **709**, 20–30 (2018)
16. Czyzowicz, J., Gąsieniec, L., Gorry, T., Kranakis, E., Martin, R., Pajak, D.: Evacuating robots via unknown exit in a disk. In: Kuhn, F. (ed.) DISC 2014. LNCS, vol. 8784, pp. 122–136. Springer, Heidelberg (2014). https://doi.org/10.1007/978-3-662-45174-8_9
17. Czyzowicz, J., et al.: Evacuation from a disc in the presence of a faulty robot. In: Das, S., Tixeuil, S. (eds.) SIROCCO 2017. LNCS, vol. 10641, pp. 158–173. Springer, Cham (2017). https://doi.org/10.1007/978-3-319-72050-0_10
18. Czyzowicz, J., et al.: God save the queen. In: 9th International Conference on Fun with Algorithms (FUN 2018) (2018)
19. Czyzowicz, J., et al.: Priority evacuation from a disk using mobile robots. In: Lotker, Z., Patt-Shamir, B. (eds.) SIROCCO 2018. LNCS, vol. 11085, pp. 392–407. Springer, Cham (2018). https://doi.org/10.1007/978-3-030-01325-7_32
20. Czyzowicz, J., et al.: Search on a line by byzantine robots. In: 27th International Symposium on Algorithms and Computation, ISAAC 2016, Sydney, Australia, 12–14 December 2016, pp. 27:1–27:12 (2016)

21. Czyzowicz, J., Georgiou, K., Kranakis, E., Narayanan, L., Opatrny, J., Vogtenhuber, B.: Evacuating robots from a disk using face-to-face communication (extended abstract). In: Paschos, V.T., Widmayer, P. (eds.) CIAC 2015. LNCS, vol. 9079, pp. 140–152. Springer, Cham (2015). https://doi.org/10.1007/978-3-319-18173-8_10
22. Deng, X., Kameda, T., Papadimitriou, C.: How to learn an unknown environment. In: FOCS, pp. 298–303. IEEE (1991)
23. Dobbie, J.: A survey of search theory. Oper. Res. **16**(3), 525–537 (1968)
24. Fekete, S., Gray, C., Kröller, A.: Evacuation of rectilinear polygons. In: Wu, W., Daescu, O. (eds.) COCOA 2010. LNCS, vol. 6508, pp. 21–30. Springer, Heidelberg (2010). https://doi.org/10.1007/978-3-642-17458-2_3
25. Fomin, F.V., Thilikos, D.M.: An annotated bibliography on guaranteed graph searching. Theoret. Comput. Sci. **399**(3), 236–245 (2008)
26. Georgiou, K., Karakostas, G., Kranakis, E.: Search-and-fetch with one robot on a disk - (track: wireless and geometry). In: Chrobak, M., Fernández Anta, A., Gąsieniec, L., Klasing, R. (eds.) ALGOSENSORS 2016. LNCS, vol. 10050, pp. 80–94. Springer, Cham (2017). https://doi.org/10.1007/978-3-319-53058-1_6
27. Georgiou, K., Karakostas, G., Kranakis, E.: Search-and-fetch with 2 robots on a disk - wireless and face-to-face communication models. In: Liberatore, F., Parlier, G.H., Demange, M. (eds.) Proceedings of the 6th International Conference on Operations Research and Enterprise Systems, ICORES 2017, Porto, Portugal, 23–25 February 2017, pp. 15–26. SciTePress (2017)
28. Georgiou, K., Kranakis, E., Steau, A.: Searching with advice: robot fence-jumping. J. Inf. Process. **25**, 559–571 (2017)
29. Hoffmann, F., Icking, C., Klein, R., Kriegel, K.: The polygon exploration problem. SIAM J. Comput. **31**(2), 577–600 (2001)
30. Kleinberg, J.: On-line search in a simple polygon. In: SODA, p. 8. SIAM (1994)
31. Lidbetter, T.: Hide-and-seek and other search games. Ph.D. thesis, The London School of Ecoomics and Political Science (LSE) (2013)
32. Mitchell, J.S.B.: Geometric shortest paths and network optimization. Handb. Comput. Geom. **334**, 633–702 (2000)
33. Nahin, P.: Chases and Escapes: The Mathematics of Pursuit and Evasion. Princeton University Press, Princeton (2012)
34. Papadimitriou, C.H., Yannakakis, M.: Shortest paths without a map. In: Ausiello, G., Dezani-Ciancaglini, M., Della Rocca, S.R. (eds.) ICALP 1989. LNCS, vol. 372, pp. 610–620. Springer, Heidelberg (1989). https://doi.org/10.1007/BFb0035787
35. Stone, L.: Theory of Optimal Search. Academic Press, New York (1975)
36. Thrun, S.: A probabilistic on-line mapping algorithm for teams of mobile robots. Int. J. Rob. Res. **20**(5), 335–363 (2001)
37. Yamauchi, B.: Frontier-based exploration using multiple robots. In: Proceedings of the Second International Conference on Autonomous Agents, pp. 47–53. ACM (1998)

Mutual Visibility by Asynchronous Robots on Infinite Grid

Ranendu Adhikary$^{(\boxtimes)}$ ⓘ, Kaustav Bose ⓘ, Manash Kumar Kundu ⓘ,
and Buddhadeb Sau

Department of Mathematics, Jadavpur University, Kolkata, India
{ranenduadhikary.rs,kaustavbose.rs,
manashkrkundu.rs}@jadavpuruniversity.in, bsau@math.jdvu.ac.in

Abstract. Consider a set of autonomous, identical, opaque point robots in the Euclidean plane. The Mutual Visibility problem asks the robots to reposition themselves, without colliding, to a configuration where they all see each other, i.e., no three of them are collinear. In this paper, we consider the problem in a grid based terrain where the movements of the robots are restricted only along grid lines and only by a unit distance in each step. We consider the luminous robots model, in which each robot is equipped with an externally visible light which can assume a constant number of predefined colors. These colors serve both as internal memory and as a form of communication. The robots operate in Look-Compute-Move cycles under a fully asynchronous scheduler. The robots do not have any common global coordinate system or chirality and do not have the knowledge of the total number of robots. Our proposed distributed algorithm solves the problem for any arbitrary initial configuration and guarantees collision-free movements.

Keywords: Distributed computing · Autonomous robots ·
Mutual visibility · Robots with lights · Asynchronous ·
Look-Compute-Move cycle · Grid

1 Introduction

Robot swarms are a distributed system of autonomous mobile robots that collaboratively execute some complex tasks. Swarms of low-cost, weak, simple robots are emerging as a viable alternative to using a single powerful and expensive robot. In the traditional model of robot swarms, the mobile robots are assumed to be *autonomous* (there is no central control), *homogeneous* (they execute the same distributed algorithm), *anonymous* (they have no unique identifiers), *identical* (they are indistinguishable by their appearance) and *disoriented* (they do not agree on any global coordinate system). The robots do not have any direct means of communication. Each robot is equipped with sensor capabilities (i.e., vision) to perceive the positions of the other robots. The robots operate in Look-Compute-Move (LCM) cycles: when a robot becomes active it takes a snapshot of the positions of the other robots, then computes a destination based on the

© Springer Nature Switzerland AG 2019
S. Gilbert et al. (Eds.): ALGOSENSORS 2018, LNCS 11410, pp. 83–101, 2019.
https://doi.org/10.1007/978-3-030-14094-6_6

snapshot using a deterministic algorithm (Compute), and then moves towards the destination along a straight line (Move).

The *opaque robots* or *obstructed visibility* model assumes that visibility can be obstructed by the presence of other robots: that is, two robots can see each other if and only if no other robot lies on the line segment joining them. The fundamental problem in this model is the MUTUAL VISIBILITY problem: starting from arbitrary distinct positions in the plane, the robots have to reposition themselves, within finite time and without colliding, to a configuration in which they are in distinct locations and no three of them are collinear. The problem is important as it provides a basis for any subsequent task requiring complete visibility. We consider this problem in the *robots with lights* or *luminous robots* model [8,10]. In this model, each robot is equipped with an externally visible light which can assume a constant number of predefined colors.

1.1   Our Contribution

In this paper, we have considered the MUTUAL VISIBILITY problem in a grid based terrain. The infinite grid is a natural discretization of the Euclidean plane. Traditional spatial representation methods in robot navigation commonly represent the world as a two dimensional grid around the robot. Grid type floor layouts are also commonly implemented in real life robot navigation systems, e.g., industrial Automated Guided Vehicles (AGV), using magnets or optical guidances on the floor [3]. The simple model of movement along grid lines from one grid point to another can be easier to implement for robots with weak mechanical capabilities as they may not be able to execute accurate movements in arbitrary directions or by arbitrarily small amounts. Although the simple model of movement may be easier to physically execute, the restrictions imposed on the movements of the robots pose the main difficulty of the algorithmic problem. Our proposed distributed algorithm solves the MUTUAL VISIBILITY problem on infinite grid for any arbitrary initial configuration. We have solved the problem in the luminous robots model using 11 colors.

1.2   Earlier Works

While fundamental problems in autonomous mobile robots like GATHERING have been studied in grid environments [5–7,11,17], the MUTUAL VISIBILITY problem has only been studied in continuous Euclidean plane. The first distributed algorithm for the MUTUAL VISIBILITY problem was presented by Di Luna et al. [9] for oblivious and semi-synchronous robots. Later, Sharma et al. [13] analyzed and modified the round complexities of the algorithm in the fully synchronous model. The MUTUAL VISIBILITY problem under the luminous robots model was first studied by Di Luna in [8]. They solved the problem with semi-synchronous scheduler using 3 colors and with asynchronous schedulers using 3 colors under one axis agreement. Later Sharma et al. [14] attained this result using only 2 colors both for semi-synchronous and for asynchronous robots. Then a series of papers [15, 16] appeared aiming towards reducing the runtime of the algorithm. Recently Bhagat and Mukhopadhyaya [4] have solved the problem for asynchronous robots without any agreement on coordinate axes or chirality. The problem has also been considered for fat robots [12] and faulty robots [2].

2   Model and Definitions

In this section, we present the model and some basic definitions.

Robots: We consider a set of $N \geq 3$ homogeneous, autonomous, anonymous and identical robots $\mathcal{R} = \{r_1, r_2, \ldots, r_N\}$ deployed on a two dimensional infinite grid. All the robots are initially positioned on distinct grid points. The robots are assumed to be dimensionless and modeled as points on the plane. The robots do not have access to any global coordinate system. The total number of robots N is not known to them.

Movement: The movement of the robots are restricted only along grid lines from one grid point to one of its four neighboring grid points. Traditionally in discrete domains, robot movements are assumed to be instantaneous. For simplicity of analysis, we also assume the movements to be instantaneous. This implies that the robots are always seen on grid points, not on edges. However, our strategy will also work without this assumption.

Lights: Each robot is equipped with an externally visible light which can assume a constant number of predefined colors. The robots explicitly communicate with each other using these lights. The lights are persistent (i.e., the color is not erased at the end of a cycle), but otherwise the robots are oblivious. The colors used in our algorithm are $\mathcal{C} = \{Off, Boundary, RequestExpansion, Expanding, Moving1, Rectangle, Square, NextToCorner, Moving2, Moving3, Done\}$.

Visibility: The visibility range of the robots is unlimited, but can be obstructed by the presence of other robots. A robot r_i can see another robot r_j if and only if there are no robots on the straight line segment $\overline{r_i r_j}$. The set of positions of all robots visible by a robot r at time t, expressed in its local coordinate system, is denoted by $\mathcal{V}_r(t)$, or simply $\mathcal{V}_r$ when there is no ambiguity.

Look-Compute-Move Cycles: The robots, when active, operate according to the so-called LOOK-COMPUTE-MOVE cycle. In each cycle, a previously idle or inactive robot wakes up and executes the following steps. In the LOOK phase a robot takes the snapshot of the positions of the robots visible to it represented in its own coordinate system. Based on the perceived configuration, the robot performs computations according to a deterministic algorithm to decide a destination point $p \in \mathbb{Z}^2$ (either the grid point on which it currently resides or one of the four neighboring grid points) and a color $c \in \mathcal{C}$. Finally based on the outcome of the algorithm, the robot changes its light to the computed color, and either remains stationary or makes an instantaneous move to an adjacent grid point.

Scheduler: We assume that the robots are controlled by a fully asynchronous adversarial scheduler. This implies that the amount of time spent in LOOK, COMPUTE, MOVE and inactive states by different robots is finite but unbounded and unpredictable. As a result, the robots do not have a common notion of time and the configuration perceived by a robot during the LOOK phase may significantly change before it actually makes a move.

Geometric Definitions: Given a configuration of robots at time t, the *smallest enclosing rectangle* is defined as the smallest axis-aligned rectangle that contains all the robots. The boundary of the largest rectangle contained inside the smallest enclosing rectangle is called the *penultimate layer*, and each side of the rectangle is called *penultimate line segment*. The robots on the boundary of the smallest enclosing rectangle will be called *boundary robots*, and otherwise *interior robots*. The boundary of the smallest enclosing rectangle of the interior robots is called the *inner boundary*, and each side of the rectangle is called *inner boundary side*. A configuration will be called an *empty rectangle* if there are only boundary robots. A robot r on a grid line segment $\mathcal{L}$ will be called *non-terminal on $\mathcal{L}$* if it lies between two robots on $\mathcal{L}$, and otherwise it will be called *terminal on $\mathcal{L}$* (Fig. 1).

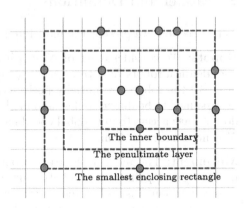

Fig. 1. Illustrations for the geometric definitions given in Sect. 2.

3   The Algorithm

The main difficulty of the problem arises from the restrictions imposed on the movements of the robots. If the four neighboring grid points of a robot are occupied, then any move made by it will lead to a collision. Our plan is to first create an empty rectangle configuration where all the robots are positioned on the boundary of the smallest enclosing rectangle. This phase is called the *Interior Depletion* phase. Our main idea is to exploit the symmetry of the empty rectangle to our advantage. In the *Symmetric Movements* phase, the robots will sequentially leave the empty rectangle and form a mutually visible configuration. During the movements, the robots may not be able to perceive the positions of other robots due to obstructed visibility, but can predict their movements from the symmetries of the empty rectangle. The robots at the corners of the empty rectangle will not move in the symmetric movements phase, but, however, will play an important role in the process. The specified lights of the corner robots will guide the movements of the other robots. From their positions relative to these corners, the robots will deduce their destination. However, the empty rectangle created in the interior depletion phase may not have robots at the corner points.

So in an intermediate phase, called *Corner Creation*, the robots terminal on the sides of the empty rectangle will reposition themselves at the corners. These phases are described in more detail in the following subsections.

3.1   Interior Depletion

The main idea of the algorithm is to sequentially move the interior robots to the boundary. In order to avoid collisions, only those robots that are on the inner boundary will move. However, there may not be an empty grid point on the boundary for the robot to position itself. In that case, the robot, using its lights, will ask the boundary robots to expand the boundary. A pseudo-code description of the procedure is presented in Algorithm 1. The geometric functions used in the algorithm are explained briefly in the following. The lights used in this phase are $\{Off, Boundary, RequestExpansion, Expanding, Moving1\}$.

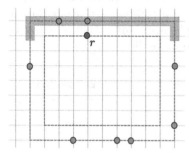

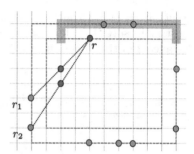

(a) Corresponding to the robot r, S_r is the shaded portion of the boundary.

(b) The robot r can not see the boundary robots r_1 and r_2, and hence is not able to gauge the full extent of the boundary. As a result its perceived S_r is different from what is actually desired.

Fig. 2. Illustrations for the function FINDSPACE().

The lights of all the robots are initially set to Off. Upon waking up, a robot r calls the function ONBOUNDARY() to decide if it is on the boundary. If it finds that there is an open-half plane, delimited by one of the grid lines passing through itself, containing no robots, then it concludes that it is a boundary robot and sets its light to *Boundary*. If r finds itself in the interior, then it calls the function ONINNERBOUNDARY(). If there is an open-half plane, delimited by one of the grid lines passing through itself, containing only robots with lights set to *Boundary*, then it is on the inner boundary. Only the robots on the inner boundary are allowed to move towards the boundary. Now, there are two cases to consider: the robot on the inner boundary is either on the penultimate layer or not.

If r is not on the penultimate layer and is terminal on an inner boundary side, then it has to move towards the boundary. But it will not move immediately. First, it will set its light to $Moving1$. Then in the next round, it will redo the same computations. An interior robot will move only if its light is already set to $Moving1$ (in some previous round).

On the other hand, if r finds itself on a penultimate line segment $\mathcal{L}$ and is terminal on $\mathcal{L}$, then it has to decide whether it is possible to move to the boundary. It does so using the function FINDSPACE(). The function FINDSPACE() works in the following way.

For the robot r, let $\mathcal{S}_r$ denote the portion of the boundary as shown in the Fig. 2a. If $\mathcal{L}$ contains more than one robot, then define $\mathcal{H}_r$ as the closed-half plane, delimited by the grid line perpendicular to $\mathcal{L}$ and passing through r, such that $\mathcal{L} \cap \mathcal{H}_r$ contains no robot other than r. If there is no other robot on $\mathcal{L}$ except r, then $\mathcal{H}_r$ is the entire plane. The robot r will scan $\mathcal{H}_r \cap \mathcal{S}_r$ for an empty grid point. If it finds an empty point, it makes sure that the shortest path to that point is not blocked by some other robot. Even if it finds an unobstructed empty point on $\mathcal{H}_r \cap \mathcal{S}_r$, a move towards it can lead to a collision. To avoid this, it must make sure that there are no robots with light $Moving1$ within Manhattan distance 2 in the direction in which it intends to move. See Fig. 3.

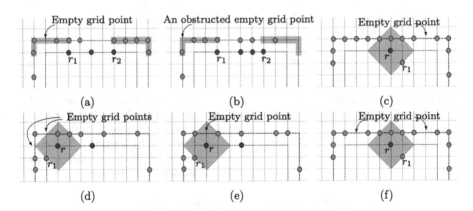

Fig. 3. Illustrations for the function FINDSPACE(). (a) The robot r_1 finds an empty grid point, while r_2 does not find any empty grid point. (b) The robot r_1 finds an empty grid point, but the shortest paths leading to it are blocked by other robots. (c)–(d) The robot r finds an empty grid point but there is a robot with light $Moving1$ within Manhattan distance 2 in the direction it should move to get there. (e) There is a robot r_1 with light $Moving1$ within Manhattan distance 2, but not in the direction towards the empty point. Hence, FINDSPACE() will return True for r. (f) r is the only terminal robot on the penultimate line segment, and hence $\mathcal{H}_r \cap \mathcal{S}_r = \mathcal{S}_r$. It has found two empty grid points in $\mathcal{S}_r$. Here, it will choose the empty grid point on the left, because on the right side there is a robot with light $Moving1$ within Manhattan distance 2.

If the function FINDSPACE() returns True, then r will move towards the empty grid point on the boundary. Again, it will move only if its light is already set to *Moving*1, otherwise it will only change its light to *Moving*1. If FIND-SPACE() returns False, r will set its light to *RequestExpansion*, requesting the boundary robots to expand the smallest enclosing rectangle so that an empty space is created on the boundary.

Algorithm 1. Interior Depletion

```
 1  Procedure INTERIORDEPLETION()
 2      r ← myself
 3      while EMPTYRECTANGLE() = FALSE do
 4          if r.color = Off then
 5              if ONBOUNDARY() = True then
 6                  r.color ← Boundary
 7              else if ONINNERBOUNDARY() = True then
 8                  if ONPENULTIMATE() = True then
 9                      if TERMINALONPENULTIMATE() = True then
10                          if FINDSPACE() = True then
11                              r.color ← Moving1
12                          else
13                              r.color ← RequestExpansion

14                  else if TERMINALONINNERBOUNDARY() = True then
15                      r.color ← Moving1

16          else if r.color = RequestExpansion then
17              if ONPENULTIMATE() = False then
18                  r.color ← Off

19          else if r.color = Moving1 then
20              if ONBOUNDARY() = True then
21                  r.color ← Boundary
22              else if ONINNERBOUNDARY() = True then
23                  if ONPENULTIMATE() = True then
24                      if TERMINALONPENULTIMATE() = True then
25                          if FINDSPACE() = True then
26                              Move towards the empty boundary point
27                          else
28                              r.color ← RequestExpansion

29                  else if ONPENULTIMATE() = False then
30                      Move towards boundary

31          else if r.color = Expanding then
32              if EXPANSIONCOMPLETED() = True then
33                  r.color ← Boundary

34          else if r.color = Boundary then
35              if ONBOUNDARY() = False then
36                  Move towards the boundary
37              else if All terminal robots on the penultimate grid line segment next to it
                        have lights set to RequestExpansion then
38                  r.color ← Expanding
39                  Move outward
```

Now consider a robot r' with light *Boundary*. It will first recheck if it is still on the boundary. It may happen that some of the boundary robots on its grid line had started the expansion earlier, leaving r' inside the smallest enclosing rectangle. However, these robots will only move at most one hop from the previous boundary. Thus, if r' finds itself in the interior, it can move to the new boundary in a single step. If r' is on the boundary, and it observes that all terminal robots on the penultimate line segment next to it have set their lights to *RequestExpansion*, then it will change its light to *Expanding* and will start moving outwards. If a robot finds its light set to *Expanding*, it checks whether all its fellow boundary robots from the previous round have completed their moves by checking the penultimate line next to it. If it finds that the expansion is completed, it will change its light to *Boundary*.

A rigorous proof of correctness of the algorithm is omitted due to space constraints. The following two lemmas address the two main issues regarding the correctness of the algorithm. The proofs of the lemmas in this subsection are briefly presented in Appendix A.

Lemma 1. *The interior depletion phase is collision free.*

Lemma 2. *If a robot r at time t_0 on the penultimate layer sets it light to RequestExpansion, then there exist a time t ($> t_0$) when the robot r reaches the boundary.*

The interior depletion phase is completed when all the robots are on the boundary of the smallest enclosing rectangle, and the lights of all the robots are set to *Boundary*. However, it may not be possible for the robots to locally detect this. This is because, if the first condition is attained, a robot r can not determine whether the second condition is satisfied, as it may not be able to see all the robots on boundary line on which it resides. We say that a robot detects the partial completion of the interior depletion phase if it can determine if the first condition is satisfied. It does so using the function EMPTYRECTANGLE(), which returns True if the following conditions are satisfied:

1. all robots in $\mathcal{V}_r$ are on the boundary of their smallest enclosing rectangle,
2. the lights of all the robots in $\mathcal{V}_r$ are set to *Boundary*.

Lemma 3. *The function EMPTYRECTANGLE() can detect the partial completion of the interior depletion phase.*

Theorem 1. *The algorithm INTERIOR DEPLETION creates an empty rectangle configuration starting from any arbitrary configuration.*

3.2 Symmetric Movements

Due to space constraints, we will not describe the corner creation phase. A brief discussion on this phase is given in Appendix C. This phase will require four different lights, namely, *Rectangle, Square, NextToCorner* and *Moving2*. The objective of the corner creation phase is to create an empty rectangle configuration with its four corners occupied by robots with specified lights. However, this may not be achievable if the empty rectangle is a square. Due to space restrictions, we will describe the symmetric movements phase assuming that the starting configuration is the generic configuration of a non-square rectangle having four corner robots with lights set to *Rectangle*. The algorithms for other configurations, like squares with possibly some missing corners or a straight line configuration, are based on the same movement strategy subject to some minor modifications. The lights that will be used in the symmetric movements phase are $\{Moving3, Done\}$. See Appendix B, for the proofs of the claims in this subsection.

In this phase, the non-terminal robots on the sides of the empty rectangle will leave the boundary and move outwards along the grid line passing through its starting position. The extent of their movement will depend on (1) the length of the sides of the empty rectangle, and (2) the starting position of the robot on the boundary.

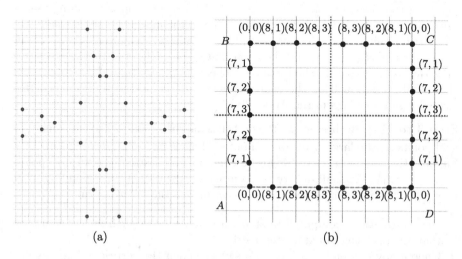

(a)   (b)

Fig. 4. (a) The final mutually visible configuration for a 7×8 empty rectangle. (b) The coordinate system of the boundary of the rectangle.

The grid points on the boundary of the rectangle will be given coordinates (p, k), where $p =$ the size of the side of the rectangle it belongs to, and $k =$ its distance from the closest corner. The coordinates of the four corners will be $(0, 0)$. This coordinate scheme is illustrated in Fig. 4b. The group of symmetries of the rectangle is generated by reflections with respect to perpendicular bisectors of its sides. The group of symmetries induces an equivalence relation on the grid points on the rectangle: $P \sim Q$ if and only if Q can be obtained from P by some reflection operations. This partitions the set of grid points on the rectangle into equivalence classes. The distance, that two robots on starting points belonging to the same equivalence class should move, have to be equal. Two points are equivalent if and only if their coordinates are equal (See Fig. 4b). We shall exploit these symmetries to design a recursive function called DESTINATION that computes the destination points of the robots. The distance that a robot starting from (p, k) should move is DESTINATION(m, n, p, k), where m, n $(m \geq n)$ are size of the sides.

The pseudocode of the function DESTINATION is omitted. We shall briefly sketch out the recursive computation of the destination points corresponding to all the points on the rectangle. At each step, the algorithm computes destinations corresponding to all the points belonging to an equivalence class. In other words, the iteration runs over the set of equivalence classes of the grid points on the rectangle. The set of all the destinations computed up to the ith step of the procedure will be denoted by C_i.

Step 0: Robots at the corners ($k = 0$) will not move. Hence, C_0 consists of the starting positions of the corner robots.

Step 1 and 2: At step 1 and 2, the destinations corresponding to the middle points of the sides, i.e., $k = \lceil \frac{m}{2} \rceil - 1$ and $\lceil \frac{n}{2} \rceil - 1$, are computed. Suppose we are computing the destinations corresponding to the middle points of a side AB. Draw two straight lines through A and B, parallel to the two diagonals of the rectangle. Let $\mathcal{H}_A$ and $\mathcal{H}_B$ be the open half-planes delimited by these straight lines that do not contain the rectangle. Then the destination of the robot(s) at the middle of AB will be the nearest grid points belonging to $\mathcal{H}_A \cap \mathcal{H}_B$ (the shaded region in Fig. 8 in the Appendix).

Step 3 to $\lceil \frac{m}{2} \rceil$: In these steps, the destinations corresponding to the remaining grid points on the larger side are computed. This is done in a recursive manner. Suppose that at the ith step, we are to compute the destinations for grid points $\{x_1, x_2, x_3, x_4\}$. We shall denote the computed destination corresponding to x_j as y_j. The destinations are computed according to the following rules.

1. The destinations computed in step i are strictly farther from the rectangle than the ones computed in step $i - 1$.
2. Choose any one of $\{x_1, x_2, x_3, x_4\}$, say x_1. Then the corresponding destination y_1 is the grid point (on the grid line passing through x_1) closest to the rectangle (respecting condition 1) such that no three points in $C_{i-1} \cup \{y_1\}$ are collinear. The destinations y_2, y_3, y_4 are obtained from y_1 from the reflectional symmetries. Since no two points in C_{i-1} are collinear with y_1, from the reflectional symmetries we can say that the same is true for y_2, y_3 and y_4. But it is still not apparent that no three points in $C_i = C_{i-1} \cup \{y_1, y_2, y_3, y_4\}$ are collinear. We prove this in Lemma 4.

Step $\lceil \frac{m}{2} \rceil + 1$ to $\lceil \frac{m}{2} \rceil + \lceil \frac{n}{2} \rceil - 2$: In these steps, the destinations corresponding to the grid points on the smaller side are computed. The procedure is the same as before.

Lemma 4. *If no three points in C_{i-1} are collinear, then the same is true for C_i.*

Theorem 2. *No three points of the destinations computed by the function DES-TINATION are collinear.*

Proof. Since no three points of C_0 are collinear, the result immediately follows from the Lemma 4.

We shall now describe the movement strategy. A pseudocode description of the procedure is given in Algorithm 2. As mentioned earlier, the algorithm is described for only non-square empty rectangle configurations with four occupied corners. In the function DESTINATION, the destinations corresponding to the middle points of the sides were computed first in the recursive process. But the movements will occur in the exactly opposite order. The robots will sequentially leave the boundary with the ones closest to the corner moving first. A boundary robot will call the function ELIGIBLETOMOVE() to determine whether it should start moving. It checks if the following conditions are satisfied:

1. It can see at least one corner robot on its boundary side, and two corner robots on the opposite side. In Fig. 5, r_1 can see c_1, c_3 and c_4. r_3 can also see c_2, c_3 and c_4. But r_5 cannot see c_1 or c_2, and so it is not eligible to move yet.
2. If there were robots initially on its boundary side between it and the corner(s) it can see, they have already completed their movements and changed their lights to *Done*. The robot r_1 checks this by scanning the shaded region A, which is empty. Hence r_1 is eligible to move. But when r_3 scans the region B, it finds r_2 with its light set to *Moving*3, and hence it will not move.

If ELIGIBLETOMOVE() returns True, the robot will change its light to *Moving*3 and leave the boundary. Note that a robot can leave the boundary even before the corner creation phase is completed. This is because the robot leaves the boundary when it sees at least three corners. The fourth corner is probably yet not created. We call this a premature move. But it can determine if it has made a premature move just after moving one hop from the boundary. This is because if the other corner is created, it will be able to see from the grid line one hop away from the boundary. If it can't, it will wait for the completion of the corner creation phase.

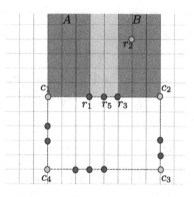

Fig. 5. Illustrations supporting the proof of Theorem 3

Note that at most one robot on a boundary line can make a premature move. Also note that PREMATURE() will always return false if the robot is more than one hop away from the boundary.

When the robot is moving, after each one hop move it has to compute DESTINATION(m, n, p, k) to find whether it has reached its destination. But that requires the knowledge of m and n. Consider the robot r_2 in Fig. 5. Since the robots closer to the corners move farther, r_2 will always be able to see c_2 and c_3. But to know the size of both the boundary sides, it also has to see another corner. If it cannot see c_1, then its view must be obstructed by some robot in the shaded region in Fig. 5. So r_2 now has to decide if its view is obstructed by a moving robot or a robot that has reached its destination. If r_2 scans the shaded region and it finds a robot with light *Moving*3 below it, then BLOCKEDBYMOV-INGROBOT() will return True. Notice that if r_2 can't see c_1, it can not identify the actual extent (on the left side) of the shaded region. But this is not a major problem as there can not be any robots (from any branch) moving in the area beyond the left boundary of the shaded region. The robots, upon reaching their destination points, will change their lights to *Done*.

Algorithm 2. SYMMETRIC MOVEMENTS

---

1 **Procedure** SYMMMOVEMENT()
2      $r \leftarrow$ myself
3   **if** *I am on the empty rectangle and* ELIGIBLETOMOVE() = *True* **then**
4          $r.color \leftarrow Moving3$
5   Move outwards
6   **else if** $r.color = Moving3$ *and* PREMATURE() = *False* **then**
7   **if** *I can see at least three robots with light Rectangle* **then**
8              $d \leftarrow$ DESTINATION(m, n, p, k)
9   **if** $r.position = d$ **then**
10                 $r.color = Done$
11   **else**
12   Move
13   **else if** BLOCKEDBYMOVINGROBOT() = *False* **then**
14   Move

Theorem 3. *Algorithm* SYMMETRIC MOVEMENTS *correctly leads all the robots to the destinations computed by the* DESTINATION *function.*

From Theorem 2 and 3, we can conclude the following.

Theorem 4. *The* MUTUAL VISIBILITY *problem on infinite grid can be solved using 11 colors.*

Note that the robots terminate the execution once their lights are set to *Rectangle* or *Done*. We say that the MUTUAL VISIBILITY problem is solved with detection if we additionally require that a robot terminates only after it detects that the mutual visibility is attained. This can be easily achieved, but will require one extra color. Each corner robot can determine if the symmetric movements have been completed in its quadrant, and then changes its light to the extra color. When all four corner robots change their colors, it implies that a mutually visible configuration is attained and all the robots in the configuration can detect this.

Theorem 5. *The* MUTUAL VISIBILITY *problem on infinite grid can be solved with detection using 12 colors.*

4 Conclusion

Our proposed distributed algorithm solves the MUTUAL VISIBILITY problem on infinite grid for any arbitrary initial configuration under the luminous robots model using 11 colors. We considered the robots as dimensionless points. A more realistic model would be to consider robots with physical extent, i.e., fat robots. The MUTUAL VISIBILITY problem for fat robots is solved as a subroutine of the gathering algorithm presented in [1]. But it is assumed that each robot knows the size of the team. Recently, Sharma et al. have solved the problem in a fully synchronous setting [12]. It would be interesting to see if our strategy can be extended to solve the problem for fat robots in asynchronous setting with less assumptions. Another direction would be to investigate if the number of colors used or the number of moves by the robots can be reduced.

Acknowledgements. The first three authors are supported by CSIR, Govt. of India, NBHM, DAE, Govt. of India and UGC, Govt. of India respectively. We would like to thank the anonymous reviewers for their valuable comments which helped us improve the quality and presentation of this paper.

Appendix A Correctness of Interior Depletion

A.1 Proof of Lemma 1

Collision can only occur when a robot r is moving along the penultimate layer towards an empty space in the boundary in the situations shown in Fig. 3c and d. Suppose that two robots r and r_1, at Manhattan distance 2 from each other, computes the same destination point. Initially none of them had their lights set to $Moving1$. This is because if one of them had its light set to $Moving1$ when the other one takes the snapshot, it would not have computed a destination point. So they will first change their lights to $Moving1$, say at time t and t_1 respectively. Let $t \leq t_1$. Now in its next LOOK phase at time $t_2(> t_1 \geq t)$, r_1 perceives that r has either already made its move or is yet to move but has set its light to $Moving1$. In either case, r_1 will not move according to our algorithm. Hence, there will be no collision.   □

A.2 Proof of Lemma 2

Assume that the robot r on a penultimate line segment $\mathcal{L}$ at time t_0 has set its light to $RequestExpansion$. If the other terminal robot on $\mathcal{L}$, say r', also sets its light to $RequestExpansion$, then the corresponding boundary robots will execute the expansion. Otherwise, r' and subsequently the other robots that will become terminal on $\mathcal{L}$ will move to the boundary. Eventually, either we have another terminal robot requesting expansion, or r is the only robot remaining on $\mathcal{L}$ still with light $RequestExpansion$. Therefore, the corresponding boundary will eventually execute expansion.

After the expansion, r is now not on the penultimate line. Then there is a time t' when it will again move to the new penultimate line $\mathcal{L}'$. Now there could be at most two robots on $\mathcal{L}'$, since only the terminal robots on the inner boundary line move. An expansion always creates at least two empty points on the boundary (See Fig. 6), and one of them is in $\mathcal{H}_r \cap \mathcal{S}_r$. If r moves to the empty point, then we are done. If not, then it implies that there is a robot r_1 with light $Moving1$ within Manhattan distance two in the direction it intends to move.

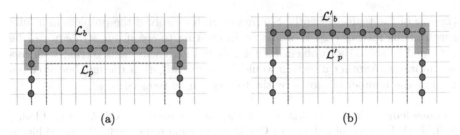

Fig. 6. (a) There is no empty grid point in the shaded region. (b) Two new empty points are created after the expansion

Case 1: Assume that r_1 is also on $\mathcal{L}'$. But since $\mathcal{H}_r \cap \mathcal{S}_r$ and $\mathcal{H}_{r_1} \cap \mathcal{S}_{r_1}$ are in opposite directions, r_1 is not in the direction in which r wants to move (and vice versa).

Case 2: Let r_1 be on the grid line below $\mathcal{L}'$, as shown in Fig. 3c and f. But since r is the only robot on $\mathcal{L}'$, it has at least two empty points available in two directions. Then it will choose the empty point which is not towards r_1.

Case 3: Now consider the situation shown in Fig. 3d, where r_1 is a robot moving on an adjacent penultimate line segment. Then r and r_1 will request another expansion. It can be seen from Fig. 7, that in the new configuration, both robots will be able to move to an empty boundary point. □

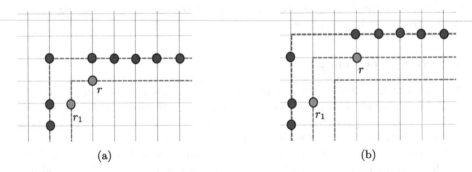

Fig. 7. (a) The situation is similar to the example shown in Fig. 3d. (b) The subsequent configuration after both boundary sides expand and both r and r_1 move to the new penultimate layer.

A.3 Proof of Lemma 3

Suppose that (1) the robots in $\mathcal{V}_r$ form an empty rectangle, (2) the lights of all the robots in $\mathcal{V}_r$ are set to *Boundary*. It may happen that some robots in $\mathcal{V}_r$ (with lights set to *Boundary*) are actually not on the boundary, but on the penultimate layer. Then these robots are in the middle of an expansion, but are yet to change their lights to *Expanding*, and are obstructing some robots on the boundary also having light *Expanding*. We argue that this is not possible.

First of all, if r itself was executing an expansion in a previous round, then all the fellow robots, with which it had previously shared a boundary side, must have also completed the expansion. This is because r has its light set to *Boundary*. Now for the remaining three boundary sides, if the robots are executing an expansion, then they must be instructed to do so by some robot in the interior with light *RequestExpansion*. But V_r has only robots with light *Boundary*. Hence, all the robots in V_r are indeed boundary robots. Hence, the empty rectangle configuration is achieved.   □

Appendix B Correctness of Symmetric Movements

B.1 Proof of Lemma 4

Suppose that there are three points in C_i, say $\{u, v, w\}$, that are collinear. No three points in C_{i-1} are collinear. So at least one of the three points is in $C_i \setminus C_{i-1}$, say u. But u is computed in such a way that it is collinear with no two points in C_{i-1}. So another one among the three points must be in C_i, say v. From the symmetries, we can say that v is one of the three possible points shown in Fig. 8 as $\{v_1, v_2, v_3\}$. Clearly the straight lines through u and v_1 or u and v_3 do not pass through any other point in C_{i-1}. If the straight line through u and v_2 passes through a point $z \in C_{i-1}$, from symmetry it will also pass through another point $z' \in C_{i-1}$ (See Fig. 8). This implies that the straight line passes through two points in C_{i-1}, which is not possible. Hence, no three points in C_i are collinear.   □

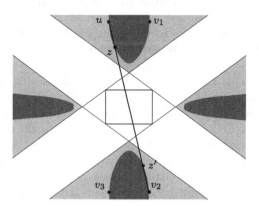

Fig. 8. Illustration supporting the proof of Lemma 4.

B.2 Proof of Theorem 3

Since the robots closer to the corners move farther, a robot r will always be able to see at least two corners. Due to obstructions, it may not see another corner. We show that this will not create any lock cases.

Case 1: (BlockedByMovingRobot() = False): If it sees no robot with light $\overline{Moving3}$, its view must be obstructed by a robot with light $Done$ that has already reached its destination. Hence, this is clearly not the destination point of r. So r will move.

Case 2 (BlockedByMovingRobot() = True): If it sees a robot with light $\overline{Moving3}$, it simply waits. In our movement strategy, at any time at most two robots can be moving in the shaded region (see Fig. 5). Hence, it sees exactly one moving robot below it, say r'. Clearly no moving robot is obstructing r''s view. Hence, r' will eventually move or turn its light to $Done$.   □

Appendix C Corner Creation

The lights that will be used in this phase are $\{Rectangle, Square, NextToCorner, Moving2\}$. The objective of this phase is to occupy the corners of the empty rectangle by robots with specified lights. For simplicity, we shall first assume that there are at least two robots on each of the four sides. Only the robots terminal on the boundary sides will move in this phase. If the configuration is a non-square rectangle, then the terminal robots on the larger side will move to the corner and set its light to $Rectangle$. Note that the robots can determine the length of the sides of the rectangle. However, if it is a square, it may not be always possible to break tie. If two terminal robots on adjacent sides of the square move to the corner, there will be a collision. If it is possible to break tie, then one of them will go to the corner and set its light to $Square$. Otherwise, the robots will move to the point adjacent to the corner and then set its light to $NextToCorner$. While moving, the terminal robots will set their lights to $Moving2$.

However, this simple scheme will not be always applicable. The initial empty rectangle configuration may have different anomalies. For example, some sides may have only a single robot, or all the robots could lie on a single straight line, or form an L-shape, etc. (See Fig. 9). While designing algorithms for these configurations, the following issues should be properly addressed. A robot may not be always able to distinguish between two configurations from their local views. In these cases, the movement specified for the robot in both configurations should not contradict each other. During the movements, the configuration may change from one case to another. Due to the asynchronous scheduler, the adversary may delay the move of a robot, which will have a pending move based on an out-dated view of the configuration. Such pending moves should not cause any inconsistencies in the algorithm. The algorithms for these different configurations are pictorially presented in Fig. 9. Proofs and other details of the algorithms are omitted due to space constraints.

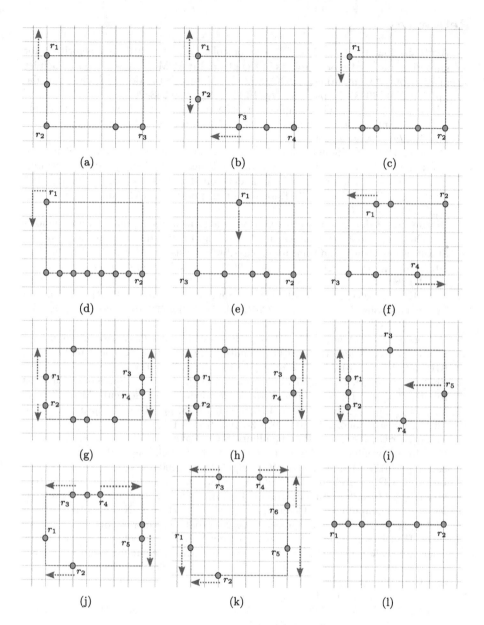

Fig. 9. Movements in the corner creation phase for atypical empty rectangle configurations.

References

1. Agathangelou, C., Georgiou, C., Mavronicolas, M.: A distributed algorithm for gathering many fat mobile robots in the plane. In: ACM Symposium on Principles of Distributed Computing, PODC 2013, 22–24 July 2013, Montreal, QC, Canada, pp. 250–259 (2013). https://doi.org/10.1145/2484239.2484266
2. Aljohani, A., Sharma, G.: Complete visibility for mobile robots with lights tolerating faults. Int. J. Netw. Comput. **8**(1), 32–52 (2018). http://www.ijnc.org/index.php/ijnc/article/view/166
3. Barberá, H.M., Quiñonero, J.P.C., Zamora-Izquierdo, M.A., Gómez-Skarmeta, A.F.: i-fork: a flexible AGV system using topological and grid maps. In: Proceedings of the 2003 IEEE International Conference on Robotics and Automation, ICRA 2003, 14–19 September 2003, Taipei, Taiwan, pp. 2147–2152 (2003). https://doi.org/10.1109/ROBOT.2003.1241911
4. Bhagat, S., Mukhopadhyaya, K.: Optimum algorithm for mutual visibility among asynchronous robots with lights. In: Proceedings of 19th International Symposium Stabilization, Safety, and Security of Distributed Systems, SSS 2017, 5–8 November 2017, Boston, MA, USA, pp. 341–355 (2017). https://doi.org/10.1007/978-3-319-69084-1_24
5. Bose, K., Adhikary, R., Chaudhuri, S.G., Sau, B.: Crash tolerant gathering on grid by asynchronous oblivious robots. CoRR abs/1709.00877 (2017). http://arxiv.org/abs/1709.00877
6. D'Angelo, G., Stefano, G.D., Klasing, R., Navarra, A.: Gathering of robots on anonymous grids and trees without multiplicity detection. Theor. Comput. Sci. **610**, 158–168 (2016). https://doi.org/10.1016/j.tcs.2014.06.045
7. Fischer, M., Jung, D., Meyer auf der Heide, F.: Gathering anonymous, oblivious robots on a grid. In: Fernández Anta, A., Jurdzinski, T., Mosteiro, M.A., Zhang, Y. (eds.) ALGOSENSORS 2017. LNCS, vol. 10718, pp. 168–181. Springer, Cham (2017). https://doi.org/10.1007/978-3-319-72751-6_13
8. Luna, G.A.D., Flocchini, P., Chaudhuri, S.G., Poloni, F., Santoro, N., Viglietta, G.: Mutual visibility by luminous robots without collisions. Inf. Comput. **254**, 392–418 (2017). https://doi.org/10.1016/j.ic.2016.09.005
9. Luna, G.A.D., Flocchini, P., Poloni, F., Santoro, N., Viglietta, G.: The mutual visibility problem for oblivious robots. In: Proceedings of the 26th Canadian Conference on Computational Geometry, CCCG 2014, Halifax, Nova Scotia, Canada (2014). http://www.cccg.ca/proceedings/2014/papers/paper51.pdf
10. Peleg, D.: Distributed coordination algorithms for mobile robot swarms: new directions and challenges. In: Pal, A., Kshemkalyani, A.D., Kumar, R., Gupta, A. (eds.) IWDC 2005. LNCS, vol. 3741, pp. 1–12. Springer, Heidelberg (2005). https://doi.org/10.1007/11603771_1
11. Poudel, P., Sharma, G.: Universally optimal gathering under limited visibility. In: Spirakis, P., Tsigas, P. (eds.) SSS 2017. LNCS, vol. 10616, pp. 323–340. Springer, Cham (2017). https://doi.org/10.1007/978-3-319-69084-1_23
12. Sharma, G., Alsaedi, R., Busch, C., Mukhopadhyay, S.: The complete visibility problem for fat robots with lights. In: Proceedings of the 19th International Conference on Distributed Computing and Networking, ICDCN 2018, 4–7 January 2018, Varanasi, India, pp. 21:1–21:4 (2018). https://doi.org/10.1145/3154273.3154319
13. Sharma, G., Busch, C., Mukhopadhyay, S.: Bounds on mutual visibility algorithms. In: Proceedings of the 27th Canadian Conference on Computational Geometry, CCCG 2015, 10–12 August 2015, Kingston, Ontario, Canada (2015). http://research.cs.queensu.ca/cccg2015/CCCG15-papers/43.pdf

14. Sharma, G., Busch, C., Mukhopadhyay, S.: Mutual visibility with an optimal number of colors. In: Bose, P., Gąsieniec, L.A., Römer, K., Wattenhofer, R. (eds.) ALGOSENSORS 2015. LNCS, vol. 9536, pp. 196–210. Springer, Cham (2015). https://doi.org/10.1007/978-3-319-28472-9_15

15. Sharma, G., Vaidyanathan, R., Trahan, J.L., Busch, C., Rai, S.: Complete visibility for robots with lights in $O(1)$ time. In: Bonakdarpour, B., Petit, F. (eds.) SSS 2016. LNCS, vol. 10083, pp. 327–345. Springer, Cham (2016). https://doi.org/10.1007/978-3-319-49259-9_26

16. Sharma, G., Vaidyanathan, R., Trahan, J.L., Busch, C., Rai, S.: O(log n)-time complete visibility for asynchronous robots with lights. In: 2017 IEEE International Parallel and Distributed Processing Symposium, IPDPS 2017, 29 May–2 June 2017, Orlando, FL, USA, pp. 513–522 (2017). https://doi.org/10.1109/IPDPS.2017.51

17. Stefano, G.D., Navarra, A.: Gathering of oblivious robots on infinite grids with minimum traveled distance. Inf. Comput. **254**, 377–391 (2017). https://doi.org/10.1016/j.ic.2016.09.004

Optimal Gathering by Asynchronous Oblivious Robots in Hypercubes

Kaustav Bose⊙, Manash Kumar Kundu$^{(\boxtimes)}$⊙, Ranendu Adhikary⊙,
and Buddhadeb Sau

Department of Mathematics, Jadavpur University, Kolkata, India
{kaustavbose.rs,manashkrkundu.rs,
ranenduadhikary.rs}@jadavpuruniversity.in, bsau@math.jdvu.ac.in

Abstract. We consider the problem of gathering a set of autonomous, identical, oblivious, asynchronous, mobile robots at a vertex of an anonymous hypercube. The robots operate in Look-Compute-Move cycles. In each cycle, a robot takes a snapshot of the current configuration (*Look*), then based on the perceived configuration, decides whether to stay idle or to move to an adjacent vertex (*Compute*), and in the later case makes an instantaneous move accordingly (*Move*). We have shown that the problem is unsolvable if the robots do not have multiplicity detection capability. With weak multiplicity detection capability, the problem is solvable in an oriented hypercube for any initial configuration of $2k + 1(k > 0)$ number of robots. For $4k(k > 0)$ number of robots, the problem is solvable under the same assumptions if and only if the group of automorphism of the configuration is trivial. Our proposed algorithms are optimal with respect to the total number of moves executed by the robots.

Keywords: Distributed computing · Autonomous robots · Gathering · Hypercube · Weber point · Asynchronous · Look-Compute-Move cycle

1 Introduction

The *gathering* problem requires a set of n mobile computational entities, usually called *robots* or *agents*, initially situated at different locations in a spatial universe, to gather at some unspecified location within finite time. When only two robots are involved, the problem is usually referred to as the *rendezvous* problem. In distributed computing, gathering has been extensively studied both in continuous and in discrete domains. In the continuous setting, the robots operate in the two-dimensional Euclidean space and in the discrete case, they operate in a network modeled as a graph. In the discrete setting, the problem has been previously studied in different graph topologies, e.g. rings [10,18,22,23], grids [8,31], trees [8] etc. The problem is particularly difficult in graphs that are highly symmetric and is solvable only in very limited cases. Hence, for characterization of gatherability, it is important to investigate the problem in highly symmetric graphs. In this paper, we investigate the problem in a hypercube graph.

© Springer Nature Switzerland AG 2019
S. Gilbert et al. (Eds.): ALGOSENSORS 2018, LNCS 11410, pp. 102–117, 2019.
https://doi.org/10.1007/978-3-030-14094-6_7

1.1   The Model

A set of autonomous mobile robots is randomly deployed on the vertices of a d-dimensional hypercube network. The *d-dimensional hypercube* Q_d is an undirected graph with vertex set $V(Q_d) = \mathbb{Z}_2^d = \{0, 1\}^d$, and two vertices are adjacent if and only if the two binary strings differ in exactly one coordinate. An *oriented hypercube* is an edge-labeled hypercube with the so-called *dimensional labeling* $\lambda : E(Q_d) \to \{1, \ldots, d\}$ where $\lambda(uv) = i$, if u and v differ in the ith coordinate. We shall denote an oriented hypercube by Q_d^O, and an unoriented hypercube by simply Q_d. The binary string labels of the vertices are for descriptive purposes, and are not known to the robots. However, in an oriented hypercube, the edge-labels are known to the robots. It is traditionally assumed that the robots can only perceive the labels of the edges adjacent to the vertex on which it resides. But since the labels of edges adjacent to a single vertex determine the dimensional labels of all the edges in a hypercube (See Theorem 3.1 in [33]), we assume without loss of generality that the robots know the labels of all the edges.

The robots are *oblivious* (they have no memory of past configurations and previous actions), *autonomous* (there is no central control), *homogeneous* (they execute the same distributed algorithm), *anonymous* (they have no unique identifiers) and *identical* (they are indistinguishable by their appearance). The robots have *global visibility*, i.e., they are able to perceive the entire graph. The robots do not agree on any global coordinate system. Furthermore, there are no means of communication between the robots.

The robots, when active, operate according to the so-called LOOK-COMPUTE-MOVE cycle. In each cycle, a previously idle or inactive robot wakes up and executes the following steps. In the LOOK phase, the robot takes the snapshot of the positions of all the robots, represented in its own coordinate system. Based on the perceived configuration, the robot performs computations according to a deterministic algorithm to decide whether to stay idle or to move to an adjacent vertex. Based on the outcome of the algorithm, the robot either remains stationary or makes an instantaneous move to an adjacent vertex. Since the moves are instantaneous, it implies that the robots are always seen on vertices, not on edges. In the fully synchronous setting (FSYNC), the activation phase of all robots can be logically divided into global rounds, where all the robots are activated in each round. The semi-synchronous (SSYNC) model coincides with the FSYNC model with the only difference that not all robots are necessarily activated in each round. The most general type of scheduler is the asynchronous scheduler (ASYNC). In ASYNC, the robots are activated independently, and the amount of time spent in LOOK, COMPUTE, MOVE and inactive states are finite but unbounded and unpredictable. As a result, the robots do not have a common notion of time.

An important capability associated to the robots is multiplicity detection. During the LOOK phase, a robot may perceive a vertex occupied by more than one robot in different ways. In *strong multiplicity detection*, the robots perceive the actual number of robots in each vertex. In *weak multiplicity detection*, the robots are only able to detect whether a vertex is occupied by more than one robot, but not the exact number. If the robots have no multiplicity detection capability, they can only decide if a vertex is occupied or empty.

1.2   Related Works

The gathering problem has been extensively studied in continuous domain under various assumptions [1,4–7,16,29,30]. In discrete domains, both gathering and rendezvous have been studied in different graph topologies [2,8,11,13–15,22,23,25,27,28,32]. The problem of gathering two robots on an anonymous ring was studied in [12,26,28]. The problem for more than two robots was studied in [15]. In [15], the robots had memory and used tokens to break symmetry. In [23], the problem was first considered in a very minimal setting with identical, asynchronous, memoryless robots without tokens or any kind of communication capability. They proved that without multiplicity detection, gathering is impossible on rings for $n \geq 2$ robots. With weak multiplicity detection capability, they solved the problem for all configurations with an odd number of robots, and all the asymmetric configurations with an even number of robots by different algorithms. In [22], symmetric configurations with an even number of robots were studied, and the problem was solved for more than 18 robots. Some of the remaining configurations were solved in [9,18,24] in separate algorithms. In [10], a single unified algorithm was proposed, that achieves gathering for all gatherable initial configurations except some potentially gatherable configurations with 4 robots. The problem was studied with weak local multiplicity detection in [19–21]. A full characterization of gatherable configurations for finite grids and trees with weak multiplicity detection was provided in [8]. Gathering in finite grids in presence of crash-faults was studied in [3]. Optimal gathering in infinite grid with strong multiplicity detection was studied in [31].

2   Theoretical Preliminaries

2.1   Group of Automorphisms

An *automorphism* of a graph $G = (V, E)$ is a bijection $\varphi : V \longrightarrow V$ such that for all $u, v \in V$, u, v are adjacent if and only if $\varphi(u), \varphi(v)$ are adjacent. The set of all automorphisms of G forms a group, called the *automorphism group* of G and is denoted by $Aut(G)$.

The automorphism group of a hypercube is generated by two types of automorphisms, namely *translation* and *rotation*.

Translation: For $a \in \mathbb{Z}_2^d$, the map $\tau_a : V(Q_d) \longrightarrow V(Q_d)$ given by $u \mapsto u \oplus a$ is called *translation by a*. Here, $u \oplus a$ is the vertex obtained by adding the binary strings u and a componentwise. The set $T = \{\tau_a \mid a \in \mathbb{Z}_2^d\}$ of all translations forms a subgroup of $Aut(Q_d)$.

Rotation: For $\sigma \in S_d$, the map $r_\sigma : V(Q_d) \longrightarrow V(Q_d)$ given by $u \mapsto \sigma(u)$ is called *rotation by* σ, where $\sigma(u)$ is the vertex obtained by permuting the binary string u by $\sigma : \{1,\ldots,d\} \longrightarrow \{1,\ldots,d\}$. The set $R = \{r_\sigma \mid \sigma \in S_d\}$ of all rotations forms a subgroup of $Aut(Q_d)$.

Theorem 1 [17]. $Aut(Q_d) = TR$.

Hence, for any automorphism $\varphi \in Aut(Q_d)$, $\exists$ a unique pair $(a, \sigma) \in (\mathbb{Z}_2^d, S_d)$, such that $\varphi : V \longrightarrow V$ can be written as $u \mapsto \sigma(u) \oplus a$.

It is easy to see that $|T| = 2^d$ and $|R| = d!$. Since $T \cap R$ is trivial, $|TR| = |T||R|/|T \cap R| = 2^d d!$. Therefore, we have the following corollary.

Corollary 1 [17]. $|Aut(Q_d)| = 2^d d!$.

The definition of automorphism of graphs can be extended to edge-labeled graphs in a natural way. Given an edge-labeled graph $G = (V, E, \lambda)$ with edge-labeling $\lambda : E \longrightarrow \mathbb{N}$, an automorphism of G is a bijection $\varphi : V \longrightarrow V$ such that for all $u, v \in V$, $\varphi(u)\varphi(v) \in E$ if and only if (1) $uv \in E$ and (2) $\lambda(\varphi(u)\varphi(v)) = \lambda(uv)$. In view of this definition, it is easy to see that the dimensional labeling of a hypercube kills all rotational automorphisms. Thus the automorphism group of an oriented hypercube consists of only translations.

Theorem 2. $Aut(Q_d^O) = T$.

2.2 Feasibility of Gathering

Consider a set of robots placed on the vertices of a simple undirected connected graph $G = (V, E)$. Define a function $f : V \longrightarrow \mathbb{N} \cup \{0\}$, where $f(v)$ is the number of robots on the vertex v. The pair (G, f) is called a *configuration of robots on G*, or simply a *configuration*. If all the robots in a configuration reside on a single vertex, then it is called *final configuration*; otherwise it is called a *non-final configuration*. Given a configuration (G, f), we define the *multiplicity function* $\tilde{f}$ in the following way. If the model assumes robots with strong multiplicity detection capability, then $\tilde{f}(v) = f(v)$ for all $v \in V$. If the robots have weak multiplicity detection capability, then $\tilde{f} : V \longrightarrow \{0, 1, 2\}$ is defined as,

$$\tilde{f}(v) = \begin{cases} 0 & \text{if } v \text{ is an empty vertex} \\ 1 & \text{if } v \text{ is occupied by exactly one robot} \\ 2 & \text{if } v \text{ is a multiplicity.} \end{cases}$$

If the robots have no multiplicity detection capability, then $\tilde{f} : V \longrightarrow \{0, 1\}$ is defined as,

$$\tilde{f}(v) = \begin{cases} 0 & \text{if } v \text{ is an empty vertex} \\ 1 & \text{if } v \text{ is occupied by at least one robot.} \end{cases}$$

Given a configuration (G, f), the pair $(G, \tilde{f})$ is called the *perceived configuration*.

An *automorphism of a perceived configuration* $(G, \tilde{f})$ is a graph automorphism $\varphi \in Aut(G)$ such that $\tilde{f}(v) = \tilde{f}(\varphi(v))$ for all $v \in V$. The set of all automorphisms of $(G, \tilde{f})$ also forms a group that will be denoted by $Aut(G, \tilde{f})$.

If $|Aut(G, \tilde{f})| = 1$, we say that $(G, \tilde{f})$ is *asymmetric*, otherwise it is said to be *symmetric*.

For an automorphism $\varphi \in Aut(G, \tilde{f})$, let $<\varphi> \subseteq Aut(G, \tilde{f})$ be the cyclic subgroup generated by φ. Elements of this group are $\{\varphi^0, \varphi^1, \varphi^2, \ldots, \varphi^{p-1}\}$, where φ^0 is the identity, $\varphi^k = \underbrace{\varphi \circ \varphi \circ \cdots \circ \varphi}_{k \text{ times}}$ and p is the order of φ.

For any subgroup H of $Aut(G, \tilde{f})$, define the equivalence relation on V given by: $x \sim y$ if and only if $x = \varphi(y)$ for some $\varphi \in H$. This equivalence relation induces a partition on V. The *orbit of a vertex $v \in V$ under the action of H* is the set $H_v = \{\sigma(v) | \sigma \in H\}$, which is the corresponding equivalence class containing v.

Partitive Automorphism: Let $\mathcal{C} = ((V, E), \tilde{f})$ be a perceived configuration. A non-trivial automorphism $\varphi \in Aut(\mathcal{C})$ is said to be *partitive* on V if $|H_v| = p$ for all $v \in V$, where $p > 1$ is the order of φ and $H = <\varphi>$.

Lemma 1. *In Q_d, any non-trivial translation is partitive.*

Theorem 2 in [32], stated for configurations of robots with strong multiplicity detection capability can be easily generalized to the following theorem.

Theorem 3. *Let $\mathcal{C} = ((V, E), \tilde{f})$ be a non-final perceived configuration. If there exists a $\varphi \in Aut(\mathcal{C})$ partitive on V, then $\mathcal{C}$ is not gatherable.*

Theorem 4. *Without multiplicity detection capability, gathering in (both oriented and unoriented) hypercubes is not deterministically solvable in SSYNC.*

Proof. Assume that there exists a correct gathering algorithm $\mathcal{A}$. In the SSYNC model, time can be logically divided into discrete global rounds. So, starting from some non-final initial configuration, consider a synchronous execution of algorithm $\mathcal{A}$, in which gathering is achieved in round t.

Case 1: Suppose that in round $t - 1$, exactly two vertices in Q_d are occupied. Hence, the perceived configuration in round $t - 1$ is $(Q_d, \tilde{f})$ where $\tilde{f}(v) = \tilde{f}(w) = 1$ for two distinct vertices $v, w \in V(Q_d)$, and $\tilde{f}(u) = 0 \ \forall u \in V(Q_d) \backslash \{v, w\}$. But, then the perceived configuration $(Q_d, \tilde{f})$ admits a partitive automorphism given by $x \mapsto x \oplus v \oplus w$. Hence by Theorem 3, gathering can not be deterministically achieved from this configuration.

Case 2: Assume that at least three vertices in Q_d are occupied in round $t - 1$. Then algorithm $\mathcal{A}$ brings all the robots to a common vertex, say u, in one step. But the adversary can choose to activate all the robots except one that is not placed at u. Then all but one robot will reach u. This will create a configuration with exactly two vertices occupied. Since this configuration admits a partitive automorphism, gathering can not be deterministically achieved from here by Theorem 3. □

Corollary 2. *Without multiplicity detection capability, gathering in (both oriented and unoriented) hypercubes is not deterministically solvable in ASYNC.*

2.3   Weber Point

Given a configuration (G, f), with $G = (V, E)$, the centrality of $v \in V$ is defined as $c_{G,f}(v) = \sum_{u \in V} d(u, v) \cdot f(u)$. When there is no ambiguity, we shall write $c_f(v)$, or simply $c(v)$.

Weber Point: Given a configuration $\mathcal{C} = (G = (V, E), f)$, a vertex $v \in V$ is a *Weber point* of $\mathcal{C}$, if $c(v) = min\{c(u)|u \in V\}$.

By definition, a Weber point is a vertex with minimum centrality. In other words, a vertex $w \in V$ is a Weber point if the sum of the lengths of the shortest paths from all robots to w is minimum. Therefore, an algorithm that gathers all the robots at a Weber point via the shortest paths is optimal with respect to the total number of moves performed by the robots. However, a configuration may have more than one Weber point. Given a configuration (G, f), we shall denote the set of Weber points by $\mathcal{W}_{G,f}$, or simply $\mathcal{W}_f$ or $\mathcal{W}$ when there is no ambiguity.

Theorem 5 [32]. *Let (G, f) be a configuration with Weber points $\mathcal{W}_f$. If a robot moves towards a Weber point $w \in \mathcal{W}_f$, resulting in a new configuration (G, f'), then*

1. $c_{f'}(v) = c_f(v) - 1$ *for each $v \in \mathcal{W}_{f'}$*
2. $w \in \mathcal{W}_{f'}$
3. $\mathcal{W}_{f'} \subseteq \mathcal{W}_f$.

We shall now discuss about the Weber points of configurations on a hypercube. Consider a set of n robots $\{r_1, r_2, \ldots, r_n\}$ on a d-dimensional hypercube Q_d. Suppose that the robots $r_1, r_2, \ldots, r_n$ are placed on the vertices $v_1, v_2, \ldots, v_n$ respectively. For $i = 1, 2, \ldots, n$, let the binary string representation of v_i be $b_{i1}b_{i2} \ldots b_{id}$, where $b_{ij} \in \{0, 1\}$. For $j = 1, 2, \ldots, d$, let us define sets $[b]_j \subseteq \{0, 1\}$ in the following way.

$$[b]_j = \begin{cases} \{0\}, & \text{if the number of 0's in the multiset } \{b_{1j}, b_{2j}, \ldots, b_{nj}\} \text{ is more} \\ & \text{than the number of 1's} \\ \{1\}, & \text{if the number of 1's in the multiset } \{b_{1j}, b_{2j}, \ldots, b_{nj}\} \text{ is more} \\ & \text{than the number of 0's} \\ \{0, 1\}, & \text{if the multiset } \{b_{1j}, b_{2j}, \ldots, b_{nj}\} \text{ has equal number of 0's and 1's.} \end{cases}$$

In Theorem 6, we show that the set of Weber points of the configuration is

$$\mathcal{W} = [b]_1 \times [b]_2 \times \ldots \times [b]_d.$$

For instance, consider a configuration of a set of 8 robots $\{r_1, r_2, \ldots, r_8\}$ on a 4-dimensional hypercube Q_4. Suppose that the robots are positioned on the following vertices respectively: 0110, 0111, 1000, 1110, 0000, 0001, 1110, 1101. Then the set of Weber points of this configuration is given by

$$\mathcal{W} = \{0, 1\} \times \{1\} \times \{0, 1\} \times \{0\}$$
$$= \{0100, 0110, 1100, 1110\}.$$

Theorem 6. *Let $\{r_i\}_{i=1}^n$ be a set of robots placed on the vertices of Q_d with binary string representations $\{b_{i1}b_{i2}\ldots b_{id}\}_{i=1}^n$ respectively. Then the set of Weber points of the configuration is $\mathcal{W} = [b]_1 \times [b]_2 \times \ldots \times [b]_d$.*

Proof. The distance between any two vertices in Q_d is the number of positions in which their binary string representations differ. Then it can be easily seen that (1) the centrality of all $w \in [b]_1 \times [b]_2 \times \ldots \times [b]_d$ are equal, (2) the centrality of any $v \in V(Q_d)\backslash[b]_1 \times [b]_2 \times \ldots \times [b]_d$ is strictly greater than the centrality of $w \in [b]_1 \times [b]_2 \times \ldots \times [b]_d$.   □

Corollary 3. *The subgraph induced by the set of Weber points $\mathcal{W}$ of a configuration on a hypercube Q_d is also a hypercube.*

Corollary 4. *The number of Weber points of a configuration on a hypercube Q_d is 2^k, where $0 \leq k \leq d$.*

2.4   Leading Weber Point

A configuration of robots on a hypercube can have more than one Weber point. We want to devise an algorithm that gathers all the robots at one of the Weber points via the shortest paths. Our proposed algorithm requires to solve a subproblem called LEADINGWEBERPOINT. Let us formally define the problem LEADINGWEBERPOINT. Consider a configuration in which no vertex contains more than one robot, and that has no partitive automorphism. Let $\mathcal{W}$ be the set of Weber points of this configuration. The problem LEADINGWEBERPOINT asks to devise an algorithm so that every robot that perceives this configuration in its local view, deterministically elects a unique Weber point $w_\ell \in \mathcal{W}$. We shall call the vertex w_ℓ the *leading Weber point*.

Since we have assumed that the robots are positioned at distinct vertices, there is no distinction between the configuration and the perceived configuration. In other words, given such a configuration (G, f), we have $\tilde{f} = f$. A vertex $v \in V$ is called a *fixed vertex* if $\varphi(v) = v, \forall\varphi \in Aut(G, f)$.

Theorem 7. LEADINGWEBERPOINT *can be deterministically solved only if $\mathcal{W}$ has at least one fixed vertex.*

Proof. See Appendix A.

Theorem 8. LEADINGWEBERPOINT *may not be deterministically solvable in an unoriented hypercube.*

Proof. Consider a configuration (Q_5, f) of a set of 14 robots on the 5-dimensional unoriented hypercube Q_5. The robots are placed on the following vertices: 00100, 00001, 11000, 10010, 01100, 01010, 00101, 00011, 11010, 10110, 11001, 10101, 01111, 11111. It is easy to see that $\mathcal{W}_f = V(Q_5)$. It can be shown that $Aut(Q_5, f) = \{e, \varphi_1, \varphi_2, \varphi_3\}$, with each φ_i given by $u \mapsto \sigma_i(u) \oplus a_i$, where $a_i \in \mathbb{Z}_2^5, \sigma_i \in S_5$ are the following: $a_1 = 00000, \sigma_1 = (1)(24)(35), a_2 = 10000, \sigma_2 = (1)(2543), a_3 = 10000, \sigma_3 = (1)(2345)$. Then it can be easily verified that (1) there is no partitive automorphism in $Aut(Q_5, f)$, (2) there is no fixed vertex in $\mathcal{W}_f = V(Q_5)$. So by Theorem 7, LEADINGWEBERPOINT is deterministically unsolvable. □

Now we show that LEADINGWEBERPOINT can be deterministically solved in an oriented hypercube.

Lemma 2. *Given a vertex u_0 in an oriented hypercube $Q_d^\mathcal{O}$, $\exists$ exactly one coordinate assignment (bijection) $\Psi : V(Q_d^\mathcal{O}) \longrightarrow \{0,1\}^d$ such that*

1. $\Psi(u_0) = 00 \ldots 0 \in \{0,1\}^d$
2. u, v are adjacent if and only if $\Psi(u), \Psi(v)$ differ in exactly one bit
3. for $uv \in E(Q_d^\mathcal{O})$, $\lambda(uv) = i$ if and only if $\Psi(u), \Psi(v)$ differ in the ith position.

Proof. The coordinates given to u_0 are $00 \ldots 0$. Then by rule (2) and (3), the coordinates of all its neighbors are uniquely determined. If the coordinates of all vertices at distance $i(< d)$ from u_0 are uniquely determined, then again by rule (2) and (3), the coordinates of all vertices at distance $i + 1$ can be determined uniquely. Hence by induction, the coordinates assigned to all the vertices are unique. □

Now for any $w \in V(Q_d^\mathcal{O})$, we define a binary string $\zeta(w)$ of length 2^d in the following the way:

1. First, give $Q_d^\mathcal{O}$ the unique coordinate assignment $\Psi : V(Q_d^\mathcal{O}) \longrightarrow \{0,1\}^d$ with $\Psi(w) = 00 \ldots 0$.
2. Now we define a total ordering $\prec$ on $V(Q_d^\mathcal{O})$ as: $u \prec v$

$$u \prec v \Leftrightarrow \begin{cases} d(u,w) < d(v,w) \\ Or, \\ d(u,w) = d(v,w), \text{and } \Psi(u) \text{ is lexicographically larger that } \Psi(v), \end{cases}$$

where $d(u,w)$ is the distance of u from w. For example, when $d = 4$, the assigned coordinates of the vertices written in increasing order will be: $\underbrace{0000}_{\text{distance 0}}$, $\underbrace{1000, 0100, 0010, 0001,}_{\text{distance 1}}$ $\underbrace{1100, 1010, 1001, 0110, 0101, 0011,}_{\text{distance 2}}$ $\underbrace{1110, 1101, 1011, 0111,}_{\text{distance 3}}$ $\underbrace{1111}_{\text{distance 4}}$.

3. Finally, scan the vertices of the hypercube according to the above ordering. For each vertex, put a 0 if it is empty, or 1 if it is occupied by a robot. Recall that any vertex can be occupied by at most one robot. The string of length 2^d thus obtained is $\zeta(w)$. In the previous example, if the occupied vertices are $0000, 1000, 0010, 1001, 0011, 1011, 1111$, then $\zeta(w) = 1101000100100101$.

Lemma 3. *For any two distinct vertices $u, v \in V(Q_d^\mathcal{O})$, if $\zeta(u) = \zeta(v)$, then the configuration has a partitive automorphism.*

Proof. It can be easily seen that if $\zeta(u) = \zeta(v)$, then the configuration has the automorphism (translation) given by $x \mapsto x \oplus u \oplus v$.   □

Theorem 9. LEADINGWEBERPOINT *is solvable in an oriented hypercube.*

Proof. Since the configuration has no partitive automorphism, $\zeta(w_1) \neq \zeta(w_2)$ for any distinct $w_1, w_2 \in \mathcal{W}$. Hence the robots can unanimously elect $w \in \mathcal{W}$ with the lexicographically (strictly) largest $\zeta(w)$ as the leading Weber point.   □

3   The Algorithm

Our plan is to solve the problem in two stages. In stage 1, we create a multiplicity at a Weber point and then in stage 2, we sequentially bring the remaining robots to that vertex. Before describing the algorithm, we first give two definitions.

Anchor: Let $(Q_d^{\mathcal{O}}, \tilde{f})$ be a non-final perceived configuration on an oriented hypercube with at most one multiplicity and no partitive automorphism. The *anchor* of $(Q_d^{\mathcal{O}}, \tilde{f})$ is a vertex $\alpha \in V(Q_d^{\mathcal{O}})$ defined as the following. If $(Q_d^{\mathcal{O}}, \tilde{f})$ has no multiplicity, then α is the leading Weber point; otherwise α is the unique vertex with multiplicity. Note that all the robots observing the configuration $(Q_d^{\mathcal{O}}, \tilde{f})$ agree on which vertex is the anchor.

Leader: Since all the robots observing the configuration $(Q_d^{\mathcal{O}}, \tilde{f})$ agree on the anchor α, they also agree on a common coordinate system, which is the unique coordinate system Ψ described in Lemma 2 with $\Psi(\alpha) = 00\ldots0$. This also allows the robots to order the vertices of the hypercube as described in the previous section. In this ordering, the first robot appearing on a non-anchor vertex will be called the *leader*.

3.1   2k + 1 (k > 0) Robots

Theorem 10. *Any configuration on a hypercube with odd number of robots has exactly one Weber point.*

Proof. Using the same notations as in Theorem 6, the set of Weber points is given by $\mathcal{W} = [b]_1 \times [b]_2 \times \ldots \times [b]_d$. Since there are odd number of robots, the multiset $\{b_{1j}, b_{2j}, \ldots, b_{nj}\}$ can never have equal number of 0's and 1's. Hence, $[b]_j = \{0\}$ or $\{1\}$, $\forall j \in \{1, \ldots, d\}$. Thus $|\mathcal{W}| = 1$.   □

Theorem 11. *Gathering in $Q_d^{\mathcal{O}}$ is optimally solvable in ASYNC with weak multiplicity detection for any configuration of odd number of robots.*

Proof. We simply ask only the leader to move towards the anchor. The anchor α is the unique Weber point of the configuration. As the leader moves towards it, the Weber point remains invariant by Theorem 5. After one or two robots reach α, a multiplicity is created at α. Throughout stage 2, α remains the unique multiplicity in the configuration, since only the leader moves. Thus, all the remaining robots will sequentially reach α. The algorithm is clearly optimal with respect to the total number of moves executed by the robots.   □

Algorithm 1. Gathering for $4k$ $(k > 0)$ robots

```
1  Procedure GATHER()
2  │    α ← anchor
3  │    if α is a multiplicity then
4  │    │    if I am leader then
5  │    │    └    Move towards α
6  │    else
7  │    │    if I am leader then
8  │    │    │    if I am on a Type 1 vertex then
9  │    │    │    │    Move towards α via a Type 1 edge
10 │    │    │    else if I am on a Type 2 vertex then
11 │    │    │    └    Move towards α
```

3.2   4k (k > 0) Robots

In view of Theorems 2, 3 and Lemma 1, any configuration with non-trivial automorphism group is ungatherable. We show that for all the remaining configurations, gathering can be optimally solved. Again our strategy is to move the leader towards the anchor. However, in this case the leader has to judiciously choose the edge via which it should approach the anchor. Unlike the previous case, the anchor may change after a move.

Consider the first stage of the algorithm, when there is no multiplicity in the configuration. Then the anchor is the leading Weber point w_ℓ. We classify all the non-anchor vertices into two types: *type 1* and *type 2*. If the configuration has 2^m $(0 \leq m \leq d)$ Weber points, then among the d neighbors of w_ℓ, m are also Weber points. This is because of Lemma 3. Let us call these Weber points $w_1, \ldots, w_m$. Since the coordinates assigned to w_ℓ are $0 \ldots 0$, the coordinates of each $w_i \in \{w_1, \ldots, w_m\}$ have exactly one 1. For each w_i, assume that its assigned coordinates have the 1 at the p_ith place, which implies that the edge joining w_i and w_ℓ has label p_i. Also, the set of Weber points, in the assigned coordinates, is given by $\mathcal{W} = [b]_1 \times [b]_2 \times \ldots \times [b]_d$, where $[b]_l$ is $\{0,1\}$ if $l \in \{p_1, \ldots, p_m\}$, and $\{0\}$ otherwise.

Type 1 Vertex: A non-anchor vertex v will be called a *type 1 vertex* if the following holds: there is at least one $p_i \in \{p_1 \ldots p_m\}$ such that the assigned coordinates of v have 1 at p_ith place. Also the edge incident to v with label p_i will be called a *type 1 edge*.

Type 2 Vertex: If a non-anchor vertex v is not type 1, then it will be called a *type 2 vertex*. This implies that the p_1th, $\ldots p_m$th terms of the assigned coordinates of v are 0. Note that if the configuration has only one Weber point, i.e., $m = 0$, then all non-anchor vertices are vacuously type 2.

Theorem 12. *Let $(Q_d^{\mathcal{O}}, f)$ be a configuration of $4k$ $(k > 0)$ robots with no multiplicities and no partitive automorphisms. Let w_ℓ be the leading Weber point, and hence the anchor. Assume that the leader r is placed at a type 1 vertex u. Suppose that it moves via a type 1 edge with label p_i to an empty vertex v, and gives rise to configuration $(Q_d^{\mathcal{O}}, f')$. Then the following holds.*

1. $w_\ell \in \mathcal{W}_{f'}$
2. If $|\mathcal{W}_f| = 2^m (m > 0)$, then $|\mathcal{W}_{f'}| = 2^{m-1}$
3. $(Q_d^\mathcal{O}, f')$ has no partitive automorphism.

Proof

(1) Assume that r moves via a type 1 edge with label p_i. Then the assigned coordinates of u and v differ in exactly one bit at the p_ith position. At the p_ith place, u has 1, while v has 0. Then v has less 1's than u, and hence v is closer to w_ℓ than u, i.e., r has moved towards w_ℓ. Hence, by Lemma 5, $w_\ell \in \mathcal{W}_{f'}$.

(2) It is easy to see that, as r moves from u to v, (i) its distance from all Weber points whose coordinates have 1 at p_ith place (there are 2^{m-1} of them), increases by one, and (ii) its distance from all Weber points whose coordinates have 0 at p_ith place, decreases by one. Hence the move reduces the set of Weber points by half.

(3) If possible, assume that $(Q_d^\mathcal{O}, f')$ admits a partitive automorphism, i.e., a non-trivial translation τ. Assume that the translation, in the assigned coordinate system, is given by $x \mapsto x \oplus a$, for some $a \in \{0,1\}^d$. Let $\mathcal{R}$ and $\mathcal{R}'$ be the set of vertices occupied by robots in $(Q_d^\mathcal{O}, f)$ and $(Q_d^\mathcal{O}, f')$ respectively. Since τ maps any vertex of $\mathcal{R}'$ to another vertex of $\mathcal{R}'$, the group $<\tau>$ induces an equivalence relation on $\mathcal{R}'$, partitioning it into $2k$ disjoint sets of cardinality 2: $\mathcal{R}' = \bigcup_{j=1}^{2k} \{x_j, \tau(x_j)\}$. Let $\mathcal{R}_{p_i}$ and $\mathcal{R}'_{p_i}$ be the multiset containing the p_ith terms of the assigned coordinates of vertices of $\mathcal{R}$ and $\mathcal{R}'$ respectively. Clearly $\mathcal{R}_{p_i}$ contains $2k$ number of 0's and $2k$ number of 1's. In $\mathcal{R}'_{p_i}$, we have $2k + 1$ number of 0's and $2k - 1$ number of 1's.

Case 1: Let the p_ith term of a be 0. Hence, if p_ith term of x is $b \in \{0,1\}$, then the p_ith term of $\tau(x) = x \oplus a$ is also b. This implies that $\mathcal{R}'_{p_i}$ has even number of 0's and 1's. This is a contradiction, as we have shown that number of 0's and 1's in $\mathcal{R}'_{p_i}$ is $2k + 1$ and $2k - 1$ respectively.

Case 2: Let the p_ith term of a be 1. So, if p_ith term of x is $b \in \{0,1\}$, then the p_ith term of $\tau(x) = x \oplus a$ is $\bar{b}$. This implies that $\mathcal{R}'_{p_i}$ has equal number of 0's and 1's. This is again a contradiction. □

Theorem 13. *Let $(Q_d^\mathcal{O}, f)$ be a configuration of $4k$ $(k > 0)$ robots with no multiplicities and no partitive automorphisms. Let w_ℓ be the leading Weber point, and hence the anchor. Suppose that the leader r is placed at a type 2 vertex u. If it moves towards w_ℓ to an empty vertex, then*

1. *the new configuration $(Q_d^\mathcal{O}, f')$ has no partitive automorphism*
2. *$\mathcal{W}_f = \mathcal{W}_{f'}$*
3. *w_ℓ is the leading Weber point of $(Q_d^\mathcal{O}, f')$*
4. *r is the leader in $(Q_d^\mathcal{O}, f')$.*

Proof. We use the same notations as in the proof of the previous theorem. For each $w \in \mathcal{W}_f \setminus \{w_\ell\}$ the following holds: (i) among the p_1th, $\ldots$, p_mth terms of the assigned coordinates, there is at least one 1, (ii) all the terms except the p_1th, $\ldots$, p_mth ones are 0. Exactly the opposite is true for the type 2 vertex u. It implies that the distance of r from w_ℓ is strictly less than its distance from any other Weber point in $\mathcal{W}_f \setminus \{w_\ell\}$. Also, after the move, its distances from all Weber points reduce by exactly 1. Hence, after the move, $\zeta(w_\ell)$ remains lexicographically strictly largest among $\{\zeta(w) \mid w \in \mathcal{W}_{f'}\}$. All the statements of the theorem easily follow from these observations. □

Lemma 4. *Let $G = (V, E)$ be an arbitrary graph. Let $\mathcal{P} = v_0 e_0 v_1 e_1 v_2 \ldots v_{l-1} e_{l-1} v_l$ be a path from v_0 to v_l. Suppose that for any e_j, a move through it by a robot from v_j to v_{j+1} reduces its distance from v_l by 1. Then $\mathcal{P}$ is a shortest path from v_0 to v_l in G.*

Lemma 5. *Suppose that a robot moves from a vertex u to an adjacent vertex v in a hypercube Q_d. Then for any $w \in V(Q_d)$, either its distance from w reduces by 1 or increases by 1, i.e., its distance from w does not remain unchanged.*

Theorem 14. *Algorithm 1 achieves optimal gathering in ASYNC, for all asymmetric configurations of $4k$ ($k > 0$) robots with weak multiplicity detection.*

Proof. By Theorems 12 and 13, no move in the first stage creates a partitive automorphism. Since at any time, only the leader is allowed to move, a multiplicity can only be created at the anchor. Since throughout stage 2, there is a unique multiplicity, a configuration with a partitive automorphism is never created. Notice that in both stages, an anchor is always a Weber point. Hence, the robots always move towards some Weber point. So, after each move, the centrality of the surviving set of Weber points is reduced by 1. Therefore, eventually the centrality of one Weber point becomes 0, which implies that gathering is accomplished.

It remains to prove that Algorithm 1 is optimal with respect to the total number of moves executed by the robots. Suppose that the algorithm gathers all the robots at w, which was a Weber point of the initial configuration. In view of Theorem 12, it is sufficient to show that every movement executed by any robot is towards w. Since every movement executed by a robot is towards some Weber point, according to Theorem 5, the set of Weber points of the configurations starting from the initial to the final configuration form the following nested series: $\mathcal{W}_0 \supseteq \mathcal{W}_1 \supseteq \ldots \supseteq \mathcal{W}_{final} = \{w\}$. In other words, w remains a Weber point throughout the progress of the algorithm. If at some step, a move by a robot is not towards w, then by Theorem 13, it moves away from w. Then the centrality of w is increased by 1, while the centrality of some other Weber point is decreases by 1. This means that w does not remain a Weber point after the move. This is a contradiction. □

4   Concluding Remarks

This is the first paper that investigates the gathering problem on a hyper-cube graph. We have provided a complete characterization of all gatherable configurations in ASYNC for $2k + 1$ and $4k$ ($k > 0$) number of robots with weak multiplicity detection in an oriented hypercube. This leaves unset-tled only the configurations with $4k + 2$ ($k > 0$) number of robots. Note that our strategy for $4k$ robots does not work for $4k + 2$ robots. To see this, consider a configuration in Q_9 of 10 robots placed on the following vertices: 000000000, 110111000, 101111000, 011111000, 000111000, 001000111, 010000111, 100000111, 111000111, 111000000. Here, the anchor, i.e., the lead-ing Weber point is 000000000 and the leader is 111000000. It can be seen that a move by the leader towards the anchor via any edge creates a configuration with a partitive automorphism. Another challenging direction of future research would be to study the problem with limited visibility. It would also be interesting to consider randomized algorithms to bypass the impossibility results.

Acknowledgements. The first three authors are supported by NBHM, DAE, Govt. of India, UGC, Govt. of India and CSIR, Govt. of India respectively. We would like to thank the anonymous reviewers for their valuable comments which helped us improve the quality and presentation of this paper.

Appendix A   Proof of Theorem 7

Consider a configuration (Q_d, f) that has no partitive automorphism. Since we have assumed that the robots are positioned at distinct vertices, there is no distinction between the configuration and the perceived configuration. In other words, we have $\tilde{f} = f$. Assume that the configuration has no fixed Weber point. Let us assume that there is an algorithm $\mathcal{A}$ that deterministically solves LEAD-INGWEBERPOINT. Let $w_1 \in \mathcal{W}$ be the leading Weber point elected by the robots. Since w_1 is not a fixed vertex, there is a $w_2 \neq w_1$ such that $\varphi(w_1) = w_2$, for some $\varphi \in Aut(G, f)$.

Each robot observes the positions of other robots in its local coordinate sys-tem. A local coordinate system of a robot is just an assignment $\Psi : V(Q_d) \longrightarrow \{0,1\}^d$, respecting the rule that $u, v \in V(Q_d)$ are adjacent if and only if $\Psi(u), \Psi(v)$ differ in precisely one bit. Since there is no global agreement, the local coordinate system of each robot is arbitrary, and is chosen by the adver-sary. Let us formally define the view of a robot. The view of a robot is given by the triplet $\mathcal{V}_\Psi = (\Psi, \tilde{f}, me)$, where $\Psi : V(Q_d) \longrightarrow \{0,1\}^d$ is the local coordinate system, $\tilde{f} : \{0,1\}^d \longrightarrow \{0,1\}$ is the multiplicity function defined on the set of vertices expressed in local coordinates, and $me \in \{0,1\}^d$ is the coordinates of the vertex on which the robot resides. The view $\mathcal{V}_\Psi$ is the input for algorithm $\mathcal{A}$. The output $\mathcal{A}(\mathcal{V}_\Psi) \in \{0,1\}^d$ is the coordinates of the required leading Weber point, i.e., the returned leading Weber point is the vertex $\Psi^{-1}(\mathcal{A}(\mathcal{V}_\Psi))$.

Consider a robot r_1 in the configuration residing at vertex v_1. The robot r_1, using a local coordinate system $\Psi_1 : V(Q_d) \longrightarrow \{0,1\}^d$, elects w_1 as the leading

Weber point. That is, given the input in the coordinate system Ψ_1, the output of $\mathcal{A}$ is $\Psi_1(w_1)$. Now consider the following cases.

Case 1: Suppose that $\varphi(v_1) = v_1$. Consider the local coordinate system $\Psi_2 = \Psi_1 \circ \varphi^{-1}$. Note that the view of r_1 in coordinate systems Ψ_1 and Ψ_2 are exactly the same, i.e., $\mathcal{V}_{\Psi_1} = \mathcal{V}_{\Psi_2}$. Since $\mathcal{A}$ is a deterministic algorithm, $\mathcal{A}(\mathcal{V}_{\Psi_1}) = \mathcal{A}(\mathcal{V}_{\Psi_2})$. Since the elected leading Weber point in local coordinate system Ψ_1 is w_1, we have $\mathcal{A}(\mathcal{V}_{\Psi_1}) = \Psi_1(w_1)$. So we have,

$$\mathcal{A}(\mathcal{V}_{\Psi_2}) = \mathcal{A}(\mathcal{V}_{\Psi_1}) = \Psi_1(w_1)$$
$$\Rightarrow \quad \Psi_2^{-1}(\mathcal{A}(\mathcal{V}_{\Psi_2})) = \Psi_2^{-1}(\Psi_1(w_1)) = \varphi \circ \Psi_1^{-1} \circ \Psi_1(w_1) = \varphi(w_1) = w_2$$

Hence, we see that in local coordinate system Ψ_1 the robot r_1 elects w_1 as the leading Weber point, while in Ψ_2 it elects w_2. This is a contradiction.

Case 2: Now assume that $\varphi(v_1) = v_2 \neq v_1$. Then there must be a robot r_2 in v_2. Suppose that the adversary sets the local coordinate system of r_2 as $\Psi_2 = \Psi_1 \circ \varphi^{-1}$. Then the view of r_1 and r_2 will be identical, i.e., $\mathcal{V}_{\Psi_1} = \mathcal{V}_{\Psi_2}$. Again we have, $\mathcal{A}(\mathcal{V}_{\Psi_2}) = \mathcal{A}(\mathcal{V}_{\Psi_1}) = \Psi_1(w_1)$ and hence, $\Psi_2^{-1}(\mathcal{A}(\mathcal{V}_{\Psi_2})) = w_2$. Therefore, r_2 will elect w_2, while r_1 elects w_1 as the leading Weber point. This is again a contradiction. $\square$

References

1. Agathangelou, C., Georgiou, C., Mavronicolas, M.: A distributed algorithm for gathering many fat mobile robots in the plane. In: ACM Symposium on Principles of Distributed Computing, PODC 2013, Montreal, QC, Canada, 22–24 July 2013, pp. 250–259 (2013). https://doi.org/10.1145/2484239.2484266
2. Barrière, L., Flocchini, P., Fraigniaud, P., Santoro, N.: Rendezvous and election of mobile agents: impact of sense of direction. Theory Comput. Syst. **40**(2), 143–162 (2007). https://doi.org/10.1007/s00224-005-1223-5
3. Bose, K., Adhikary, R., Chaudhuri, S.G., Sau, B.: Crash tolerant gathering on grid by asynchronous oblivious robots. CoRR abs/1709.00877 (2017). http://arxiv.org/abs/1709.00877
4. Cieliebak, M., Flocchini, P., Prencipe, G., Santoro, N.: Solving the robots gathering problem. In: Baeten, J.C.M., Lenstra, J.K., Parrow, J., Woeginger, G.J. (eds.) ICALP 2003. LNCS, vol. 2719, pp. 1181–1196. Springer, Heidelberg (2003). https://doi.org/10.1007/3-540-45061-0_90
5. Cieliebak, M., Flocchini, P., Prencipe, G., Santoro, N.: Distributed computing by mobile robots: gathering. SIAM J. Comput. **41**(4), 829–879 (2012). https://doi.org/10.1137/100796534
6. Cohen, R., Peleg, D.: Convergence properties of the gravitational algorithm in asynchronous robot systems. SIAM J. Comput. **34**(6), 1516–1528 (2005). https://doi.org/10.1137/S0097539704446475
7. Czyzowicz, J., Gasieniec, L., Pelc, A.: Gathering few fat mobile robots in the plane. Theor. Comput. Sci. **410**(6–7), 481–499 (2009). https://doi.org/10.1016/j.tcs.2008.10.005

8. D'Angelo, G., Stefano, G.D., Klasing, R., Navarra, A.: Gathering of robots on anonymous grids and trees without multiplicity detection. Theor. Comput. Sci. **610**, 158–168 (2016). https://doi.org/10.1016/j.tcs.2014.06.045

9. D'Angelo, G., Di Stefano, G., Navarra, A.: Gathering of six robots on anonymous symmetric rings. In: Kosowski, A., Yamashita, M. (eds.) SIROCCO 2011. LNCS, vol. 6796, pp. 174–185. Springer, Heidelberg (2011). https://doi.org/10.1007/978-3-642-22212-2_16

10. D'Angelo, G., Stefano, G.D., Navarra, A.: Gathering on rings under the look-compute-move model. Distrib. Comput. **27**(4), 255–285 (2014). https://doi.org/10.1007/s00446-014-0212-9

11. Das, S., Mihalák, M., Šrámek, R., Vicari, E., Widmayer, P.: Rendezvous of mobile agents when tokens fail anytime. In: Baker, T.P., Bui, A., Tixeuil, S. (eds.) OPODIS 2008. LNCS, vol. 5401, pp. 463–480. Springer, Heidelberg (2008). https://doi.org/10.1007/978-3-540-92221-6_29

12. Dessmark, A., Fraigniaud, P., Kowalski, D.R., Pelc, A.: Deterministic rendezvous in graphs. Algorithmica **46**(1), 69–96 (2006). https://doi.org/10.1007/s00453-006-0074-2

13. Dobrev, S., Flocchini, P., Prencipe, G., Santoro, N.: Multiple agents rendezvous in a ring in spite of a black hole. In: Papatriantafilou, M., Hunel, P. (eds.) OPODIS 2003. LNCS, vol. 3144, pp. 34–46. Springer, Heidelberg (2004). https://doi.org/10.1007/978-3-540-27860-3_6

14. Flocchini, P., Kranakis, E., Krizanc, D., Luccio, F.L., Santoro, N., Sawchuk, C.: Mobile agents rendezvous when tokens fail. In: Královič, R., Sýkora, O. (eds.) SIROCCO 2004. LNCS, vol. 3104, pp. 161–172. Springer, Heidelberg (2004). https://doi.org/10.1007/978-3-540-27796-5_15

15. Flocchini, P., Kranakis, E., Krizanc, D., Santoro, N., Sawchuk, C.: Multiple mobile agent rendezvous in a ring. In: Farach-Colton, M. (ed.) LATIN 2004. LNCS, vol. 2976, pp. 599–608. Springer, Heidelberg (2004). https://doi.org/10.1007/978-3-540-24698-5_62

16. Flocchini, P., Prencipe, G., Santoro, N., Widmayer, P.: Gathering of asynchronous robots with limited visibility. Theor. Comput. Sci. **337**(1–3), 147–168 (2005). https://doi.org/10.1016/j.tcs.2005.01.001

17. Godsil, C., Royle, G.F.: Algebraic Graph Theory. GTM, vol. 207. Springer Science & Business Media, New York (2001). https://doi.org/10.1007/978-1-4613-0163-9

18. Haba, K., Izumi, T., Katayama, Y., Inuzuka, N., Wada, K.: On gathering problem in a ring for 2n autonomous mobile robots. In: Proceedings of the 10th International Symposium on Stabilization, Safety, and Security of Distributed Systems (SSS), poster (2008)

19. Izumi, T., Izumi, T., Kamei, S., Ooshita, F.: Mobile robots gathering algorithm with local weak multiplicity in rings. In: Patt-Shamir, B., Ekim, T. (eds.) SIROCCO 2010. LNCS, vol. 6058, pp. 101–113. Springer, Heidelberg (2010). https://doi.org/10.1007/978-3-642-13284-1_9

20. Kamei, S., Lamani, A., Ooshita, F., Tixeuil, S.: Asynchronous mobile robot gathering from symmetric configurations without global multiplicity detection. In: Kosowski, A., Yamashita, M. (eds.) SIROCCO 2011. LNCS, vol. 6796, pp. 150–161. Springer, Heidelberg (2011). https://doi.org/10.1007/978-3-642-22212-2_14

21. Kamei, S., Lamani, A., Ooshita, F., Tixeuil, S.: Gathering an even number of robots in an odd ring without global multiplicity detection. In: Rovan, B., Sassone, V., Widmayer, P. (eds.) MFCS 2012. LNCS, vol. 7464, pp. 542–553. Springer, Heidelberg (2012). https://doi.org/10.1007/978-3-642-32589-2_48

22. Klasing, R., Kosowski, A., Navarra, A.: Taking advantage of symmetries: gathering of many asynchronous oblivious robots on a ring. Theor. Comput. Sci. **411**(34–36), 3235–3246 (2010). https://doi.org/10.1016/j.tcs.2010.05.020
23. Klasing, R., Markou, E., Pelc, A.: Gathering asynchronous oblivious mobile robots in a ring. Theor. Comput. Sci. **390**(1), 27–39 (2008). https://doi.org/10.1016/j.tcs.2007.09.032
24. Koreń, M.: Gathering small number of mobile asynchronous robots on ring. Zeszyty Naukowe Wydziału ETI Politechniki Gdańskiej. Technologie Informacyjne **18**, 325–331 (2010)
25. Kranakis, E., Krizanc, D., Markou, E.: Deterministic symmetric rendezvous with tokens in a synchronous torus. Discrete Appl. Math. **159**(9), 896–923 (2011). https://doi.org/10.1016/j.dam.2011.01.020
26. Kranakis, E., Santoro, N., Sawchuk, C., Krizanc, D.: Mobile agent rendezvous in a ring. In: 23rd International Conference on Distributed Computing Systems, ICDCS 2003, Providence, RI, USA, 19–22 May 2003, pp. 592–599 (2003). https://doi.org/10.1109/ICDCS.2003.1203510
27. Di Luna, G.A., Flocchini, P., Pagli, L., Prencipe, G., Santoro, N., Viglietta, G.: Gathering in dynamic rings. In: Das, S., Tixeuil, S. (eds.) SIROCCO 2017. LNCS, vol. 10641, pp. 339–355. Springer, Cham (2017). https://doi.org/10.1007/978-3-319-72050-0_20
28. Marco, G.D., Gargano, L., Kranakis, E., Krizanc, D., Pelc, A., Vaccaro, U.: Asynchronous deterministic rendezvous in graphs. Theor. Comput. Sci. **355**(3), 315–326 (2006). https://doi.org/10.1016/j.tcs.2005.12.016
29. Pagli, L., Prencipe, G., Viglietta, G.: Getting close without touching: near-gathering for autonomous mobile robots. Distrib. Comput. **28**(5), 333–349 (2015). https://doi.org/10.1007/s00446-015-0248-5
30. Prencipe, G.: Impossibility of gathering by a set of autonomous mobile robots. Theor. Comput. Sci. **384**(2–3), 222–231 (2007). https://doi.org/10.1016/j.tcs.2007.04.023
31. Stefano, G.D., Navarra, A.: Gathering of oblivious robots on infinite grids with minimum traveled distance. Inf. Comput. **254**, 377–391 (2017). https://doi.org/10.1016/j.ic.2016.09.004
32. Stefano, G.D., Navarra, A.: Optimal gathering of oblivious robots in anonymous graphs and its application on trees and rings. Distrib. Comput. **30**(2), 75–86 (2017). https://doi.org/10.1007/s00446-016-0278-7
33. Tel, G.: Network orientation. Int. J. Found. Comput. Sci. **5**(1), 23–57 (1994). https://doi.org/10.1142/S0129054194000037

Barrier Coverage Problem in 2D

Adil Erzin[1,2]([envelope]) and Natalya Lagutkina[2]

[1] Sobolev Institute of Mathematics, Novosibirsk, Russia
adilerzin@math.nsc.ru
[2] Novosibirsk State University, Novosibirsk, Russia
lagutnat@yandex.ru

Abstract. This paper deals with the NP-hard problem of covering a line segment by n initially arbitrarily arranged circles on the plane by moving their centers to the segment in such a way that the sum of the Euclidean distances between the initial and final positions of the centers of the disks would be minimal. In the case of identical circles, a dynamic programming algorithm is known, which constructs a $\sqrt{2}$–approximate solution to the problem with $O(n^4)$–time complexity. In this paper, we propose a new algorithm that has the same accuracy, but the complexity of which is reduced by n^2 times to $O(n^2)$.

Keywords: Sensor networks · Mobile sensors · Barrier coverage

1 Introduction

The sensor network consists of devices, each of which collects data within a proximity, which is called a *coverage area*. On the plane, a coverage area most often is a circle (disk) with a sensor in its center [5,7,19,24]. Though both an ellipse [16] and a sector [17] can be a coverage area of the sensor. In the wireless sensor networks energy of the sensors is often irreplaceable because the recharge or change of the battery is either impossible or impractical. The energy of the sensors defines network's lifetime. Rational use of energy prolongs the lifetime of the sensor network [5,7]. For energy efficient operation of the sensor network, it is necessary to solve several optimization problems. One of the problems is optimal placement of sensors and determination of the values of their parameters. As sensing energy consumption is proportional to the coverage area, this problem is reduced to the classical min-density covering problem [5–7,13,24].

In barrier monitoring, it is necessary to detect an unauthorized crossing of a barrier separating the two territories. In some cases, the barrier is considered as a line [1,4,10–14,20,21], in others as a strip [17,23]. The barrier can be covered by stationary sensors [2,9,10,15,17,20,22,23,25], and by mobile sensors [1,4,11–14,18]. A coverage area is often considers as a circle [4,11–14,20], but sometimes (in the case of directed devices) it is a sector [17,22,25]. In [20] a notion of weak coverage is introduced and the critical conditions for the existence of weak barrier coverage in a randomly deployed sensor network is proposed. Later, in [9] the

© Springer Nature Switzerland AG 2019
S. Gilbert et al. (Eds.): ALGOSENSORS 2018, LNCS 11410, pp. 118–130, 2019.
https://doi.org/10.1007/978-3-030-14094-6_8

algorithm of guaranteed detection and localization of intruders, the trajectory of which is located in the area of sensors placement, is proposed. How to assess and ensure the quality of the barrier coverage is examined in [10].

If a sensor is mobile, the movement energy consumption is proportional to the distance traveled by a sensor. The important optimization problem for mobile sensor networks is minimization of the total distance traveled by sensors [1, 4, 11–14, 18]. It is necessary to move the sensors in such a way that each point of the barrier (the line segment) belongs to the coverage area of at least one sensor, and the total length of the relocations would be minimal. In [13] the NP-hardness of the problem is proved. The efficiency of the different placement strategies of sensors for a barrier coverage is studied in [21]. There was also studied the question how to improve a barrier coverage after the placement using the mobility of the sensors. The authors of [21] presented an algorithm for efficient improvement of barrier coverage with a wide range of parameters of placement the sensors. Circular barriers in the plane were studied in [4, 11]. Paper [4] presents a $O(n^2)$–time algorithm for the special case of the circular barrier covering problem (when the sensors are placed along the boundary of the region uniformly) with approximation ratio $1 + \pi$, where n is the number of sensors. Later, in [1], this result was improved by presenting an algorithm with the same running time and approximation ratio 3. A PTAS was also proposed for this problem in [4], which was improved in [8]. In [3] for the case, when arbitrary disks are lying on the line containing the segment, and the disks in the cover do not intersect, an FPTAS is proposed.

In the literature on barrier monitoring, as a rule, the problem of covering a *line segment* with *identical* circles when the centers of the circles move *to the segment*, is considered. In case of the Euclidean metric it is nothing known about the complexity of this problem, however, there is a polynomial $\sqrt{2}$–approximation algorithm [12]. A line segment coverage problem in the special case when equal disks initially lie on the segment is considered in [1] and a $O(n \log_2 n)$–time algorithm is proposed to solve this problem.

Sometimes the problem of barrier coverage is considered in 3D space [22]. We within this paper consider a problem on a plane. Let barrier be a line segment on abscissa axis, and let us number the circles according to the nondecreasing abscissas of their centers. A solution in which after moving the sensors the order preserves is called an *order-preserving covering* (OPC). In the general case may not exist an optimal OPC [12]. In [12] the authors presented a $O(n^4)$–time $\sqrt{2}$–approximation algorithm.

In this paper, we propose a dynamic programming $\sqrt{2}$–approximation algorithm that solves the problem with $O(n^2)$–time complexity. Compared with the known algorithm [12], the degree of the time complexity polynomial is halved.

This paper is organized as follows. Section 2 presents a mathematical formulation of the problem. Section 3 gives the description of new dynamic programming $\sqrt{2}$–approximation algorithm $\mathcal{A}$. In Sect. 4 it is proved that the time complexity of the algorithm $\mathcal{A}$ is $O(n^2)$. The Conclusion section contains summary and further directions of the research. In the Appendix A we describe in detail the solution of one example of a covering of a given line segment by three identical circles.

2 Problem Formulation

Let barrier is a L-length line segment on the plane. It is required to cover it by mobile sensors with circle coverage areas. We introduce a coordinate system in such a way that the barrier is a segment between the points $(0,0)$ and $(L,0)$. Let S be a set of disks (corresponding to the coverage areas of the sensors), $|S| = n$, each of which is given by initial coordinates of its center $p_i = (x_i, y_i)$ and radius $r_i > 0$, $i \in S$. We assume that the sensors are numbered from left to right according to the values x_i, $i = 1, 2, \ldots, n$.

Definition 1. *The function $\hat{p} \colon S \to R^2$ is called a covering assignment if the segment is completely covered when the final positions of the sensors are $\hat{p}_i = (\hat{x}_i, \hat{y}_i)$, $i \in S$.*

Let $d(p_i, \hat{p}_i)$ be a distance between the points p_i and $\hat{p}_i$. The problem of barrier coverage by mobile sensors is to find a covering assignment $\hat{p}^*$ of minimum cost, which is the solution of the problem

$$cost(\hat{p}^*) = \min_{\hat{p}} cost(\hat{p}) = \min_{\hat{p}} \sum_{i \in S} d(p_i, \hat{p}_i). \tag{1}$$

In the general case, the covering of a segment can be obtained not necessarily by moving the sensors *to a segment*. However, in [12–14] the special case of the problem (1) when the sensors move *on the barrier* is considered. In this paper, we also consider this case, though we can modify our algorithm in such a way that it builds the solution in a general case. However, within the framework of this paper, we do not set ourselves the goal of describing the general case.

In the case when disks have different radii, the problem (1) is known to be NP-hard even to approximate up to a constant factor [13,14]. However, if the circles are identical it is unknown whether this problem is NP-hard or it is polynomially solvable [12,14]. Paper [12] presents a dynamic programming algorithm for finding $\hat{p}$ that determines an optimal OPC under L_1 metric with $O(n^4)$–time complexity. Meanwhile, the optimal solution of the problem (1) under metric L_1 is a $\sqrt{2}$–approximate solution under the Euclidean metric [12].

3 Algorithm $\mathcal{A}$

In the following, as earlier, we shall identify the centers of the circles (disks) with the sensors. Let the circles be numbered in the nondecreasing order of the abscissas of their centers. We start with the known definitions and simple observations.

Definition 2. *A covering assignment $\hat{p}$ is order-preserving if for every $i, j \in S$ we have $\hat{x}_i < \hat{x}_j$ iff $i < j$.*

Lemma 1 [12]. *If the circles are identical, then there is an optimal order-preserving covering assignment under L_1 metric.*

Lemma 2 [12]. *If the circles are identical, then any optimal order-preserving covering under L_1 metric is a $\sqrt{2}$-approximate solution to the problem (1) under Euclidean metric.*

Further, in this section, we present a new dynamic programming algorithm $\mathcal{A}$ that constructs an OPC, which, in the case of identical circles, is an optimal solution to the problem (1) under the metric L_1. The algorithm consists of one forward recursion and one backward recursion.

3.1   Forward Recursion

Let $S_k(l)$ be a minimum sum of the distances $d(p_i, \hat{p}_i) = |x_i - \hat{x}_i| + |y_i - \hat{y}_i|$, $i = 1, \ldots, k$, for the first k, $k = 1, \ldots, n$, sensors that form an OPC of the segment $[0, l]$, $0 \leq l \leq L$. Without loss of generality, we suppose that $y_i \geq 0$, $i \in S$. Then we can calculate the cost

$$
S_1(l) = \begin{cases} d(p_1, \hat{p}_1(l)), & 2r_1 \geq l \\ +\infty, & 2r_1 < l. \end{cases}
$$

Here the point $\hat{p}_1(l)$ is lying on the segment, it is the nearest point to the point p_1, and the segment $[0, l]$ is covered by disk 1. The value $d(p_1, \hat{p}_1(l))$ is defined analytically depending on the initial position of the sensor 1. We assume that the center of the disk 1 moves to the point $(x, 0)$ on the segment (see Fig. 1). Then

$$
d(p_1, \hat{p}_1(l)) = \begin{cases} -x_1 + y_1, & x_1 \leq 0, \quad l \leq r_1, \quad x = 0 \\ l - r_1 - x_1 + y_1, & \max\{x_1 + r_1, r_1\} \leq l, \quad x = l - r_1 \\ y_1, & 0 \leq l - r_1 \leq x_1 \leq r_1, \quad x = x_1 \\ x_1 - r_1 + y_1, & x_1 > r_1, \quad r_1 \leq L, \quad x = r_1 \\ x_1 - L + y_1, & x_1 > r_1 > L, \quad x = L. \end{cases}
$$

Let now there are two disks 1 and 2 with radii r_1 and r_2. If $l \leq 2\min\{r_1, r_2\}$, then either two sensors, or one of the sensors can be used in the cover. Let's consider the following possible cases.

1. The sensor 2 is not used in the cover.
2. The sensor 1 is not used in the cover.
3. Both sensors 1 and 2 are used in the cover.

In the first case, we have $S_2(l) = S_1(l)$. In the second case, we let $S_1^2(l)$ be the minimum distance of movement of a sensor 2 for covering $[0, l]$ (suppose that $S_2(l) = +\infty$, if $2r_2 < l$). Let now both sensors are used in the cover of the

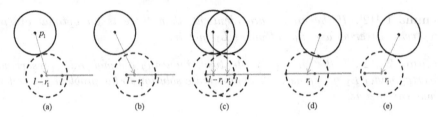

Fig. 1. Movement of disk 1 depending on r_1, p_1 and l. The original location (p_1) and the final location ($\hat{p}_1$) are connected by arrow. (a) If $l - r_1 \leq 0$, $x_1 \leq 0$, then $x = 0$. (b) If $x_1 \leq l - r_1$, $l - r_1 \geq 0$, then $x = l - r_1$. (c) If $0 \leq l - r_1 \leq x_1 \leq r_1$, then $x = x_1$. (d) If $x_1 \geq r_1$, $r_1 \leq L$, then $x = r_1$. (e) If $x_1 > r_1$, $r_1 > L$, then $x = L$.

segment $[0, l]$ and x is a point where the center of the disk 2 moves. Thus we can calculate the cost

$$S_2(l) = \begin{cases} \min\{S_1(l), S_1^2(l), \overline{S}_2(l)\}, & 0 < l \leq \min\{2(r_1 + r_2), L\} \\ +\infty, & 2(r_1 + r_2) < l, \end{cases}$$

where

$$\overline{S}_2(l) = \begin{cases} \min\limits_{x \in D_2(l)} \{|x_2 - x| + y_2 + S_1(x - r_2)\}, & l < x_2 + r_2 \\ l - r_2 - x_2 + y_2 + S_1(l - 2r_2), & l \geq x_2 + r_2, \end{cases}$$

and $D_2(l) = [\max\{r_2, l - r_2\}, \min\{2r_1 + r_2, l + r_2, L\}]$. Obviously, in the case 3, we have $x > r_2$ (see Fig. 2).

Fig. 2. Options of movement of the disk 2 in the case 3.

Let the values of all functions $S_i(l)$, $i = 1, \ldots, k - 1$, be counted, and let the segment $[0, l]$ is covering by disks $1, 2, \ldots, k$. Then, the following recursions can be used to calculate $S_k(l)$, $k = 1, \ldots, n$.

$$S_k(l) = \begin{cases} \min\left\{S_{k-1}(l), S_{k-1}^2(l), \overline{S}_k(l)\right\}, & 0 < l \leq \min\{2\sum\limits_{i=1}^{k} r_i, L\} \\ +\infty, & 2\sum\limits_{i=1}^{k} r_i < l, \end{cases}$$

where

$$\overline{S}_k(l) = \begin{cases} \min\limits_{x \in D_k(l)} \{|x_k - x| + y_k + S_{k-1}(x - r_k)\}, & l < x_k + r_k \\ l - r_k - x_k + y_k + S_{k-1}(l - 2r_k), & l \geq x_k + r_k, \end{cases}$$

$S_{k-1}^2(l)$ is the cost function if only one disk k is covering the segment, and

$$D_k(l) = \left[\max\{r_k, l - r_k\}, \min\{2\sum_{i=1}^{k-1} r_i + r_k, l + r_k, L\} \right].$$

After computing $S_n(L)$, the optimal position of the last disk n is found.

3.2   Backward Recursion

If the sensor n is used in the constructed cover then the position of its center $(\hat{x}_n, 0)$ is known, so the segment $[0, L - \hat{x}_n - r_n]$ is covered by the first $n - 1$ disks. The formulas for calculating the value $S_{n-1}(l)$ is found for any l and hence for the argument $l = L - \hat{x}_n - r_n$ too. If the sensor n is not used in the optimal cover, then we consider the sensor $n - 1$. If the sensor $n - 1$ is used in the cover, then we know the position of its center $(\hat{x}_{n-1}, 0)$ and the segment $[0, L - \hat{x}_{n-1} - r_{n-1}]$ is covered by the first $n - 2$ disks. Continuing the backward recursion, we find the covering of the whole segment $[0, L]$.

4   Time Complexity

In this section, we will prove that the proposed algorithm $\mathcal{A}$ can be implemented within the time complexity $O(n^2)$.

Definition 3. *We call $l_i \in [0, l]$ the switching points for the function $S_k(l)$ if in the segment $(l_i, l_{i+1}) \subseteq [0, l]$ the function is defined by one analytical expression $F_i(l)$ and $F_i(l_{i+1}) = F_{i+1}(l_{i+1})$, $F_i(l_i) = F_{i-1}(l_i)$, or if $S_k(l)$ is unlimited (equals $+\infty$).*

Lemma 3. *When adding the next disk k, the number of switching points for the function $S_k(l)$, $k = 1, \ldots, n$, increases by $O(1)$ with respect to the number of switching points for the function $S_{k-1}(l)$.*

Proof. Let first $k = 1$. If $2r_1 < l$ then $S_1(l) = +\infty$. Otherwise if $2r_1 \geq l$, then depending on p_1 and r_1, we have one of the three options for calculation $S_1(l)$.

1. If $x_1 < 0$, then

$$S_1(l) = \begin{cases} -x_1 + y_1, & l \leq r_1, \quad x = 0 \\ l - r_1 - x_1 + y_1, & l > r_1, \quad x = l - r_1. \end{cases}$$

2. If $0 \leq x_1 \leq r_1$, then

$$S_1(l) = \begin{cases} y_1, & l \leq r_1 + x_1, \quad x = x_1 \\ l - r_1 - x_1 + y_1, & l > r_1 + x_1, \quad x = l - r_1. \end{cases}$$

3. If $x_1 > r_1$, then

$$S_1(l) = \begin{cases} x_1 - r_1 + y_1, & r_1 \leq L, \quad x = r_1 \\ x_1 - L + y_1, & r_1 > L, \quad x = L. \end{cases}$$

Thus, for any value of x_1 the number of switching points for function $S_1(l)$ is bounded by $O(1)$ (constant).

Let now the two first disks can be used in the covering. It is necessary to consider all cases for the calculation of $S_2(l)$. Due to the limitations of the space, we will consider in detail only two cases from the set of cases.

Let's consider, for example, the case when $0 < l < x_2 + r_2$. As both sensors are used in the cover of the segment $[0, l]$, then $x_2 > \max\{l - r_2, r_2\}$. Note that the sensor 1 does not cover the point $(0, l)$, and the sensor 2 does not cover the point $(0, 0)$. Therefore,

$$S_2(l) = \begin{cases} x_2 - x + y_2 + S_1(x - r_2), & x \in X_1 \\ x - x_2 + y_2 + S_1(x - r_2), & x \in X_2, \end{cases}$$

where $X_1 = [\max\{r_2, l - r_2\}, \min\{2r_1 + r_2, l + r_2, L, x_2\}]$, and $X_2 = [\max\{r_2, l - r_2, x_2\}, \min\{2r_1 + r_2, l + r_2, L\}]$.

The function S_1 depends on the x_1, and it is computed as follows:

– if $x_1 < 0$, then

$$S_1(x - r_2) = \begin{cases} -x_1 + y_1, & x \leq r_1 + r_2 \\ x - r_1 - r_2 - x_1 + y_1, & x > r_1 + r_2; \end{cases}$$

– if $0 \leq x_1 \leq r_1$, then

$$S_1(x - r_2) = \begin{cases} y_1, & x \leq x_1 + r_1 + r_2 \\ x - r_1 - r_2 - x_1 + y_1, & x > x_1 + r_1 + r_2; \end{cases}$$

– if $x_1 > r_1$, then $S_1(x - r_2) = x_1 - r_1 + y_1$, $r_1 \leq L$.

Assume that $x_1 < 0$, $x \leq x_2$ and $x \leq r_1 + r_2$. Then we have the formula:

$$S_2(l) = x_2 - x + y_2 + S_1(x - r_2),$$

where $\max\{r_2, l - r_2\} \leq x \leq \min\{2r_1 + r_2, l + r_2, L, x_2\}$ and $S_1(x - r_2) = -x_1 + y_1$. As a result, we have the following analytical expression

$$S_2(l) = x_2 - x + y_2 - x_1 + y_1,$$

where $\max\{r_2, l - r_2\} \leq x \leq \min\{2r_1 + r_2, l + r_2, L, x_2\}$ and $x = \min\{r_1 + r_2, x_2, l + r_2, L\}$.

Let's consider one more case when $l \geq x_2 + r_2$ and $x_1 < 0$. Then we have the formula $S_2(l) = l - r_2 - x_2 + y_2 + S_1(l - 2r_2)$, where

$$
S_1(l - 2r_2) = \begin{cases} -x_1 + y_1, & x_1 \leq 0,\, l - 2r_2 \leq r_1 \\ l - r_1 - 2r_2 - x_1 + y_1, & x_1 \leq 0,\, l - 2r_2 > r_1. \end{cases}
$$

If $l - 2r_2 \leq r_1$, then in order to cover the segment $[0, l - 2r_2]$ the disk 1 moves to the point $(0,0)$. Otherwise, if $l - 2r_2 > r_1$, the disk 1 moves to the point $(l - 2r_2 - r_1, 0)$.

Assume that $l - 2r_2 \leq r_1$, then the function $S_2(l) = l - r_2 - x_2 + y_2 + y_1 - x_1$.

Other cases are considered similarly. Thereby the number of switching points for the function $S_2(l)$ is upper bounded by constant.

For an arbitrary number of sensors $k = 1, 2, \ldots, n$, we can calculate the cost as follows.

$$
S_k(l) = \begin{cases} \min\left\{ S_{k-1}(l), S_{k-1}^2(l), \overline{S}_k(l) \right\}, & 0 < l \leq \min\{2 \sum\limits_{i=1}^{k} r_i, L\} \\ +\infty, & 2 \sum\limits_{i=1}^{k} r_i < l, \end{cases}
$$

where

$$
\overline{S}_k(l) = \begin{cases} \min\limits_{x \in D_k(l)} \{|x_k - x| + y_k + S_{k-1}(x - r_k)\}, & l < x_k + r_k \\ l - r_k - x_k + y_k + S_{k-1}(l - 2r_k), & l \geq x_k + r_k, \end{cases}
$$

$S_{k-1}^2(l)$ is the cost function if only one disk k is covering the segment, and
$$
D_k(l) = \left[\max\{r_k, l - r_k\}, \min\{2 \sum\limits_{i=1}^{k-1} r_i + r_k, l + r_k, L\} \right].
$$

When calculating the value of $\overline{S}_k(l)$ it is considered two cases $l < x_k + r_k$ and $l \geq x_k + r_k$, and the number of switching points increases by constant. Hence, for calculation of the next value of $S_k(l)$, $k = 1, \ldots, n$, the constant number of switching points is added, that completes the proof.

Corollary 1. *When calculating the function $S_k(l)$ the optimal position of the center of the disk k, $k = 1, \ldots, n$ can be computed with time complexity equals $O(n)$.*

Remark 1. In the case of different disks may not exist an optimal order-preserving assignment under L_1 metric (see Fig. 3). Therefore, we can apply the proposed algorithm $\mathcal{A}$, but we cannot obtain a $\sqrt{2}$-approximate solution.

The main result of this paper is the

Theorem 1. *In the case of identical disks the algorithm $\mathcal{A}$ constructs a $\sqrt{2}-$ approximate solution to the problem (1) with time complexity equals $O(n^2)$.*

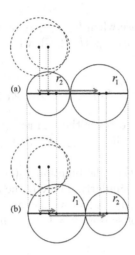

Fig. 3. (a) The optimal cover under L_1 metric. (b) The order-preserving cover which is worse by $2(r_1 - r_2)$ than the optimal cover.

Proof. It is known that in the considered case an optimal order-preserving covering under L_1 metric is a $\sqrt{2}$-approximate solution to the problem (1) under Euclidean metric [12]. Taking into account that the functions $S_k(l)$ are calculated n times and Corollary 1, we find that the complexity of the algorithm $\mathcal{A}$ is $O(n^2)$. The theorem is proved.

To illustrate the operation of the algorithm, in the Appendix A an example is given.

5   Conclusion

The paper deals with the problem of moving the centers of n circles located at arbitrary position on a plane on a given line segment of length L so that the line is completely covered by the circles while minimizing the cumulative Euclidean distance between the initial position of centers and their position on the segment. It is known that this problem is NP-hard in the case of a non-identical disks [13,14]. When the disks are identical the complexity of the problem is unknown, but there is a $O(n^4)$-time $\sqrt{2}$-approximation algorithm. In this paper, we propose a $O(n^2)$-time algorithm that is applicable in general case and constructs a $\sqrt{2}$-approximate solution to the problem in the case of n identical circles.

In the further research, we plan to clarify the complexity of the problem in the case of identical disks and to give up a requirement of movement the sensors *on the segment*. Moreover, we are planning to design an FPTAS for the case when each barrier point is covered by exactly one disk.

Acknowledgements. The research is partly supported by the Russian Foundation for Basic Research (Projects 16-07-00552 and 17-51-45125) and by the Ministry of Science and Higher Education of the Russian Federation under the 5-100 Excellence Programme.

A   Appendix

Example. Let it is required to cover the line segment $[0, 4.5]$ by three identical disks with radii equal to 1, which initial positions of the centers are $p_1 = (-0.5, 1)$, $p_2 = (2.5, 2)$ and $p_3 = (5.5, 0)$ (Fig. 4(a)).

Since $x_1 < 0$, then

$$S_1(l) = \begin{cases} -x_1 + y_1 = 1.5, & l \leq 1 \\ l - 1 - x_1 + y_1 = l + 0.5, & l > 1 \\ +\infty, & l > 2. \end{cases}$$

The disk 1 moves to the point $(0, 0)$, if $l \leq 1$ (Fig. 4(b)) and it moves to the point $(l - 1, 0)$, if $l > 1$ (Fig. 4(c)). Thus, we have the switching points $0, 1, 2$ and 4.5.

Let now two circles participate in the covering. If $l \leq 1$, then it is easy to see, that only disk 1 covers the segment $[0, l]$ and $S_2(l) = 1.5$ (Fig. 4(d)).

If $1 < l \leq 2$, then the segment $[0, l]$ can be covered ether by one disk 1 or by one disk 2. We have that $d(p_1, \hat{p}_1) = l + 0.5 \leq d(p_2, \hat{p}_2)$. So, in this case only disk 1 covers the segment $[0, l]$. Suppose that both disks 1 and 2 participate in

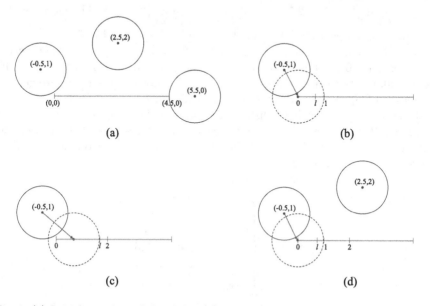

Fig. 4. (a) Initial position of the disks; (b) one disk in the case when $l \leq 1$; (c) one disk in the case when $1 < l \leq 2$; (d) two disks in the case when $l \leq 1$.

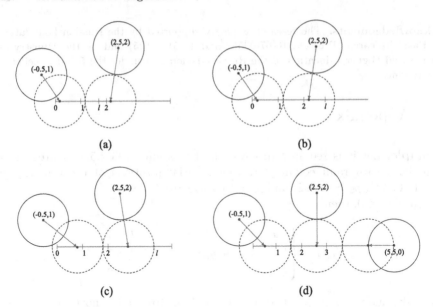

Fig. 5. (a) Two disks in the case when $1 < l \leq 2$; (b) two disks in the case when $2 < l \leq 3.5$; (c) two disks in the case when $3.5 < l \leq 4$; (d) the optimal OPC.

the covering of the segment $[0, l]$. Let us denote by $x \in (1, 3)$ the point at which the center of disk 2 moves. Then the segment $[0, x - 1]$ must be covered by disk 1. If $x \leq 2.5$ then $S_2(l) = 2 + 2.5 - x + S_1(x - 1) = 4$. If $2.5 < x \leq 3$ then $S_2(l) = \min_{x \in [2.5,3]} \{2 + x - 2.5 + S_1(x - 1)\} = \min_{x \in [2.5,3]} \{2x - 1\}$. Therefore, in this case only the center of disk 1 moves to the point $(l - 1, 0)$ (Fig. 5(a)).

If $2 < l \leq 3.5$, then the both disks 1 and 2 must participate in the covering of the segment $[0, l]$. If x is a point where the center of disk 2 moves, then the segment $[0, x - 1]$ must be covered by disk 1. For any $x \in [l - 1, 2.5]$, we get the same value of $S_2(l) = 4$ and set $x = 2.5$ (Fig. 5(b)).

If $3.5 < l \leq 4$, then both disks participate in the covering of the segment $[0, l]$. If $x \in [l - 1, 3]$ is a point where the center of disk 2 moves, then the segment $[0, x - 1]$ must be covered by disk 1. In this case $2.5 \leq x \leq 3$. Moreover, $x = l - 1$ and $S_2(l) = l - 1 - 2.5 + 2 + 1 + l - 2 + 0.5 = 2l - 2$ (Fig. 5(c)).

Therefore, the following formula holds

$$S_2(l) = \begin{cases} 1.5, & 0 < l \leq 1 \\ l + 0.5, & 1 < l \leq 2 \\ 4, & 2 < l \leq 3.5 \\ 2l - 3, & 3.5 < l \leq 4 \\ +\infty, & l > 4, \end{cases}$$

where the switching points are $0, 1, 2, 3.5, 4, 4.5$.

Let now all three sensors participate in the covering. The center of disk 3 can move to the point $x \in [3.5, 4.5]$. Then the segment $[0, x - 1]$ must be covered by disks 1 and 2 and

$$S_3(4.5) = \min_{x \in [3.5, 4.5]} \{5.5 - x + S_2(x - 1)\} = \min_{x \in [3.5, 4.5]} \{9.5 - x\} = 5.$$

Then the center of disk 3 moves to the point $(4.5, 0)$.

The backward recursion allows us to restore the optimal coverage, which is shown in the Fig. 5(d).

References

1. Andrews, A.M., Wang, H.: Minimizing the aggregate movements for interval coverage. Algorithmica **78**(1), 47–85 (2017)
2. Astrakov, S.N., Erzin, A.I.: Efficient band monitoring with sensors outer positioning. Optim. A J. Math. Program. Oper. Res. **62**(10), 1367–1378 (2013)
3. Benkoczi, R., Friggstad, Z., Gaur, D., Thom, M.: Minimizing total sensor movement for barrier coverage by non-uniform sensors on a line. In: Bose, P., Gąsieniec, L.A., Römer, K., Wattenhofer, R. (eds.) ALGOSENSORS 2015. LNCS, vol. 9536, pp. 98–111. Springer, Cham (2015). https://doi.org/10.1007/978-3-319-28472-9_8
4. Bhattacharya, B.K., Burmester, M., Hu, Y., Kranakis, E., Shi, Q., Wiese, A.: Optimal movement of mobile sensors for barrier coverage of a planar region. Theor. Comput. Sci. **410**(52), 5515–5528 (2009)
5. Cardei, M., Du, D.-Z.: Improving wireless sensor network lifetime through power aware organization. ACM Wirel. Netw. **11**(3), 333–340 (2005)
6. Cardei, M., Wu, J.: Energy-efficient coverage problems in wireless ad-hoc sensor networks. Comput. Commun. **29**, 413–420 (2006)
7. Carle, J., Simplot, D.: Energy-efficient area monitoring by sensor networks. IEEE Comput. **37**(2), 40–46 (2004)
8. Carmi, P., Katz, M.J., Saban, R., Stein, Y.: Improved PTASs for convex barrier coverage. In: Solis-Oba, R., Fleischer, R. (eds.) WAOA 2017. LNCS, vol. 10787, pp. 26–40. Springer, Cham (2018). https://doi.org/10.1007/978-3-319-89441-6_3
9. Chen, A., Kumar, S., Lai, T.H.: Designing localized algorithms for barrier coverage. In: Proceedings of the 13th Annual ACM International Conference on Mobile Computing and Networking, pp. 63–74 (2007)
10. Chen, A., Lai, T.H., Xuan, D.: Measuring and guaranteeing quality of barrier coverage in wireless sensor networks. In: Proceedings of the ACM International Symposium on Mobile Ad Hoc Networking and Computing (MobiHoc), pp. 421–430 (2008)
11. Chen, D.Z., Tan, X., Wang, H., Wu, G.: Optimal point movement for covering circular regions. Algorithmica **72**(2), 379–399 (2015)
12. Cherry, A., Gudmundsson, J., Mestre, J.: Barrier coverage with uniform Radii in 2D. In: Fernández Anta, A., Jurdzinski, T., Mosteiro, M.A., Zhang, Y. (eds.) ALGOSENSORS 2017. LNCS, vol. 10718, pp. 57–69. Springer, Cham (2017). https://doi.org/10.1007/978-3-319-72751-6_5
13. Czyzowicz, J., et al.: On minimizing the sum of sensor movements for barrier coverage of a line segment. In: Nikolaidis, I., Wu, K. (eds.) ADHOC-NOW 2010. LNCS, vol. 6288, pp. 29–42. Springer, Heidelberg (2010). https://doi.org/10.1007/978-3-642-14785-2_3

14. Dobrev, S., et al.: Complexity of barrier coverage with relocatable sensors in the plane. Theor. Comput. Sci. **579**, 64–73 (2015)
15. Erzin, A.I., Astrakov, S.N.: Min-density stripe covering and applications in sensor networks. In: Murgante, B., Gervasi, O., Iglesias, A., Taniar, D., Apduhan, B.O. (eds.) ICCSA 2011. LNCS, vol. 6784, pp. 152–162. Springer, Heidelberg (2011). https://doi.org/10.1007/978-3-642-21931-3_13
16. Erzin, A.I., Astrakov, S.N.: Covering a plane with ellipses. Optim.: A J. Math. Program. Oper. Res. **62**(10), 1357–1366 (2013)
17. Erzin, A.I., Shabelnikova, N.A.: On the density of a strip covering with identical sectors. J. Appl. Ind. Math. **9**(4), 461–468 (2015)
18. He, S., Chen, J., Li, X., Shen, X., Sun, Y.: Mobility and intruder prior information improving the barrier coverage of sparse sensor networks. IEEE Trans. Mob. Comput. **13**(6), 1268–1282 (2014)
19. Kershner, R.: The number of circles covering a set. Am. J. Math. **61**(3), 665–671 (1939)
20. Kumar, S., Lai, T.H., Arora, A.: Barrier coverage with wireless sensors. In: Proceedings of the 11th Annual International Conference on Mobile Computing and Networking, pp. 284–298 (2005)
21. Saipulla, A., Westphal, C., Liu, B., Wang, J.: Barrier coverage with line-based deployed mobile sensors. Ad Hoc Netw. **11**, 1381–1391 (2013)
22. Si, P., Wu, C., Zhang, Y., Jia, Z., Ji, P., Chu, H.: Barrier coverage for 3D camera sensor networks. Sensors **17**(8), 1771 (2017)
23. Wu, F., Gui, Y., Wang, Z., Gao, X., Chen, G.: A survey on barrier coverage with sensors. Front. Comput. Sci. **10**(6), 968–984 (2016)
24. Zalyubovskiy, V., Erzin, A., Astrakov, S., Choo, H.: Energy-efficient area coverage by sensors with adjustable ranges. Sensors **9**(4), 2446–2460 (2009)
25. Zhao, L., Bai, G., Shen, H., Tang, Z.: Strong barrier coverage of directional sensor networks with mobile sensors. Int. J. Distrib. Sens. Netw. **14**(2) (2018). https://doi.org/10.1177/1550147718761582

Time- and Energy-Aware Task Scheduling in Environmentally-Powered Sensor Networks

Lars Hanschke[✉] and Christian Renner

Research Group smartPORT, Hamburg University of Technology, Hamburg, Germany
{lars.hanschke,christian.renner}@tuhh.de

Abstract. In the past years, the capabilities and thus application scenarios of Wireless Sensor Networks (WSNs) increased: higher computational power and miniaturization of complex sensors, e.g. fine dust, offer a plethora of new directions. However, energy supply still remains a tough challenge because the use of batteries is neither environmentally-friendly nor maintenance-free. Although energy harvesting promises uninterrupted operation, it requires adaption of the consumption—which becomes even more complex with increased capabilities of WSNs. In existing literature, adaption to the available energy is typically rate-based. This ignores that the underlying physical phenomena are typically related in time and thus the corresponding sensor tasks cannot be scheduled independently. We close this gap by defining task graphs, allowing arbitrary task relations while including time constraints. To ensure uninterrupted operation of the sensor node, we include energy constraints obtained from a common energy-prediction algorithm. Using a standard Integer Linear Programming (ILP) solver, we generate a schedule for task execution satisfying both time and energy constraints. We exemplarily show, how varying energy resources influence the schedule of a fine dust sensor. Furthermore, we assess the overhead introduced by schedule computation and investigate how the size of the task graph and the available energy affect this overhead. Finally, we present indications for efficiently implementing our approach on sensor nodes.

Keywords: Energy harvesting · Task scheduling ·
Integer linear programming

1 Introduction

The popularity of Wireless Sensor Networks (WSNs), based on still increasing computational power, new sensing capabilities [2] or sharing of WSNs [1,16], offers a plethora of new application scenarios. Sensor nodes are equipped with multiple sensors, e.g. fine dust, humidity or ozone, or different radio interfaces, e.g. LoRa [6], WiFi [10] or IEEE 802.15.4. With an increasing number of different peripherals, the complexity of the underlying program structure grows steadily.

© Springer Nature Switzerland AG 2019
S. Gilbert et al. (Eds.): ALGOSENSORS 2018, LNCS 11410, pp. 131–144, 2019.
https://doi.org/10.1007/978-3-030-14094-6_9

Each of these radio interfaces or peripherals imposes a new task on the sensor node. Since these tasks are influenced by physical phenomena, e.g. detecting a change or reacting on a control command, they are not independent from each other. The sampling rate has to be adapted to the physical phenomenon but additional information of the surroundings has to be considered, e.g. a gas sensor typically needs to be compensated for temperature or humidity. If power supply and energy constraints are neglected, this complexity is manageable with today's approaches: generating a schedule for all envisaged tasks and performing these at a constant rate is sufficient.

Supplying sensor nodes with ambient energy from renewable resources, e.g. solar energy, allows for reducing the environmental footprint of WSNs and also decreases costs. In many scenarios, energy harvesting allows perpetual operation, but if the energy budget is restricted, e.g. due to physical size limitation of node and energy storage, the consumption of the sensor node has to be adjusted carefully. Since power is only available sporadically and the power consumption of the sensor node often drastically exceeds source power, scheduling the tasks of a node correspondingly is key. We also highlight this aspect in Sect. 2. Over the past years, many classes of energy-driven devices evolved [17]: continuously operating, as presented in [15] and [21], or even intermittently-powered, e.g. as shown in [12] and [7]. While these devices support simple program structures, e.g. sample-and-transmit-to-sink, it is still unclear how they perform with more complex program structures imposed of today's capabilities.

Recent work [13] emphasizes to include time constraints between tasks in scheduling for sensor networks. While this approach concentrates on intermittently-powered devices, we strongly agree with it. Especially more complex sensors, as gas sensors, need up-to-date information of their surroundings, e.g. the temperature or humidity, before a meaningful sensor reading can be obtained. Thus, sampling a gas sensor without querying related sensors in-time wastes energy.

We argue that future sensor nodes powered by ambient energy need a new scheduling approach: without consideration of time constraints and dependencies between tasks, energy is wasted and thus operation is inefficient. In this paper, we show how dependencies, time and energy constraints can be used for task graph generation and scheduling.

1.1 Contributions

An important aspect of task scheduling is still missing in existing literature: tasks are not independent from each other and are strongly related in time. Thus, we present our approach, which ensures the following:

- Description for a program structure satisfying time constraints.
- Definition of the mathematical problem.
- Automated scheduling for dependent tasks.
- Evaluation of the performance and indications for solving the problem on low-power microcontrollers used in WSNs.

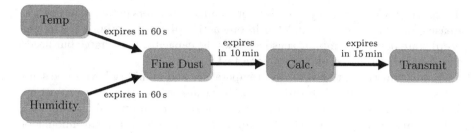

Fig. 1. Process of fine dust sampling; limited energy storage prevents consecutive execution of all tasks at once.

Our approach allows complex sensors to operate with ambient energy, especially with size-restricted energy storage units. This paves the ground for running multi-purpose sensors, with different sensors and radio interfaces, perpetually for fine-grained environmental monitoring.

2 Example

In order to illustrate the shortcomings of existing approaches and to show the benefits of our solution, we give a practical example depicted in Fig. 1. High aerial fine dust concentration in European cities is a major influencing factor of human health [25]. Responsible authorities for inner cities, residential areas but also larger production plants need to monitor their air quality finely grained to take countermeasures. Especially in port areas, where ships are the main contributor to fine dust pollution, flexible placement of fine dust sensors is desired. Interruption of production processes for sensor maintenance is time-consuming and cost-intensive; thus, supplying these sensors by ambient energy is a promising approach.

Low-cost and small sensors to measure fine dust particles, i.e. $PM_{2.5}$ and PM_{10}[1], have been studied in [22] and [4]. However, solely supplying the sensor nodes with ambient energy requires careful adaption of the sensing activities, because of the high power consumption compared to the harvest. A cheap fine dust sensor as the Nova SDS011 draws 70 mA at 5 V, which depletes the energy stored in a 100 F supercapacitor in just 9 min. Furthermore, fine dust is cross-sensitive to the relative humidity in ambient air. This requires a humidity and temperature sensor to be sampled before a valid fine dust measurement can be processed. Commercial fine dust sensors offer internal humidity compensation, but they are orders of magnitudes more expensive.

As recent work in [12] and [7] shows, it is not desirable to supply energy harvesting sensor nodes with large capacitors. Due to non-deterministic events, the voltage level of the capacitor might drop below the minimum voltage. Since recharge time is much longer with large capacitors, the system is unavailable for

---

[1] $PM_{2.5}$ and PM_{10} are air particles with diameter less than 2.5 μm and 10 μm respectively.

a longer time. Unfortunately, the need for smaller capacitors entails that energy-intensive tasks cannot be completed in one active phase. Entering a low-power (sleep) state between sampling sensors allows the capacitor to recharge but needs consideration of time constraints among tasks.

Figure 1 illustrates this behavior: temperature and humidity have to be sampled shortly before the fine dust sensor to ensure correlation between sensor values. However, once sensor values are gathered, calculation and transmission of these values can be delayed. Most applications require a certain amount of information per time interval, e.g. two times per hour. Thus, the margin for calculation and transmission allows for recharging of the capacitor, ensuring perpetual operation in the future.

3   Related Work

A plethora of literature exists for scheduling tasks with energy constraints in sensor networks. The declared goal of all scheduling techniques is to arrange and serve tasks considering limited energy storage and varying energy harvesting conditions. The principle of ensuring that the power consumption of a sensor nodes stays below the energy income and simultaneously ensuring a minimum battery level is better known as Energy-Neutral Operation (ENO) [15]. While all presented work satisfy this principle, their strategies vary.

One of the first approaches presented in [14] solely relies on adjusting the duty cycle based on future energy income to ensure ENO. Neglecting the variety of different power consumption states of a sensor node and its connected peripherals, apart from *sleep* and *active* state, the approach aims to maximize the duty cycle. However, for sensor nodes with different tasks and more complex program structures, e.g. as presented in Sect. 2, this approach is limited. Furthermore, a greedy behavior is pointless for a variety of scenarios, e.g. Delay Tolerant Networking (DTN).

The work presented in [18] resolves the shortcomings of real-time Earliest Deadline First (EDF) algorithms by presenting two classes of Lazy Scheduling Algorithms (LSAs). Arriving tasks are scheduled as late as possible but with a margin that allows to schedule potentially arriving new tasks without deadline misses. Unfortunately, LSA relies on independent and preemptive tasks, which is unsuitable for the application presented in Sect. 2. Furthermore, the approach assumes the power consumption of the node to be continuously adaptable. Since the power consumption has discrete levels, determined by the state of the microcontroller and peripherals, the approach is not directly applicable in practice.

By using Directed Acyclic Graphs (DAGs) and two corresponding ILP approaches, [23] describes the tasks of the used platform. The approach tries to find one path in the graph which minimizes the sum of execution times or maximizes the usefulness of tasks, respectively. Again, a rate-based duty-cycling ensures ENO. Their graph-based task model does not include timely distances between tasks: each task is executed directly after its predecessor although this might not be valuable in practice. Furthermore, energy savings in practice remain unclear, since their rate adaption requires re-evaluating the task graph before each new iteration. Since this can be time-consuming on low-power microcontrollers, it is not clear if this leads to energy savings in real-world experiments.

A controller-based approach for adjusting the activity rate of the sensor node is presented in [19]. The authors formally describe their controller design which is solved by Integer Linear Programmings (ILPs). While controller generation is done offline, they reduce the complexity of their online approach by only determining the control region of the actual system state. Although their approach allows for rate adaption of different tasks, they do not take dependencies among tasks into account which might lead to energy wastage in more complex applications, i.e. as described in Sect. 2.

Two approaches for tackling the scheduling problem of energy harvesting are presented in [3]. By introducing the concept of *virtual tasks*, the authors opt for smoothing the average power consumption or achieving full utilization. While they show that their concept outperforms LSA and EDF in simulations with static schedules, it is unclear how their approach is influenced by dynamically changing weather conditions and thus a dynamically changing schedule. Furthermore, they lack the support for dependencies between tasks.

A real-time scheduling approach called Earliest Deadline with Energy Guarantees (EDeg) is presented in [9] as a variation of the classical EDF algorithm. By simulations, the authors show that EDeg outperforms energy-oblivious scheduling approaches in terms of system runtime and deadline miss rate. One drawback of EDeg is that it assumes tasks to be preemptive without loss of energy. In practice, this is hard to justify when communicating with external devices, e.g. sensors or other nodes in the network. Restarting a task, e.g. quiring an external sensor, involves repetition and thus wastes energy.

The authors of [26] identified that sampling rates of sensors are typically not known in beforehand in research experiments. Thus, they develop an interactive algorithm which allows change of sampling rate upon user request at runtime considering energy limitations. They ensure ENO by keeping track of the saved energy due to under sampling phases and allowing to spend saved energy at later point in time. However, they do not check for task dependencies while varying the different sampling rates.

Initially developed for intermittently-powered systems, the approach presented in [13] provides a clear description for time constraints of sensor tasks. As it focuses on systems which face power outages frequently, the approach uses time constraints for checking validity of sensor values after execution. However, for ensuring ENO, integrating these time constraints into task planning – so before execution – is key to ensure both ENO and time constraints.

4   Task Model

As existing scheduling techniques lack the support for describing relations between tasks for WSNs, e.g. sensing, actuating, processing or sending, we describe the underlying mathematical model of our approach. Our model aims at

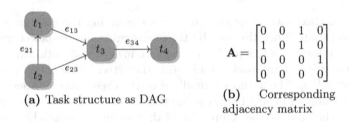

(a) Task structure as DAG

$$\mathbf{A} = \begin{bmatrix} 0 & 0 & 1 & 0 \\ 1 & 0 & 1 & 0 \\ 0 & 0 & 0 & 1 \\ 0 & 0 & 0 & 0 \end{bmatrix}$$

(b)   Corresponding adjacency matrix

Fig. 2. Example for describing complex program structures by the use of DAGs.

task planning, so tasks are considered as non-preemptive and known in beforehand. First, we describe the program structure of nodes as a DAG and introduce the definitions of nodes and edges of the graph. Second, we define the time and third, the energy constraints which are solved via ILP. Fourth, we elaborate on the problem definition and how to choose the objective function.

4.1   Program Flow

The time constraints between different tasks of a sensor node are also described in [13] by the Mayfly language. The authors of [13] claim that tasks can be divided into three categories or can be described with three properties:

- `misd`: minimum inter-sample distance; the time after which new data is valuable
- `expires`: the time after which a sensor value looses its meaning
- `collect`: describes a distinct **number** of samples to be collected within a certain `timeframe`.

We strongly agree with theses definitions, as they allow to describe the program flow presented in Sect. 2. `misd` and `expires` describe the relation between consecutive activities, i.e. the time difference. Additionally, `collect` embraces numerous activities, which all have to be scheduled into a certain `timeframe`.

Beyond Mayfly. However, the schedule constructed from the compiler of [13] only ensures a sequence of task execution. Since intermittently-powered devices may experience power outages between two consecutive tasks, they cannot ensure that execution of every task maintains its meaning. If task information is outdated, is only checked *after* execution. We believe that devices ensuring ENO have the benefit of deciding the meaningfulness of tasks *before* execution by accurate task planning. The basis for our task scheduling is the program flow described as DAG; see also Fig. 2.

The task graph $\mathcal{G} = (\mathcal{T}, \mathcal{E})$ is defined by the set of vertices, here tasks, $\mathcal{T} = \{t_1, t_2, ..., t_N\}, N \in \mathbb{N}^+$ and edges $\mathcal{E} \in \mathcal{T} \times \mathcal{T}$. Since $\mathcal{G}$ is a directed graph, each edge $e_{vw} \in \mathcal{E}$ is represented by a 2-tuple of starting vertice v and end vertice w. Tasks which are connected by edges are called *adjacent*; thus, the

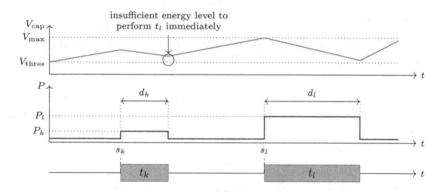

Fig. 3. To prevent depletion of the capacitor, i.e. a voltage below V_{thres}, sleeping before execution of t_l is needed.

relationships in $\mathcal{G}$ can be described by an adjacency matrix **A**. The entries of $\mathbf{A} = [a_{i,j}]$ are defined as

$$a_{i,j} = \begin{cases} 1 & \text{if } (i,j) \in \mathcal{E} \\ 0 & \text{else} \end{cases} \tag{1}$$

Tasks. To develop an energy-efficient schedule, more information about the tasks and the edges of the graph is needed. Typical microcontrollers, e.g. the ARM Cortex series or the low-cost ESP 8266, have different hardware states, which vary in computing power, usable internal components and consequently power consumption. Additionally, the nodes power external peripherals, such as sensors and actuators, to interact with their surroundings. Thus, a task t_i is described by its power consumption $P_i, i \in \mathbb{N}^+$.

To query a sensor or to perform calculations, a microcontroller needs a time duration. For task t_i, this duration d_i can be obtained or updated at runtime by internal timers or external devices [8]. Together with the power consumption, this allows to compute the energy consumption of the node but also enables overlapping-free scheduling of tasks. To develop the schedule, we calculate all s_i, defined as the starting time of t_i.

Edges. Apart from its start vertice v and end vertice w, each edge comprises the constraints for the relation between v and w. Following the definitions in [13], edges are described by a 4-tuple $e_{vw} = (\mathtt{misd}, \mathtt{expires}, \mathtt{timeframe}, \mathtt{number})$. This 4-tuple describes the constraints which have to be fulfilled by a valid schedule so that a following task w can work with up-to-date data. Not every task has to fulfill all parameters; e.g. if only one sensor value has to be sampled, no value for $\mathtt{timeframe}$ is defined.

4.2   Constraints

To compute a valid but also energy efficient schedule, numerous definitions are needed for satisfying the constraints. We also visualize the constraints in Fig. 3.

Time. In the following, k and l denote indices of their corresponding tasks which are connected by an edge, so that $a_{k,l} = 1$. A mandatory requirement is that connected tasks of the program flow are consecutive and that their execution times do not overlap:

$$s_l \geq s_k + d_k. \tag{2}$$

Furthermore, the time constraints described by the connecting edges have to be fulfilled. Thus, the difference between starting times has to be:

$$\texttt{misd}_{kl} \leq s_l - s_k \leq \texttt{expires}_{kl}. \tag{3}$$

If one task requests a certain **number** of samples to be collected within **timeframe**, the scheduler ensures a greater time distance towards the following tasks, i.e.

$$s_k + \texttt{timeframe}_k \leq s_l. \tag{4}$$

Energy. Sensor nodes, which are powered from ambient solar energy, mostly run energy prediction algorithms, e.g. [20] or [5], which give an approximation of the harvest within the next prediction horizon (usually 24 h in timeslots of 1 h). This can be used to obtain, e.g. as presented in [21], an approximation of the maximum allowed energy consumption per timeslot within the prediction horizon. Typically, this approximation prevents depletion of the energy storage, e.g. the voltage of a supercapacitor does not drop below a threshold. Within the prediction horizon, not all tasks can be executed subsequently without breaks and thus require sleep times in between to ensure recharging of the capacitor. If less energy is available, the time difference of our tasks are spread further to the boundaries of **expires**; if more energy is available, further to **misd**.

The energy consumption of the sensor node within one timeslot of the prediction horizon is given by the energy consumed by all executed tasks. If there is a budget b^2 which can be taken from the energy storage without depletion within the next slot of the prediction horizon h, the energy consumption of the sensor node has to satisfy:

$$\sum_{i=1}^{N} n_i \cdot d_i \cdot P_i + \left(h - \sum_{i=1}^{N} n_i \cdot d_i\right) \cdot P_q \leq b \cdot V_{cc} \cdot h. \tag{5}$$

The budget b describes a constant current drawn by the sensor node at its supply voltage V_{cc}. Here, n_i denotes the number of executions of task t_i—if the budget is

---

[2] In compliance with [21], the budget is a current; the energy follows directly with constant supply voltage and known time.

high, the task graph may be executed more than once. Additionally, P_q denotes the quiescent power consumption in sleep state. For simplicity, we only consider one sleep state but plan to investigate the use of different sleep states in future work.

4.3   Problem Definition and Objective Function

For the input of the ILP solver, we need the time and energy constraints, but also the objective function f for optimization. For our concrete scheduling problem, we use the definition of the canonical form:

$$
\begin{aligned}
\text{maximize} \quad & f^T s \\
\text{subject to} \quad & Cs \le c \\
\text{and} \quad & s \ge 0
\end{aligned}
\tag{6}
$$

The starting times of all instances of the tasks are contained in s, while the constraints presented in Sect. 4.2 are listed and C and c. We use Eq. (5) to determine the number of instances n_i for every task which can be executed without violating energy constraints. Microcontrollers offer a plethora of timers of different resolutions, typically limited by the clock frequency of the Microcontroller Unit (MCU). As we do not explicitly aim on real-time systems, we limit our resolution to milliseconds since ENO systems typically do not benefit from higher time resolution. Still, it requires the variables, i.e. the starting times, to be strictly positive integers. Since ILPs are in general more complex to solve, there is room for decreasing the complexity of our approach in future work.

The choice of f is typically non-trivial. Commonly used objectives for scheduling problems are minimizing lateness, delay, or a serving time. Other objectives embrace maximizing the usefulness of a schedule by assigning a utility or priority to each task. While we agree with opting for a utility-based objective, it is very application-specific, has to be provided by the user and varies with scenarios.

As our approach is aiming at ENO, we choose a different objective function. Assuming constant harvest during one timeslot, e.g. 1 h, is simplifying the real world, as shown in [20]. The harvest can be below average at the start of the timeslot and above average at the end. A constant energy consumption computed based on the average hence risks depletion of energy storage at the beginning of the time slot, even if the average harvest prediction is met. Consequently, we argue that tasks should be preferably executed at the end of the timeslot to decrease the risk of depletion—but only if time constraints are still satisfied. The following objective function is constant, which favors starting times at the end of a timeslot. However, due to the time constraints, the majority of tasks is still distributed evenly across the timeslot. As choosing different objective functions offers potential for increasing the usefulness of our schedule, we plan to investigate this in future work.

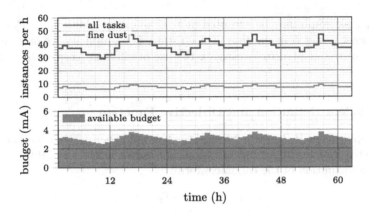

Fig. 4. Energy-adapting task scheduling of the fine dust example; number of instances per h varies upon budget; budget trace obtained from real-world tests.

5   Implementation

For assessing the performance of our approach, we implement the problem definition, i.e. the constraints as defined in Sects. 4.2 and 4.3, and feed it into a MATLAB ILP solver. For the problem definition, we only need two inputs: the task graph and the budget of the timeslot we are generating the schedule for. Our goal is to find the number of entries in **s** as well as matrix **C** and its corresponding vector **c** to solve Eq. (6). To obtain the number of entries in **s**, we incrementally add tasks to the list of tasks to be executed within this timeslot as long as the budget is not exceeded. Once the number of starting times is known, we loop through them incrementally. First, we check for the task to be executed at the current starting time. Second, we query the adjacency matrix **A** if the task of the current starting time has neighbors. Neighbors can be predecessors or successors in the task graph — for both of them, we add the constraints as defined in Eqs. (2) to (4) to **C** and **c**. If tasks are executed more than once, we also add a constraint to ensure the first task of the next repetition of the task graph is executed after the last task of the current execution. Last, we ensure that starting times are upper bounded by the length of a timeslot.

To speed up computation and reduce memory footprint, we use the sparse matrix representation for **C**. We use `intlinprog` from the optimization toolbox of MATLAB. As `intlinprog` uses several pre-processing techniques, e.g. reducing the size of the problem or solve relaxed problems first, the time to solve the problem only gives a hint on complexity-influencing factors, not on the execution time itself. Still, we are able to show if our approach generates feasible schedules and to highlight influencing factors.

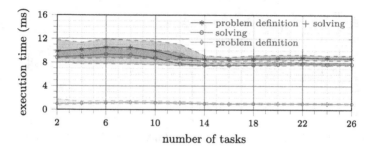

Fig. 5. Median of execution times influenced by the number of tasks in task graph; dashed lines indicate upper and lower quartiles respectively; constant budget of 5 mA; since budget stays constant, number of instances in total and thus complexity of the problem remains relatively steady.

6   Results

First, we show how an energy-aware schedule is generated for the example of Sect. 2. Second, we analyze the performance of our approach and show how the number of tasks in the graph and an increasing budget influences execution time. Third, we discuss what we can learn for our microcontroller implementation. We use an Intel Core i7-6700 with 16 GB of RAM for our simulations in MATLAB. Execution times are measured by the provided cputime function.

6.1   Example

We use the task graph shown in Fig. 1 for generating a schedule. Additionally, we feed real-world budget traces obtained from a long-term experiment with a solar energy harvester [11] into simulation. For simplicity, we only depict the first 64 h in Fig. 4.

Depending on the budget, we can see that instances per hour of the fine dust sampling task ranges between 6 and 9. The number of instances of all tasks shows a similar behavior, ranging between 29 and 47. This shows that even more complex program structures defined by a task graph and with time constraints can be scheduled being energy-aware simultaneously.

6.2   Performance

Our main metric for comparing the performance is the execution time, i.e. the time to define and solve the problem. Although energy-aware scheduling is necessary for uninterrupted operation, it is still only a tool for performing the designated task. Consequently, the energy spent for calculating a schedule means an energy overhead and thus the time spent for scheduling should be as short as possible. Figure 5 depicts how the number of tasks in the task graph influences the execution time. We randomly generate fully connected task graphs with fixed number of tasks, varying number of edges and a constant budget.

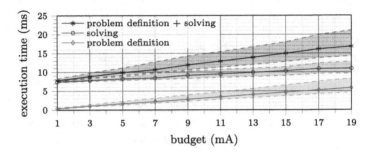

Fig. 6. Median of execution times influenced by increasing budget. Dashed lines indicate upper and lower quartiles respectively; increased complexity influences both problem definition and solving; however, increase in execution time primarily influenced by increasing time for problem definition.

For each number of tasks, we generate 500 different task graphs. Although the complexity of the task graph increases due to increased number of tasks and edges, the execution time remains relatively constant. Due to the fixed budget, an increasing number of tasks does not increase the number of variables—the number of variables is equal if 15 tasks are executed four times each or if 20 tasks are executed three times each.

However, if the budget increases, also the execution time rises because of more variables to determine. Figure 6 shows how the increased budget influences the performance. Again, we randomly generate a task graph and compute a schedule with increasing budget each. We repeat this process 500 times. The execution time strongly increases with budget, e.g. increasing the budget from 3mA to 9mA demands 30% more execution time. Although this seems noteworthy, the increase is good-natured: an increasing budget means good energy harvesting conditions and thus tolerates a larger overhead for scheduling purposes. However, it is important to mind this aspect for microcontroller implementations.

6.3   Discussion

Our investigations show that a high-performance PC can generate a schedule for 10 tasks and 20 edges for a typical budget of 3 mA in 8.8 ms. However, the computational power of microcontrollers is orders of magnitudes lower. To have an idea on the energy overhead, we compare the DMIPS[3] of a common microcontroller, the ARM Cortex M0+ and our simulation PC. We measure 11.28 DMIPS/MHz the with Intel Core i7, while a Cortex M0+ offers just 0.95 DMIPS/MHz [24]. Including the clock frequencies in the calculation, the approximated time for executing the schedule for one timeslot on a Cortex M0+ is roughly 11 s. While this sounds relatively long, the energy overhead is still only 0.7%, assuming a current consumption of 6.85 mA [24], a budget of 3 mA and a timeslot length of 1 h. Surely, these numbers will differ in practice but they indicate that our approach is feasible.

---

[3] DMIPS = Dhrystone Million Instruction per Seconds; common performance measure generated by the Dhrystone benchmark.

7 Conclusion

The increasing capabilities of WSNs offer a plethora of new applications but also pose new demands on energy-aware task scheduling. We showed that using task graphs, which describe dependencies between tasks, and defining time and energy constraints, allows for calculation of schedules for sensor nodes. We plan to investigate the use of different objective functions and to implement our approach for microcontrollers to perform real-world-experiments in future work.

References

1. Adkins, J., Campbell, B., Ghena, B., Jackson, N., Pannuto, P., Dutta, P.: Energy isolation required for multi-tenant energy harvesting platforms. In: Proceedings of the 5th ACM International Workshop on Energy Harvesting and Energy-Neutral Sensing Systems, ENSsys 2017, pp. 27–30. ACM (2017)
2. Arora, C., Arora, N., Choudhary, A., Sinha, A.: Intelligent vehicular monitoring system integrated with automated remote proctoring. In: Hu, Y.-C., Tiwari, S., Mishra, K.K., Trivedi, M.C. (eds.) Intelligent Communication and Computational Technologies. LNNS, vol. 19, pp. 325–332. Springer, Singapore (2018). https://doi.org/10.1007/978-981-10-5523-2_30
3. Audet, D., MacMillan, N., Marinakis, D., Wu, K.: Scheduling recurring tasks in energy harvesting sensors. In: 2011 IEEE Conference on Computer Communications Workshops, INFOCOM WKSHPS 2011, pp. 277–282. IEEE (2011)
4. Budde, M., El Masri, R., Riedel, T., Beigl, M.: Enabling low-cost particulate matter measurement for participatory sensing scenarios. In: Proceedings of the 12th International Conference on Mobile and Ubiquitous Multimedia, MUM 2013, p. 19. ACM (2013)
5. Cammarano, A., Petrioli, C., Spenza, D.: Pro-energy: a novel energy prediction model for solar and wind energy-harvesting wireless sensor networks. In: IEEE 9th International Conference on Mobile Adhoc and Sensor Systems, MASS 2012, pp. 75–83. IEEE (2012)
6. Cattani, M., Boano, C.A., Römer, K.: An experimental evaluation of the reliability of LoRa long-range low-power wireless communication. J. Sens. Actuator Netw. **6**(2), 7 (2017)
7. Colin, A., Ruppel, E., Lucia, B.: A reconfigurable energy storage architecture for energy-harvesting devices. In: Proceedings of the Twenty-Third International Conference on Architectural Support for Programming Languages and Operating Systems, ASPLOS 2018, pp. 767–781. ACM (2018)
8. Dutta, P., Feldmeier, M., Paradiso, J., Culler, D.: Energy metering for free: augmenting switching regulators for real-time monitoring. In: Proceedings of the 7th International Conference on Information Processing in Sensor Networks, IPSN 2008, pp. 283–294. IEEE (2008)
9. Ghor, H.E., Chetto, M., Chehade, R.H.: A real-time scheduling framework for embedded systems with environmental energy harvesting. Comput. Electr. Eng. **37**(4), 498–510 (2011)
10. Hanschke, L., Heitmann, J., Renner, C.: Challenges of WiFi-enabled and solar-powered sensors for smart ports. In: Proceedings of the 4th ACM International Workshop on Energy Neutral Sensing Systems, ENSsys 2016. ACM (2016)

11. Hanschke, L., Heitmann, J., Renner, C.: Stop waiting: mitigating varying connecting times for infrastructure WiFi nodes. In: Proceedings of the 16th GI/ITG KuVS Fachgespräch "Sensornetze", FGSN 2017 (2017)

12. Hester, J., Sitanayah, L., Sorber, J.: Tragedy of the coulombs: federating energy storage for tiny, intermittently-powered sensors. In: Proceedings of the 13th ACM Conference on Embedded Networked Sensor Systems, SenSys 2015, pp. 5–16. ACM (2015)

13. Hester, J., Storer, K., Sorber, J.: Timely execution on intermittently powered batteryless sensors. In: Proceedings of the 15th ACM Conference on Embedded Network Sensor Systems, SenSys 2017, pp. 17:1–17:13. ACM (2017)

14. Hsu, J., Zahedi, S., Kansal, A., Srivastava, M., Raghunathan, V.: Adaptive duty cycling for energy harvesting systems. In: Proceedings of the 2006 International Symposium on Low Power Electronics and Design, ISLPED 2006, pp. 180–185. ACM (2006)

15. Kansal, A., Hsu, J., Zahedi, S., Srivastava, M.B.: Power management in energy harvesting sensor networks. ACM Trans. Embed. Comput. Syst. (TECS) 6(4), 32 (2007)

16. La Porta, T., Petrioli, C., Spenza, D.: Sensor-mission assignment in wireless sensor networks with energy harvesting. In: 2011 8th Annual IEEE Communications Society Conference on Sensor, Mesh and Ad Hoc Communications and Networks, SECON 2011, pp. 413–421. IEEE (2011)

17. Merrett, G.V., Al-Hashimi, B.M.: Energy-driven computing: rethinking the design of energy harvesting systems. In: Design, Automation and Test in Europe Conference and Exhibition, DATE 2017, pp. 960–965. IEEE (2017)

18. Moser, C., Brunelli, D., Thiele, L., Benini, L.: Real-time scheduling with regenerative energy. In: 18th Euromicro Conference on Real-Time Systems, ECRTS 2006, IEEE (2006)

19. Moser, C., Thiele, L., Brunelli, D., Benini, L.: Adaptive power management for environmentally powered systems. IEEE Trans. Comput. 59(4), 478–491 (2010)

20. Renner, C.: Solar harvest prediction supported by cloud cover forecasts. In: Proceedings of the 1st ACM International Workshop on Energy Neutral Sensing Systems, ENSsys 2013, ACM (2013)

21. Renner, C., Meier, F., Turau, V.: Policies for predictive energy management with supercapacitors. In: International Conference on Pervasive Computing and Communications Workshops, PERCOM Workshops 2012 (2012)

22. Ruprecht, A.A., et al.: Mass calibration and relative humidity compensation requirements for optical portable particulate matter monitors: the IMPASHS (impact of smoke-free policies in EU member states) Wp2 preliminary results. Epidemiology 22(1), S206 (2011)

23. Steck, J.B., Rosing, T.S.: Adapting task utility in externally triggered energy harvesting wireless sensing systems. In: 2009 Sixth International Conference on Networked Sensing Systems, INSS 2009, pp. 1–8. IEEE (2009)

24. STMicroelectronics: Datasheet STM32L072x8, September 2017. rev. 4

25. World Health Organization (WHO): Health Risks of Air Pollution in Europe - HRAPIE Project: Recommendations for Concentration-response Functions for Cost-benefit Analysis of Particulate Matter, Ozone and Nitrogen Dioxide. UN City, Copenhagen, Denmark (2013)

26. Yang, J., Tilak, S., Rosing, T.S.: An interactive context-aware power management technique for optimizing sensor network lifetime. In: Proceedings of the 5th International Conference on Sensor Networks, SENSORNETS 2016, vol. 1, pp. 69–76. SciTePress (2016)

Mobility-Aware, Adaptive Algorithms for Wireless Power Transfer in Ad Hoc Networks

Adelina Madhja[1,2], Sotiris Nikoletseas[1,2], and Alexandros A. Voudouris[3(✉)]

[1] Department of Computer Engineering and Informatics, University of Patras,
Patras, Greece
madia@ceid.upatras.gr
[2] Computer Technology Institute and Press "Diophantus" (CTI), Rion, Greece
nikole@cti.gr
[3] Department of Computer Science, University of Oxford, Oxford, UK
alexandros.voudouris@cs.ox.ac.uk

Abstract. We investigate the interesting impact of mobility on the problem of efficient wireless power transfer in ad hoc networks. We consider a set of mobile agents (consuming energy to perform certain sensing and communication tasks), and a single static charger (with finite energy) which can recharge the agents when they get in its range. In particular, we focus on the problem of efficiently computing the appropriate range of the charger with the goal of prolonging the network lifetime. We first demonstrate (under the realistic assumption of fixed energy supplies) the limitations of any fixed charging range and, therefore, the need for (and power of) a dynamic selection of the charging range, by adapting to the behavior of the mobile agents which is revealed in an online manner. We investigate the complexity of optimizing the selection of such an adaptive charging range, by showing that two simplified offline optimization problems (closely related to the online one) are NP-hard. To effectively address the involved performance trade-offs, we finally present a variety of adaptive heuristics, assuming different levels of agent information regarding their mobility and energy.

1 Introduction

Over the last decade, the continuously increasing development and excessive use of energy-hungry mobile devices (like smartphones, tablets, or even electric vehicles; see [3, 12]) in ad hoc networks, has given rise to the problem of efficient power management under various objectives. A viable solution to this critical problem, that has been extensively studied in the recent related literature due to its efficiency and wide applicability, is the Wireless Power Transfer (WPT)

This work was supported by the Greek State Scholarships Foundation (IKY), and by a PhD scholarship from the Onassis Foundation. The third author would like to thank Ioannis Caragiannis for fruitful discussions at early stages of this work.

© Springer Nature Switzerland AG 2019
S. Gilbert et al. (Eds.): ALGOSENSORS 2018, LNCS 11410, pp. 145–158, 2019.
https://doi.org/10.1007/978-3-030-14094-6_10

technology using magnetic resonant coupling [11] combined together with ultra-fast rechargeable batteries [13]. By exploiting such a technology, it is possible to recharge the network devices as required and prolong their lifetime.

In rechargeable ad hoc networks, there are two main types of entities that are distributed in the network area, called *chargers* and *agents*, respectively. Usually, a charger is a special device that has high energy supplies and acts as a transmitter, while an agent has significantly lower battery capacity and acts as a receiver. The charger is responsible for the energy management in the network, by effectively transferring parts of its energy to the agents. In contrast, the agents are the actual network devices which consume energy while performing communication and sensing tasks (like collecting and routing data) and are, therefore, in need of energy replenishment to sustain their normal operation.

There are generally many different assumptions regarding the charging process, whether there is a single or multiple chargers that are mobile or not, as well as the information that is available about the energy levels and the locations of the (possibly mobile) agents. As the survey of all these different settings are not the main focus of this paper, we refer the interested reader to the book [20].

Our Contribution. We consider ad hoc networks that consist of mobile agents and a single static charger. The agents move around following a mobility model and consume energy for communication purposes. The charger is assumed to have initial finite energy that can be used to replenish the battery of the agents that get in its charging range. The finite energy assumption here is well motivated in scenarios where we would like to cover isolated areas (for instance, mountains where people go hiking) and there are simply no wired sources capable to provide unlimited energy to the charger. See Sect. 2 for a description of our model.

As the mobility and energy consumption characteristics of the agents become available online, the charger must respond to the behavior of the agents by dynamically changing its transmission power which, in turn, defines the charging range. The main objective of this adaptive selection of the charging range is to extend the network lifetime, which can be defined as the time period during which there is at least one agent with non-zero energy or the time period during which a percentage of agents have non-zero energy; of course, this is not the only objective that one may be interested in. To the best of our knowledge, this is the first paper that systematically studies the setting where the charging range is dynamically selected adaptively to the agents status.

We theoretically and experimentally showcase the need for adaptiveness. In particular, for every possible fixed range that the charger may have, we identify worst-case scenarios where there is always an adaptive solution that performs better (see Sect. 3). In addition, we define two simplified offline optimization problems that are closely related to the online multi-objective one, and prove their computational intractability (see Sect. 4). Furthermore, we design three adaptive algorithms and compare them to each other in terms of various metrics using a non-trivial simulation setup, where we consider probability distributions over randomized mobility and energy consumption scenarios that are designed to test our methods in highly heterogeneous instances (see Sect. 5).

Related Work. Mobility in ad hoc networks has been thoroughly studied and many models have been proposed over the years. Generally, such mobility models assume that the agents perform different kinds of random walks that may depend on many different parameters [2,4], and even be influenced by social network attributes that attempt to capture human behavior [10,16,21]. In this work, we slightly deviate from previous work and adopt a mobility model that allows us to construct many interesting mobility patterns for the agents, that can also simulate cases where human mobility may be arbitrary, greedy or even irrational.

Recharging in mobile ad hoc networks has been the focus of many research papers. Indicatively, Nikoletseas et al. [19] considered mobile ad hoc networks with multiple static chargers of finite energy supplies, and evaluated (using real devices) two algorithms that decide which chargers must be active during each round, in order to maximize charging efficiency and achieve energy balance, respectively. Angelopoulos et al. [1] also considered mobile ad hoc networks, with the difference that there exists a single mobile charger that has infinite energy and traverses the network in order to recharge the agents as needed. They focused on designing optimal traversal strategies for the mobile charger with the goal of prolonging the network lifetime.

He et al. [9] studied the energy provisioning problem to minimize the number of chargers and compute where they should be located in the network area so that all (possibly) mobile agents are always active (have or get enough energy to complete their tasks). By taking into account an agent's velocity and battery capacity, Dai et al. [5] showed that the agent's continuous operation cannot be guaranteed, and introduced the Quality of Energy Provisioning (QoEP) metric to characterize the expected time that the agent is actually active.

Dai et al. [6] considered static networks and studied the safe charging problem to maximize charging utility, while simultaneously ensuring that the electromagnetic radiation (EMR) does not exceed a threshold value at any point of the network. In [7], the authors studied a variation, where the power of each charger can be adjusted once at the beginning of time. Nikoletseas et al. [17] studied the low radiation efficient wireless charging problem as well, but they defined a different charging model that takes into account hardware constraints for the chargers and the agents. The last two papers are the most related ones to ours, in the sense that the power of each charger is adjustable. However, observe that since the agents are static in both models considered in [7,17], each charger adjusts its power only once, at the beginning of the time horizon. In contrast, the power of the charger in our setting constantly changes over time, adaptively to the behavior of the mobile agents which is revealed in an online manner. Therefore, even though our setting and that of [7,17] are seemingly similar, they are fundamentally different and uncomparable to each other.

There are several studies that deviate from the above modeling assumptions. In particular, Zhang et al. [22] introduced the notion of collaborative charging, where the chargers are able to transfer energy to each other as well. This feature was extended by Madhja et al. [14] in a hierarchical structure. Furthermore, recent studies do not even use chargers, but they assume that the agents

themselves are able to both receive and send power wirelessly [15,18]. Another
research direction deals with the simultaneous energy transfer and data collection by the charger (e.g. [23]). In this setting, practically, the charger acts as an
energy transmitter as well as a sink.

2   Model

There are n agents that move around in a bounded network area $\mathcal{A}$, and a
single static charger that is positioned at the center of $\mathcal{A}$. For simplicity, we
assume that $\mathcal{A}$ is represented by a rectangle defined by the points $(0,0)$ and
$(x_{\max}, y_{\max})$ on the Euclidean space. Hence, the position p_{charger} of the charger
is given by the coordinates $(\frac{1}{2}x_{\max}, \frac{1}{2}y_{\max})$. Further, we assume that there is a
discrete time horizon $T \in \mathbb{N}_{\geq 0}$ consisting of a number of distinct rounds each
of which runs for a constant period of time τ. For every agent i, we denote by
$p_i(t) = (x_i(t), y_i(t)) \in \mathcal{A}$ its position at the beginning of round t. The positions of
the agents are updated as they move around in $\mathcal{A}$. For the charger, we denote by
$R(t) \in [R_{\min}, R_{\max}]$ its range during round t. $R(t)$ is decided by the transmission
power of the charger and defines a circle of radius $R(t)$ around p_{charger}; let
$\mathcal{C}_{R(t)} \subseteq \mathcal{A}$ denote this circle on the plane. All agents that pass through $\mathcal{C}_{R(t)}$
during round t can get recharged (if they need to).

Mobility Model. At the beginning of each round t, every agent i randomly selects
a *speed mode* $\mu_i(t) \in [3]$ which indicates whether its velocity takes random values
in the intervals $I_1 = [0, \frac{1}{4}v_{\max}]$, $I_2 = (\frac{1}{4}v_{\max}, \frac{1}{2}v_{\max}]$, or $I_3 = (\frac{1}{2}v_{\max}, v_{\max}]$,
where $v_{\max}$ is the maximum possible velocity. Each agent i performs a *random
walk* as follows. At round t, it starts from position $p_i(t) \in \mathcal{A}$, and randomly
chooses a new direction $\theta_i(t) \in [0, 2\pi)$ and a new velocity $v_i(t) \in I_{\mu_i(t)}$. The
direction $\theta_i(t)$ together with $p_i(t)$, define a line along which the agent travels
with the chosen velocity $v_i(t)$ until it reaches its final position at the end of the
round, which is the position $p_i(t+1) \in \mathcal{A}$ at the beginning of the next round.
In particular, $p_i(t+1)$ has coordinates $x_i(t+1) = x_i(t) + v_i(t) \cdot \tau \cdot \cos\theta_i(t)$ and
$y_i(t+1) = y_i(t) + v_i(t) \cdot \tau \cdot \sin\theta_i(t)$. We remark that if these equations do not
define a point in $\mathcal{A}$, then the movement is redefined accordingly. Starting from
$t = 1$ and the initial deployment of the agents in $\mathcal{A}$, the above process is repeated
for all rounds $t \in [T]$.[1]

Energy Model. Let $E_i(t)$ be the energy of agent i at the beginning of round t. All
agents have the same battery characteristics in the sense that they have the same
battery capacity B. We assume that initially all agents are fully charged, i.e.,

---

[1] Notice that the mobility model we consider here is similar to the random way-point
model, but we also allow for special restrictions in the movements of the agents that
give birth to many interesting and extreme scenarios. We identify such worst-case
scenarios in Sect. 3 and utilize them in our experimental evaluation in Sect. 5, where
we consider probability distributions over both general and special mobility scenarios
to test our algorithms in highly heterogeneous settings.

$E_i(1) = B$ for every agent i. During round t, each agent i consumes an amount of energy $E_i^c(t)$ for communication purposes which depends on random sensing and routing events. Since the thorough study of such events are out of the scope of this paper, following previous work (e.g., see [1]), we simply assume that $E_i^c(t)$ follows a poisson probability distribution with expected value $\gamma_i \in [\gamma_{min}^i, \gamma_{max}^i]$. The energy of agent i at the beginning of the next round $t+1$ (assuming no recharging takes place), is equal to $E_i(t+1) = \max\{0, E_i(t) - E_i^c(t)\}$. We remark that the agents are assumed to *not* consume any energy due to movement as the necessary energy can be supplied by different sources. For example, in any crowdsensing scenario it is supplied by the humans that carry around their smart devices.

Charging Model. Let $E_{charger}(t)$ denote the energy that the charger has at the beginning of round t. We assume that the charger initially has some *finite* amount of energy $E_{charger}(1) = C$ that can be used to replenish the energy that the agents consume. In particular, if the charger has the appropriate amount of energy, then all agents that get in its range receive a positive amount of energy. Let $f_i(t)$ and $\ell_i(t)$ be the first and last position of agent i that are in range. These may or may not be defined depending on whether the agent travels through $\mathcal{C}_{R(t)}$ or not; Fig. 1 depicts an example of all possible cases about the relations between $p_i(t)$, $p_i(t+1)$, $f_i(t)$ and $\ell_i(t)$. The time that agent i spends in the charger's range is then equal to

$$T_i^{in}(t) = \begin{cases} \frac{\|f_i(t) - \ell_i(t)\|}{v_i(t)}, & \text{if } f_i(t) \neq \ell_i(t), v_i(t) \neq 0 \\ \tau, & \text{if } f_i(t) = \ell_i(t), v_i(t) = 0 \\ 0, & \text{otherwise,} \end{cases}$$

where $\|f_i(t) - \ell_i(t)\|$ denotes the Euclidean distance between points $f_i(t)$ and $\ell_i(t)$. We assume that agent i receives energy according to a simplified version of the well-known Friis transmission equation. In particular,

$$E_i^r(t) = \frac{\alpha \cdot R(t)^2 \cdot T_i^{in}(t)}{(\|p_{charger} - f_i(t)\| + \beta)^2}, \tag{1}$$

where α and β are environmental and technological constants. The energy of agent i at the beginning of round $t+1$ (accounting for both energy consumption and recharging), is equal to $E_i(t+1) = \min\{B, \max\{0, E_i(t) - E_i^c(t) + E_i^r(t)\}\}$. Observe that the amount of energy that the agent receives must respect its battery limit. Of course, the energy of the charger is also decreased accordingly.

3   The Need for Adaptiveness

Here, we aim to justify the need for algorithms that dynamically change the charging range over time by adapting to the behavior of the agents. The simplest algorithm that one could come up with, is to have the range *fixed* during the whole period of time; this is the typical algorithm that has been used in most of

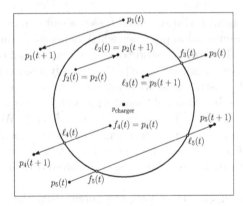

Fig. 1. An example of all possible cases regarding the relation between the line along which an agent may travel and $\mathcal{C}_{R(t)}$. Here, agent 1 does not get in range and, hence, $f_1(t)$ and $\ell_1(t)$ are undefined. Agent 2 starts and ends in range, agent 3 starts out of range but ends up inside, agent 4 starts inside but ends up out of range, and finally agent 5 travels through $\mathcal{C}_{R(t)}$.

the related literature so far. However, observe that there are essentially infinitely many different fixed values. Therefore, finding the one that works efficiently (with respect to the various objectives that we could be interested in) for *every* possible instance is improbable. In fact, in the following we will prove that this is actually impossible.

We begin by showing that for any fixed non-max range value there exists an instantiation of the agents' movements for which no recharging will take place.

Proposition 1. *For any range value $R < R_{\max}$, there exists a scenario for which fixing the range equal to R is equivalent to not using a charger at all.*

Proof. Consider the mobility scenario according to which no agent ever passes through the circle $\mathcal{C}_R$. Then, if the range is set to R for the whole period of time, no agent will ever get recharged. □

A scenario similar to the one in the proof of Proposition 1 exists even for the maximum possible range $R_{\max}$. However, in such a case there exists *no* algorithm that can do any better. Hence, we need to make the critical assumption that all agents will pass through the circle $\mathcal{C}_{R_{\max}}$ at least once. Next, we prove a stronger statement that holds true even when we consider the maximum range value.

Proposition 2. *There exists a scenario for which setting the charger's range equal to any fixed value R is not optimal.*

Proof. Consider the scenario according to which the agents get in range only when their energy levels are below a threshold. Assume that the agents have the following energy consumption characteristics. There are $n - 1$ agents with small energy consumption and a *single* greedy agent that consumes all of its available energy, at every round.

If the charger's range is fixed to any R during the whole time horizon, this single greedy agent can choose its in-range position so that it gets its battery fully recharged. As a result, the charger's energy can be quickly drained out (if the initial energy is small enough), before the other agents have a chance to get recharged. In contrast, consider the algorithm that adapts to the behavior of this greedy agent and, in each round, sets the range such that this agent gets a minimum amount of energy. For example, it can set the range equal to the distance between the agent and the charger so that, according to Eq. (1), it gives to the agent only a small amount of energy every time. This way, the charger conserves energy for the rest of the agents and the network's lifetime can be expanded.   □

4   Optimization Problems

In this section, we define two simplified offline optimization problems and prove their computational intractability. These two problems are closely related to the online one that we defined in the previous sections, and each of them focuses on a particular objective goal, the number of charges that are performed by the charger during a given time horizon, and the number of rounds during which the network is active, respectively. The hardness of these problems is only indicative of the hardness of the actual online multi-objective problem.

As input, we are given all information about the behavior of the agents during a time horizon T. The charger has initial energy C and its range can be chosen from a set of k distinct values $\{R_1, ..., R_k\}$ such that $0 \leq R_1 < ... < R_k$. All non-fully charged agents that are in the specified charging range receive appropriate energy according to the adopted charging model. For any $t \in [T]$, the objective of MNC (standing for MAXIMIZE NUMBER OF CHARGES) is to set the range $R(t)$ of the charger in order to maximize the total number of recharged agents until the charger is left out of energy. The objective of MNL (standing for MAXIMIZE NETWORK LIFETIME) is to set the range $R(t)$ in order to maximize the total rounds during which there exists at least one agent with non-zero (strictly positive) energy.

Theorem 1. *MNC is NP-hard, even for two range values.*

Proof. We use a reduction from the KNAPSACK PROBLEM (KP, for short) which is known to be NP-hard [8]. Its formal description is as follows.

KP: Consider a collection of q items $a_1, ..., a_q$ such that item a_i has value $v(a_i) \in \mathbb{R}_{\geq 0}$ and weight $w(a_i) \in \mathbb{R}_{\geq 0}$. We are given a knapsack of capacity $W \in \mathbb{R}_{\geq 0}$, and the goal is to select a set of items of total weight at most $W \in \mathbb{R}_{\geq 0}$ in order to maximize the total value of these items.

Given an instance of KP we will design an instance of MNC. First, without loss of generality, we assume that the values of the items as well as the weight W of the knapsack in the instance of KP are rescaled so that they are integer numbers (for example they are all multiplied by some large number). Second, there are no items with zero value (as such items can be discarded) and no items with zero weight (as such items are for free).

Now, our MNC instance is as follows:

- There are $n = \max_t v(a_t)$ agents with battery $B = \max_t \frac{w(a_t)}{v(a_t)}$.
- The initial energy of the charger is $C = W$ (the knapsack corresponds to the charger).
- There are $T = q$ rounds (every item corresponds to a round) and each of them lasts for a unit of time.
- The range of the charger can either be set to 0 or $R = \max_t \sqrt{\frac{w(a_t)}{v(a_t)}}$; essentially, the charger is either *inactive* or *active* (and its range is R).
- For each round t, the movement and energy consumption characteristics of the agents are as follows. At the beginning of the round, all agents are fully charged. There is a set A_t of exactly $v(a_t)$ agents at distance $d_t = R\sqrt{\frac{v(a_t)}{w(a_t)}}$ each of whom travels along the circle $\mathcal{C}_{d_t}$, and consumes energy equal to $\frac{w(a_t)}{v(a_t)} \leq B$ in case the charger is active, and 0 otherwise; such an energy consumption may be due to the communication of the agents with the charger itself. All other agents (if there are any) do not have any energy consumption during round t and move arbitrarily (but consistently to future positioning requirements).

Now, let us focus on an arbitrary round $t \in [q]$. If the charger is inactive during this round, then of course no agent gets recharged. However, according to the above specified energy consumption characteristics, all agents remain fully charged in such a case. On the other hand, if the charger is active during round t, then according to Eq. (1) with $\alpha = 1$ and $\beta = 0$, every agent in A_t receives energy equal to

$$\frac{R^2}{d_t^2} = \frac{R^2}{R^2 \frac{v(a_t)}{w(a_t)}} = \frac{w(a_t)}{v(a_t)},$$

which is exactly its energy consumption during this round. Therefore, the charger needs to spend $w(a_t)$ units of energy in total in order to fully recharge these $v(a_t)$ agents during round t. In other words, the number of charges corresponds to the total value of the selected items and the total needed energy corresponds to the total weight of these items. Consequently, any set of items with maximum total value satisfying the knapsack capacity corresponds to a set of rounds during which the charger is active with maximum number of charges satisfying the initial energy of the charger, and vice versa. The proof is complete.   □

Theorem 2. *MNL is NP-hard, even for one agent and two range values.*

Proof. We again use a reduction from KP (see the proof of Theorem 1 for its formal definition). Given an instance of KP, we define an instance of MNL:

- There is a single agent with battery $B = \max_i w(a_i)$.

- The initial energy of the charger is $C = W$ (the knapsack corresponds to the charger).
- Every round lasts for a unit of time.
- The charger can either be inactive with zero range or active with range $R = \max_i \sqrt{\frac{w(a_i)}{v(a_i)}}$.
- During the first round t_1, the agent is out of the range of the charger and consumes all of its battery. For each item $i \in [q]$, there is time horizon T_i consisting of $v(a_i)$ rounds. During the first of these rounds the agent is in range at fixed distance $d_i = R\sqrt{\frac{1}{w(a_i)}}$ (for example it travels along the circle $\mathcal{C}_{d_i}$ or is static), while for the remaining $v(a_i) - 1$ rounds, the agent moves out of range and has a total energy consumption of $w(a_i) \leq B$ so that during all these rounds it has non-zero energy. These q time horizons are continuous, given a permutation of the items: $T = t_1 \cup_i T_i$.

If the charger is inactive during the first round of any time horizon T_i, then the agent does not get and does not have any energy during T_i (a total of $v(a_i)$ rounds). On the other hand, if the charger is active during the first round of T_i, since the agent is at distance d_i from the charger, and using Eq. (1) with $\alpha = 1$ and $\beta = 0$, the energy that the agent receives by the charger is equal to

$$\frac{R^2}{d_i^2} = \frac{R^2}{R^2 \frac{1}{w(a_i)}} = w(a_i),$$

which is exactly the energy that it consumes during T_i. Therefore, if the charger is active during the first round of T_i, the agent is active for $v(a_i)$ rounds and the charger spends $w(a_i)$ units of energy. As a result, the number of rounds that the agent is active is equal to the total value of the selected items. Hence, any set of items with maximum total value satisfying the knapsack capacity corresponds to a set of time horizons with maximum number of rounds (during which the agent is active) satisfying the energy capacity of the charger, and vice versa. The proof is complete.   □

5   Adaptive Algorithms and Experimental Evaluation

We propose three adaptive algorithms and experimentally compare them to each other. The algorithms are presented in an increasing order in terms of the knowledge they require in order to decide the charging range during any round t.

Least Distant Agent or Maximum Range (LdMax). The LdMax algorithm uses a parameter $q \in [0, 1]$ and works as follows. At the beginning of each round t, it sets $R(t) := \max\{R_{\min}, \min_{i:p_i(t) \in \mathcal{C}_{R_{\max}}} \|p_{\text{charger}} - p_i(t)\|\}$ with probability q, and $R(t) := R_{\max}$ otherwise (with probability $1 - q$).

Maintain Working Agents (MWA). The MWA algorithm uses a parameter $\mu \in [n]$ and, during each round t, sets the range $R(t)$ in an attempt to guarantee that there are at least μ working agents in the network (i.e. agents that either have positive energy at the beginning of the round or get recharged during it). To find the appropriate range $R(t)$ it works as follows. First, it counts the number $k_1(t)$ of agents that are in $\mathcal{C}_{R_{\max}}$ and have positive energy at the beginning of the round. If $k_1(t) \geq \mu$, then it sets $R(t) := R_{\min}$ since the requirement is already satisfied. Otherwise, it counts the number $k_2(t)$ of agents that have zero energy at the beginning of the round and $p_i(t) \in \mathcal{C}_{R_{\max}}$ or $p_i(t+1) \in \mathcal{C}_{R_{\max}}$. If $k_1(t) + k_2(t) < \mu$, then it sets $R(t) := R_{\max}$ since the requirement cannot be satisfied. Otherwise, it searches for the smallest R^* such that the circle $\mathcal{C}_{R^*}$ covers at least $\mu - k_1(t)$ agents, and sets $R(t) := R^*$.

Maximize Charges over Energy Ratio (MCER). Let $\mathcal{R}$ be a set of discrete range values in $[R_{\min}, R_{\max}]$. Let $\nu_j(t)$ be the number of agents that get recharged when the range is equal to $R_j \in \mathcal{R}$ during round t, and let $\varepsilon_j(t)$ be the total given energy in this case. The MCER algorithm uses a parameter $\lambda \geq 1$ and sets $R(t) := \arg\max_{R_j \in \mathcal{R}} \frac{\nu_j(t)^\lambda}{\varepsilon_j(t)}$. This algorithm attempts to strike a balance between the number of charges and the energy that it has to give in order to perform these charges. However, observe that it needs to perform many heavy computations as, in order to choose the best range, it has to simulate the whole recharging process multiple times.

Simulation Setup. We now experimentally compare these adaptive algorithms. We consider a simulation setup[2] with $n = 100$ agents that move around in a 25×25 network area $\mathcal{A}$. The charger is positioned at the center of $\mathcal{A}$, has initial energy $C = 10^5$, and its range can take values in $[1, 5]$. Each agent has battery $B = 1000$, maximum velocity $v_{\max} = 3$, and its speed mode is redefined with probability $1/4$ in each round. Also, the agents are randomly partitioned into 4 groups, namely, (S_1, S_2, S_3, S_4) of expected sizes $\left(\frac{n}{2}, \frac{n}{4}, \frac{n}{8}, \frac{n}{8}\right)$. Then, agent i consumes energy following a poisson distribution with randomly chosen expected value γ_i such that $\gamma_i \in [0, 10 \cdot 2^{j-1}]$ if $i \in S_j$. We remark that the expected values are chosen non-uniformly from the corresponding intervals so that there is heterogeneous energy consumption among the agents.

For the agent mobility behavior we consider three randomized scenarios: (S1) All agents randomly move around in $\mathcal{A}$; (S2) Choose $R \in [R_{\min}, \frac{1}{2}R_{\max}]$ uniformly at random so that no agent is allowed to enter circle $\mathcal{C}_R$; (S3) Choose $\delta \in \left[\lfloor \frac{n}{10} \rfloor\right]$, $R_\ell \in [R_{\min}, \frac{1}{4}(R_{\min} + R_{\max}))$ and $R_h \in \left[\frac{1}{4}(R_{\min} + R_{\max}), R_{\max}\right]$ uniformly at random so that δ agents live in the ring $\mathcal{C}_{R_h} \setminus \mathcal{C}_{R_\ell}$, while the remaining $(n - \delta)$ agents randomly move around in $\mathcal{A}$. We create a probability distribution over these three mobility scenarios by repeating our simulation

---

[2] We remark that the setup that we present here is only indicative. Actually, we have experimented with many different setups that differ on the number of agents and their battery capacity, the network size, and the initial energy of the charger. For all such setups, the relative performance of our algorithms is similar.

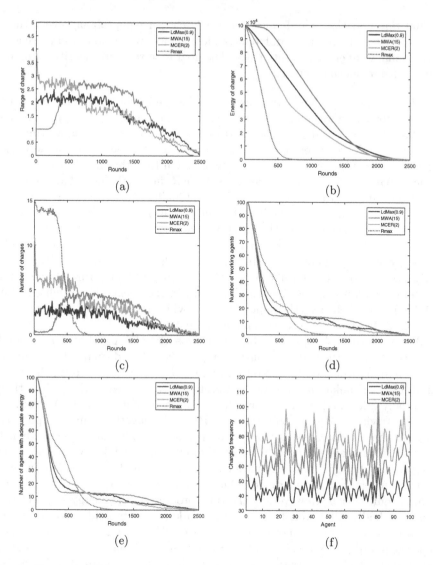

Fig. 2. Comparison between the three adaptive algorithms LdMax(0.9), MWA(15) and MCER(2) as well as the fixed $R_{\max}$ value algorithm. Figure (a) depicts the evolution of the charging range over time. Figure (b) depicts the decrease of the charger's energy over time. Figure (c) depicts the number of charges that were performed over time. Figure (d) depicts the number of working agents over time. Figure (e) depicts the number of agents with adequate energy over time. Figure (f) depicts the charging frequency of the agents (the number of times they were recharged). The simulated data presented here are averages over 100 executions. The performance of each algorithm in the different executions is robust and sharply concentrated around the depicted average value.

for 100 times so that a different scenario is chosen equiprobably every time. Observe that there are many different random choices to be made and these give birth to many different instantiations. The goal is to test our algorithms under a highly heterogeneous setting.

Results and Interpretation. After extensive fine-tuning of the parameters used by our adaptive algorithms, we have concluded that setting $q = 0.9$, $\mu = 15$ and $\lambda = 2$ are the best values for the particular simulation setup that we consider here. In general, we expect q to depend heavily on the density of the network; it should be smaller for more sparse networks. On the other hand, $\lambda = 2$ seems to nicely balance the ratio considered by MCER due to the fact that the given energy is of square order according to Eq. (1). Finally, parameter μ can be picked by the designer to maintain a sufficient number of agents, depending on the needs of the network, the energy of the charger, etc.

Due to its definition, MWA guarantees for a long period of time a stable number of working agents (as well as agents with adequate energy). However, MCER seems to outperform the other two algorithms in terms of the total number of charges and the charging frequency of the agents. Essentially, MWA and MCER work in exactly opposite ways, while LdMax lies somewhere in-between of these two, due to its randomized nature.

To interpret this data, we will briefly analyze how MWA and MCER respond to the behavior of the agents by inspecting Fig. 2a which displays the evolution of the charging range over time depending on the algorithm. During the early rounds of the simulation, most of the agents are considered working since they are initially fully charged. Therefore, the requirement of maintaining 15 working agents is trivially satisfied and MWA starts by having the minimum possible range, so that it stores energy for future use (see Fig. 2b). In contrast, MCER chooses a higher range in order to perform more charges while giving away little energy; since the agents already have energy, they request only a small amount of energy when they get in range, which means that the cost (in energy) per charge is quite small. However, as the time progresses, the energy levels of the agents gradually get lower, there are less working agents, and when an agent gets in range requests for more energy. As a result, MWA is forced to increase the range in order to keep satisfying the requirement of maintaining 15 working agents, while MCER decreases its range as the cost per charge has increased substantially.

6   Conclusion and Possible Extensions

In this paper, we studied the problem of dynamically selecting the appropriate charging range of a single static charger to prolong the lifetime of a network of mobile agents. We proved the hardness of the problem, and presented three interesting heuristics that perform fairly well in the simulation setups that we considered. Of course, there are multiple interesting future directions.

Can we design better adaptive algorithms that perform well under any possible scenario regarding the agents' characteristics? An interesting way to try to tackle this, would be to consider a machine learning like approach. In particular, given statistical information (a prior probability distribution) about the behavior of the agents, is it possible to learn the "correct" sequence of values for the charging range in order to prolong the network lifetime as much as possible, while maintaining a fair amount of working agents? We remark that our algorithms do not exploit such training information, and function based only on the online behavior of the agents. Another possible direction could be to consider the natural generalization of using multiple chargers that can move around in the network, and even be able to charge each other. This, couples (in a nontrivial way) our work together with that of Angelopoulos et al. [1], and definitely deserves investigation.

References

1. Angelopoulos, C.M., Buwaya, J., Evangelatos, O., Rolim, J.D.P.: Traversal strategies for wireless power transfer in mobile ad-hoc networks. In: Proceedings of the 18th ACM International Conference on Modeling, Analysis and Simulation of Wireless and Mobile Systems (MSWiM), pp. 31–40 (2015)
2. Bettstetter, C., Resta, G., Santi, P.: The node distribution of the random waypoint mobility model for wireless ad hoc networks. IEEE Trans. Mob. Comput. **2**(3), 257–269 (2003)
3. Bi, Z., Kan, T., Mi, C.C., Zhang, Y., Zhao, Z., Keoleian, G.A.: A review of wireless power transfer for electric vehicles: prospects to enhance sustainable mobility. Appl. Energy **179**, 413–425 (2016)
4. Camp, T., Boleng, J., Davies, V.: A survey of mobility models for ad hoc network research. Wirel. Commun. Mob. Comput. **2**(5), 483–502 (2002)
5. Dai, H., Chen, G., Wang, C., Wang, S., Wu, X., Wu, F.: Quality of energy provisioning for wireless power transfer. IEEE Trans. Parallel Distrib. Syst. **26**(2), 527–537 (2015)
6. Dai, H., et al.: Safe charging for wireless power transfer. IEEE/ACM Trans. Netw. **25**(6), 3531–3544 (2017)
7. Dai, H., et al.: SCAPE: safe charging with adjustable power. IEEE/ACM Trans. Netw. **26**(1), 520–533 (2018)
8. Garey, M.R., Johnson, D.S.: Computers and Intractability: A Guide to the Theory of NP-Completeness. W. H. Freeman, San Francisco (1979)
9. He, S., Chen, J., Jiang, F., Yau, D.K.Y., Xing, G., Sun, Y.: Energy provisioning in wireless rechargeable sensor networks. IEEE Trans. Mob. Comput. **12**(10), 1931–1942 (2013)
10. Hrabčák, D., Matis, M., Dobos, L., Papaj, J.: Students social based mobility model for MANET-DTN networks. Mob. Inf. Syst. **2017**, 2714595:1–2714595:13 (2017)
11. Kurs, A., Karalis, A., Moffatt, R., Joannopoulos, J.D., Fisher, P., Soljačić, M.: Wireless power transfer via strongly coupled magnetic resonances. Science **317**(5834), 83–86 (2007)
12. Li, S., Mi, C.: Wireless power transfer for electric vehicle applications. IEEE J. Emerg. Sel. Top. Power Electron. **3**, 4–17 (2015)
13. Lin, M., et al.: An ultrafast rechargeable aluminium-ion battery. Nature **520**, 324 (2015)

14. Madhja, A., Nikoletseas, S.E., Raptis, T.P.: Hierarchical, collaborative wireless energy transfer in sensor networks with multiple mobile chargers. Comput. Netw. **97**, 98–112 (2016)

15. Madhja, A., Nikoletseas, S.E., Raptopoulos, C., Tsolovos, D.: Energy aware network formation in peer-to-peer wireless power transfer. In: Proceedings of the 19th ACM International Conference on Modeling, Analysis and Simulation of Wireless and Mobile Systems (MSWiM), pp. 43–50 (2016)

16. Musolesi, M., Hailes, S., Mascolo, C.: An ad hoc mobility model founded on social network theory. In: Proceedings of the 7th International Symposium on Modeling Analysis and Simulation of Wireless and Mobile Systems (MSWiM), pp. 20–24 (2004)

17. Nikoletseas, S.E., Raptis, T.P., Raptopoulos, C.: Radiation-constrained algorithms for wireless energy transfer in ad hoc networks. Comput. Netw. **124**, 1–10 (2017)

18. Nikoletseas, S.E., Raptis, T.P., Raptopoulos, C.: Wireless charging for weighted energy balance in populations of mobile peers. Ad Hoc Netw. **60**, 1–10 (2017)

19. Nikoletseas, S.E., Raptis, T.P., Souroulagkas, A., Tsolovos, D.: Wireless power transfer protocols in sensor networks: experiments and simulations. J. Sens. Actuator Netw. **6**(2), 4 (2017)

20. Nikoletseas, S.E., Yang, Y., Georgiadis, A. (eds.): Wireless Power Transfer Algorithms, Technologies and Applications in Ad Hoc Communication Networks. Springer, Heidelberg (2016). https://doi.org/10.1007/978-3-319-46810-5

21. Vastardis, N., Yang, K.: An enhanced community-based mobility model for distributed mobile social networks. J. Ambient Intell. Humaniz. Comput. **5**(1), 65–75 (2014)

22. Zhang, S., Wu, J., Lu, S.: Collaborative mobile charging. IEEE Trans. Comput. **64**(3), 654–667 (2015)

23. Zhao, M., Li, J., Yang, Y.: Joint mobile energy replenishment with wireless power transfer and mobile data gathering in wireless rechargeable sensor networks. In: Nikoletseas, S., Yang, Y., Georgiadis, A. (eds.) Wireless Power Transfer Algorithms, Technologies and Applications in Ad Hoc Communication Networks, pp. 667–700. Springer, Cham (2016). https://doi.org/10.1007/978-3-319-46810-5_25

Distributed Leader Election
and Computation of Local Identifiers
for Programmable Matter

Nicolas Gastineau[⊠], Wahabou Abdou, Nader Mbarek, and Olivier Togni

LE2I, Université Bourgogne Franche-Comté, Dijon, France
nicolas.gastineau@u-bourgogne.fr

Abstract. The context of this paper is programmable matter, which consists of a set of computational elements, called *particles*, in an infinite graph. The considered infinite graphs are the square, triangular and king grids. Each particle occupies one vertex, can communicate with the adjacent particles, has the same clockwise direction and knows the local positions of neighborhood particles. Under these assumptions, we describe a new leader election algorithm affecting a variable to the particles, called the k-local identifier, in such a way that particles at close distance have each a different k-local identifier. For all the presented algorithms, the particles only need a $O(1)$-memory space.

Keywords: Programmable matter · Leader election · Identifier · Graph coloring

1 Introduction

Programmable matter can be seen as modular robots (called modules or particles) able to fix to adjacent modules and send (receive) messages to (from) other modules fixed to the entity. Thus, the different modules form a geometric shape which is a network. Usually, a module can fix to another module using a finite number of ports (see Fig. 1 for an example of spherical modules). Also, the modules know the ports that are in contact with other modules and have a knowledge about the geographic position of their ports. Moreover, the ports are supposed to be homogeneously distributed along the surface of each module. Such assumptions imply that the way how the modules are on a plane can be modeled by a grid. In this paper, we only consider modules on a plane surface, i.e. two dimensional grids. In this context, the geometric amoebot model [6–11] aims to model the properties of a network for programmable matter.

Distributed algorithms aim to give a theoretical algorithmic framework in order to model the execution of an algorithm that runs on a network of computational elements that can cooperate in order to solve network problems. In distributed algorithm frameworks, it is often supposed that the different elements of the network do not have a unique identity, i.e., the network is anonymous. In

© Springer Nature Switzerland AG 2019
S. Gilbert et al. (Eds.): ALGOSENSORS 2018, LNCS 11410, pp. 159–179, 2019.
https://doi.org/10.1007/978-3-030-14094-6_11

anonymous networks, a natural question is how to perform a leader election, i.e., how to determine a singular element in an anonymous network. It is well known that for some network structures, the ring for example, there is no deterministic leader election algorithm [1].

In 1999, Mazurkiewicz [19] has presented a deterministic general algorithm to determine a leader (in the case it is possible to do so). In the situation where the elements have access to a random source, then it is also proven that no algorithm can correctly determine a leader in a ring with any probability $\alpha > 0$ [15]. Due to the assumption we make about the ports of the particles in the context of programmable matter (a particle knows the ports which are in contacts with other particles and knows the geographic position of its ports), the leader election problem becomes different than in the classical system. In particular, in the field of programmable matter, there exists a probabilistic algorithm that determine a leader (and in particular for a ring) with probability 1 [5].

Several projects aim to build programmable matter prototypes. One of such projects [20,23], financed by the french National Agency for Research, aims to build cuboctahedral particles able to deform them-selves in order to move. This project can be split in two phases, one consists in manufacturing the hardware of prototype matters, the second consists in proposing algorithms for programmable matter. The final goal of this project is to sculpt a shape-memory polymer sheet with programmable matter. In the continuity of the algorithm phase of this project [20], we propose algorithms for the self-configuration, i.e., in order to create identifiers and spanning trees.

In the context of programmable matter [3,4,14,18,23,24], it is supposed that a network can contain several millions of modules and that each module has possibly a nano-centimeter size. These two facts lead us to believe that even a $O(\log(n))$-space memory for each module, n being the number of modules, is not technically possible. Also, because of the large number of modules, it can be very challenging and time consuming to implement a unique identity to the modules when they are created. In this context, we suppose that the modules can not store a unique identity, i.e., that the network is anonymous. In this paper we propose deterministic $O(1)$-space memory algorithms to determine a leader in the network and to create k-local identifiers of the particles. A k-local identifier is a variable affected to each module of the network which is different for every two modules at distance at most k. Note that leader election [5,13] plays a significant role in numerous problems of programmable matter.

Our contribution is the following: we introduce a leader election algorithm based on local computations and simple to implement. This algorithm works when the structure the particles form has no hole (see Sect. 3). Also, since the algorithm can be described as a sequence of local computations, its limits (message complexity, required memory-space, etc.) are easy to analyze. We present a distributed algorithm to construct a spanning tree in the context of programmable matter and, also, a distributed algorithm to re-organize the port numbers of the particles. Finally, we present an algorithm to assign a k-local identifier to each particle. In order to compute k-local identifiers, we suppose

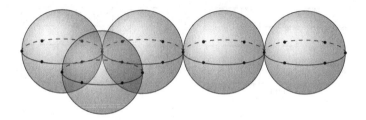

Fig. 1. Five spherical particles forming a simple structure (circle: port of the particles).

that we have done a leader election before. The k-local identifiers are determined using graph theoretical results about the coloring of the kth power of the grids. An advantage of the given k-local identifiers is that they are really simple to update in case the particles move and, consequently, the structure that the particles form changes.

This paper is organized as follows: in Sect. 2, we present our algorithmic framework in the context of distributed algorithms for programmable network. In the third section, we present our leader election algorithm. Finally, in Sect. 4, we present our algorithm to assign k-local identifiers to the particles (using the colorings from Appendix D).

2   Notation, Definitions and Our Programmable Matter Algorithmic Framework

The geometric amoebot model [6–11] aims to model the computations that can occur in the context of programmable matter. In this paper, we use an algorithmic framework inspired by the geometric amoebot model. We assume that any structure the different particles can form is a subgraph of an infinite graph G. In this graph, $V(G)$ represents all possible positions the particles can occupy and $E(G)$ represents possible connections between particles. The set $E(G)$ also represents the possible movements from a position to another position (for a particle). We suppose that two particles can bond each other, i.e., can communicate only in the case they are on adjacent positions. The two following paragraphs are dedicated to the notation and definitions we use for graphs.

For a graph G, we denote by $V(G)$ the *vertex set* of G and by $E(G) \subseteq V(G) \times V(G)$ the *edge set* of G. We denote by $d_G(u, v)$, the usual distance between two vertices u and v in G. If we consider the distance in a subgraph H of G, the distance will be denoted by $d_H(u, v)$. The *diameter* of G, denoted by $\text{diam}(G)$, is $\max(\{d_G(u,v)|\ u, v \in V(G)\})$. The set $N_G(u) = \{v \in V(G)|\ uv \in E(G)\}$ is the set of *neighbors* of u. By $\Delta(G)$, we denote the *maximum degree* in G, i.e., the maximum cardinality of $N_G(u)$, for $u \in V(G)$. Finally, we denote by $G[S]$, for $S \subseteq V(G)$, the *subgaph* induced by the vertices from S and by $G - S$ the subgraph of G induced by the vertices from $V(G) \setminus S$.

In the remaining part of this paper, the graphs considered will be the infinite *square*, *triangular* and *king* grids. We denote by $\mathcal{S}$ the square grid, by $\mathcal{T}$ the

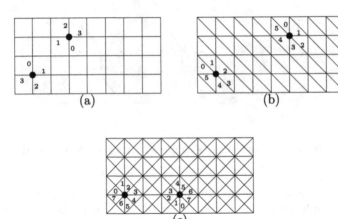

Fig. 2. Subgraphs of the square (a), triangular (b) and king (c) grids, with the port numbers of two particles.

triangular grid and by $\mathcal{K}$ the king grid. A subgraph of each of these three infinite graphs is represented in Fig. 2. Moreover, we suppose that these three grids are represented on a plane as in Fig. 2. For these grids, the considered vertex set is $\{(i,j)|\ i,j \in \mathbb{Z}\}$ and the edge sets are the following:

- $E(\mathcal{S}) = \{(i,j)(i \pm 1, j)|\ i, j \in \mathbb{Z}\} \cup \{(i,j)(i, j \pm 1)|\ i, j \in \mathbb{Z}\};$
- $E(\mathcal{T}) = E(\mathcal{S}) \cup \{(i,j)(i + 1, j - 1)|\ i, j \in \mathbb{Z}\} \cup \{(i,j)(i - 1, j + 1)|\ i, j \in \mathbb{Z}\};$
- $E(\mathcal{K}) = E(\mathcal{T}) \cup \{(i,j)(i + 1, j + 1)|\ i, j \in \mathbb{Z}\} \cup \{(i,j)(i - 1, j - 1)|\ i, j \in \mathbb{Z}\}.$

We also remind the distance between two vertices (i,j) and (i',j') in the three different grids:

- $d_{\mathcal{S}}((i,j),(i',j')) = |i - i'| + |j - j'|;$
- $d_{\mathcal{T}}((i,j),(i',j')) = \begin{cases} \max(|i - i'|, |j - j'|), & \text{if } (i \geq i' \wedge j \leq j') \vee (i \leq i' \wedge j \geq j'); \\ |i - i'| + |j - j'|, & \text{otherwise}; \end{cases}$
- $d_{\mathcal{K}}((i,j),(i',j')) = \max(|i - i'|, |j - j'|).$

Note that there is a way to draw the triangular grid in which each triangle is equilateral. However, we prefer to draw it as a subgraph of the king grid (see Fig. 2) in order to have illustrations for which the vertex set $\{(i,j)|\ i,j \in \mathbb{Z}\}$ corresponds to the position of the vertices in the plane. In both representation, the notion of distance coincide but is easier to observe in our chosen representation. However, note that the representation of the triangular grid in which each triangle is equilateral corresponds to the optimal way to pack unit disks in the plane (the position of the vertices in this representation corresponds to the center of the unit disk and an edge represents a contact between two disks).

We also denote by $i \pmod{p}$ or $i_{\pmod{p}}$, depending on the context, the integer j such that $j \equiv i \pmod{p}$ and $0 \leq j < p$. The remaining part of this subsection is dedicated to our programmable matter algorithmic framework.

We give the following properties about the particles and vertices of the graph:

- each particle occupies a single vertex and each vertex is occupied by at most one particle;
- the subgraph induced by the occupied vertices is supposed to be connected.

The subgraph induced by the occupied vertices of $V(G)$ is called the *particle graph* and is denoted by P. The vertex occupied by a particle p is denoted by $s(p)$. For a particle p, $N_G(p) = \{u \in V(G)| \ u \in N_G(s(p))\}$. The *ports* of a particle are the endpoints of communication. Each particle has $\Delta(G)$ ports in a regular grid G ($\Delta(G) = 4$ for $G = \mathcal{S}$, $\Delta(G) = 6$ for $G = \mathcal{T}$ and $\Delta(G) = 8$ for $G = \mathcal{K}$). The ports of a particle occupying a vertex u are represented by the edges incident with u. An edge between two vertices represents a possible communication between two particles p_1 and p_2 occupying these two vertices using each one a different port. A particle has the following properties:

- each particle is anonymous, i.e., it does not have an identifier;
- each particle has a collection of ports, each labeled by a different integer from $\{0, \ldots, \Delta(G) - 1\}$;
- the port numbers are given as a function of the position of the edges on a plane representation of the grids (see Fig. 2);
- each particle knows the labels of the ports that can communicate with particles from the neighborhood;
- each particle knows the state of the neighbors.

In our algorithmic framework, we suppose that the particles have their ports labeled following the same clockwise order. Thus, consecutive port numbers correspond to consecutive edges around a vertex (as in the representation on the plane from Fig. 2). Note that the particles do not have the same notion of orientation, i.e., there is possibly not a unique label for ports that correspond to edges going in the same cardinal direction. In the presented algorithms, the state of a particle will contain a variable corresponding to the status of the particle in the leader election algorithm and the information regarding its parents and childs for a constructed spanning tree.

The proposed algorithms in our algorithmic framework are results of successive local computations [2,21]. In particular, the first presented leader election algorithm from Sect. 3 can be described by a graph relabeling system [2] which is a local computation system. In this paper, the correct execution of the different algorithms is only guaranteed if the algorithms are ran in the order depicted in Fig. 3.

We suppose the following:

- each particle contains the same program and begins in the same state;
- the computation process is represented by successive local computations;
- no local computation occurs simultaneously on two particles at distance at most 2;
- during a local computation, a particle can perform a bounded number of computations and can send messages to its neighbors;

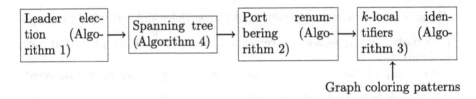

Fig. 3. An illustration of the algorithm dependency (arrow between algorithms/results: dependency of one algorithm to another algorithm/result).

- a *round* is a sequence of successive local computations for which each particle does at least one local computation;
- an algorithm finishes in k rounds if after any k successive rounds the algorithm is finished.

Note that the concept of rounds is used to bound the running time of the algorithms. In our algorithm framework we suppose that no two particles at distance at most 2 perform computations simultaneously in order to simplify the presentation of our results. However, this supposition can be removed by implementing, for example, a probabilistic leader election algorithm on the vertices at distance at most 2 of one of the two vertices, i.e., by computing a random value on the vertices at distance 2 and doing the local computation following the increasing order of the values. In order to compute the running time of an algorithm in case of a specific programmable matter prototype, the complexity of the algorithm should be computed using the required number of rounds and the required running time in order to avoid that two particles at distance at most 2 perform computations simultaneously.

3   Leader Election

In this section we present a new leader election algorithm. This algorithm is very easy to implement but requires that the particle graph has a specific structure. In this algorithm, the required memory space is constant, the messages have constant size, the required computation power of the particle has been optimized and the required number of rounds is less than $2n$ (n being the number of particles).

A *hole* in a subgraph G' of a graph G among the three grids is a subgraph H of G satisfying three properties:

(i) $V(H)$ is finite, H is connected and $|V(H)| \geq 1$;
(ii) $V(H) \cap V(G') = \emptyset$;
(iii) every vertex $u \in V(H)$ satisfies $N_G(u) \subseteq V(H) \cup V(G')$.

Less formally, a subgraph G' of one of the three grids contains a hole if there is a finite space only containing vertices from $V(G) \setminus V(G')$ which are surrounded by vertices of G'. A hole containing three vertices is illustrated in the left part of Fig. 4. We call G' *hole-free*, when G' has no holes.

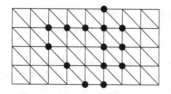 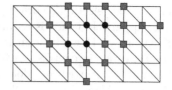

Fig. 4. A hole in P (on the left) and the border of P in the case P is hole-free (on the right; square: particle on the border of P).

If the particle graph P on G is hole-free, then every particle p which satisfies $|N_G(p) \cap V(P))| < \Delta(G)$ is at the geographical border of the shape of P. Moreover, we call the set of particles p which satisfy $|N_G(p) \cap V(P))| < \Delta(G)$ and such that the vertices $N_G(p) - V(P)$ are not all in a hole of P, the *border* of P. The right part of Fig. 4 illustrates the border of P.

In addition, for a particle p occupying a vertex (i, j) of the square grid, the four vertices $(i+1, j+1)$, $(i-1, j+1)$, $(i+1, j-1)$ and $(i-1, j-1)$ are the *corners* of p and the set of corners is denoted by $C(p)$. The *extended neighborhood* of a particle p, denoted by $M_G(p)$, is the set $N_G(p)$ if G is the triangular grid or king grid or the set $N_G(p) \cup C(p)$ if G is the square grid. Note that we define the extended neighborhood differently for the square grid in order to be able to present a generic algorithm (Algorithm 1) that works for all the three grids.

We give the following definition of S-contractible particle (see Fig. 5) that will be used in our leader election algorithm.

Definition 1. *Let G be an infinite grid among $\mathcal{S}$, $\mathcal{T}$ and $\mathcal{K}$ and let $S \subseteq V(P)$, for P the particle graph on G. A particle p is said to be S-contractible if it satisfies the following properties:*

(I) $G[M_G(p) \cap S]$ is connected;
(II) $|N_G(p) \cap S| < \Delta(G)$, i.e., there exists a neighbor of p in G which is not occupied by a particle from S.

A particle p is an *articulation* of a connected subgraph G' of one of the three grids if $G' - \{s(p)\}$ is not connected. Derakhshandeh et al. [8] proposed

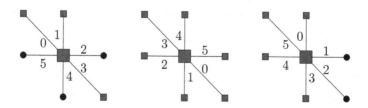

Fig. 5. Two non S-contractible particles (at the center of the left and the middle drawing) and an S-contractible particle (at the center of the right drawing) in the triangular grid (square: particle in S; circle: particle not in S).

a randomized leader election algorithm in the geometric amoebot model in the case there is no particle which is an articulation. Our proposed leader election algorithm (Algorithm 1) works even if $V(P)$ contains a particle p which is an articulation. However, in contrast with the leader election algorithm from Derakhshandeh et al. [8], Algorithm 1 does not work if P has holes. In the remaining part of this paper, Algorithm 1 is called the S-contraction algorithm.

Recently, Daymude et al. [5] have improved the algorithm from Derakhshandeh et al. [8] in order that it works when $V(P)$ contains an articulation. However, it remains challenging to implement it.

Also, very recently, Di Luna et al. [13] have introduced a leader election algorithm called consumption algorithm. The consumption and the S-contraction algorithms both consist in successively removing the candidacy of the particles on the border of P. However, one can easily notice that, in our algorithm, we possibly remove the candidacy of particles having four or five neighbors (which is not considered in the consumption algorithm). Also, the consumption algorithm does not work on square and king grids and the considered theoretical frameworks for the two algorithms are different.

In the S-contraction algorithm (Algorithm 1), the particles can be in three different states: **C** (candidate), **N** (not elected) and **L** (leader). We suppose that every particle begins in the state **C**.

Let S be the particle in state **C**. Algorithm 1 consists in removing from S the particles which are both on the border of $G[S]$ and not articulations of $G[S]$. An example of the execution of Algorithm 1 is illustrated by Fig. 6. Note that, depending on the order in which the local computations occur, the result of the execution of the algorithm could be different. For example, between the configuration of Fig. 6c and that of Fig. 6d, we suppose that the local computations occur in this order: first a local computation occurs for the bottom left particle, second it occurs for the upper left particle, third it occurs for the upper right particle and fourth it occurs for the last particle (we only consider the particles which are in state **C**).

Algorithm 1. The S-contraction algorithm for a particle p and S the set of particles in state **C**.

Case 1: State **C**.
if the particle is S-contractible **then**
    if the particle has no neighbor in S **then**
        set the state to **L**.
    else
        set the state to **N**.
    end if
else
    stay in state **C**.
end if
Case 2: States **L** or **N**.
Perform no further actions.

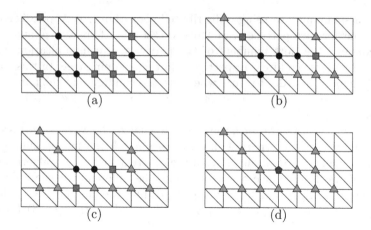

Fig. 6. An example of the execution of S-contraction algorithm after one round (a), two rounds (b), three rounds (c) and after four rounds (d; circle: non S-contractible particle; square: S-contractible particle; triangle: particle in state **N**; pentagon: particle in state **L**; S being the set of particles in state **C**).

Theorem 1. *Let S be the set of particles in state C and P be the particle graph on G. If P is hole-free, then at the end of the execution of the S-contraction algorithm, there will be exactly one particle in the state L.*

In Appendix A, the proof of Theorem 1 is given. Also, a bound on the complexity of the S-contraction algorithm is given. In Appendix C, it is explained how to combine the S-contraction algorithm with a general leader election algorithm.

4   Assigning k-Local Identifiers to Particles

In this section, we combine the results from Sect. 3 and Appendix D in order to correctly compute a k-local identifier. In a first subsection, we describe a way to create a spanning tree of particles and a way to change the ports numbering of the different particles. In a second subsection, we describe how to compute k-local identifiers based on the coloring functions from Appendix D.

We suppose that Algorithms 2 and 3 are preceded by a leader election algorithm (which could be Algorithm 1). Then it follows that there is a single particle in a specific state (the leader) and all the remaining particles are in the same state (non elected).

4.1   Re-organizing the Particles

By $N_G^+(u)$ we denote the set of port numbers which can communicate with particles occupying vertices from $N_G(u)$. When there is a leader, we can easily compute a spanning tree using a distributed algorithm (see Appendix B). Now suppose that for each particle p, we have two set of ports parent(p) and child(p) which contains the port numbers of the particles in communication with its parent and with its children, respectively, in the spanning tree. In this way, the required memory in order to store where are the children and the parent of the particle in a spanning tree is constant (since the maximum degree is bounded in the considered grids).

In our proposed Algorithm 2, the goal is to change the way the port are numbered in order that every particle has its ports numbered by the same number going in the same cardinal direction in the different grids. This algorithm does not work if we do not have a leader among the different particles. The function r_G used in Algorithm 2 is defined, depending the choice of G, as follows: $r_S(i) = (i+2) \pmod 4$, $r_{\mathcal{T}}(i) = (i+3) \pmod 6$ and $r_{\mathcal{X}}(i) = (i+3) \pmod 6$.

Algorithm 2. The port renumbering algorithm for a particle p.

Case 1: State **L**.

for each port a from child(p) send a message m_a, containing a, through port a.

Case 2: State **N**.

if   the particle receives the message m_b, containing b, through the port a **then**

change the port number a to $r_G(b)$ and changes the port numbers of the other ports following the clockwise order;

update both parent(p) and child(p).

end if

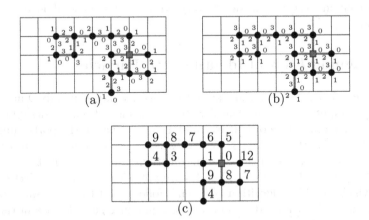

Fig. 7. One spanning tree of particles, a possible numbering of the ports of the particles before (a) and after the execution of Algorithm 2 (b) and the 4-identifier obtained by executing Algorithm 3 (c) in the square grid (square: leader; thick line: edge of the spanning tree; small number: port number of a particle; big number: 4-identifier of a particle).

The idea behind Algorithm 2 is to reproduce, in each particle, the way the ports are numbered in the leader particle. To achieve this goal, each particle p receives a message from its parent containing the port number of the parent connected to p and p renumbers its own ports in order that its port numbers are coherent with the sent number. Figures 7a and b illustrate the port numbers of particles before and after the execution of Algorithm 2.

4.2   The k-Local Identifiers

Now, we aim to give to each particle a variable id, called its k-local identifier, such that every two particles p_1 and p_2 with the same identifier satisfy $d_G(s(p_1), s(p_2)) > k$. If we suppose that the particles have not a memory of at least $\log_2(n)$ bits, for $n = |P|$, then it is not possible to record a unique variable for each particle. However, it is possible to have a k-local identifier in the three considered grids only using at most $\log_2((k+1)^2)$ bits where k is a parameter given by the user. Our proposed Algorithm 3 presents an optimal way (in term of memory) to compute k-local identifiers. We suppose that the port renumbering algorithm (Algorithm 2) has been done before executing Algorithm 3.

Algorithm 3. The k-local identifier algorithm for a particle p.

Case 1: State L.
set $i = 0$, $j = 0$, $id = 0$;
send i and j through each port from child(p).
Case 2: State N.
if the particle receives the integers i' and j' through the port a **then**
    set $i = I_G^k(i', a)$, $j = I_G^k(j', a)$, $id = f_G^k(i, j)$;
    send i and j through each port from child(p).
end if

Algorithm 3 consists in assigning a variable which corresponds to a color in a coloring of the kth power on the grid. More precisely, the function f_G^k consists in assigning a color depending the Cartesian coordinate of the vertices. Since the colors are given following a pattern, the Cartesian coordinate can be stored relatively to the size of the patterns. In Algorithm 3, the leader affects to itself the color 0 and following the direction where the messages are transmitted, the particles reproduce the coloring patterns given in Appendix D. The functions f_G^k and I_G^k, J_G^k, used in Algorithm 3 are defined, depending on the choice of G, as follows: $f_{\mathcal{S}}^k(i, j) = (i + kj) \pmod{m_k}$, $f_{\mathcal{K}}^k(i, j) = i_{\pmod{k+1}} + (k+1)j_{\pmod{k+1}}$ and $f_{\mathcal{T}}^k(i, j) = (i_{\pmod{3(k+1)/2}} + j(3(k+1)/2) + \lfloor 2j/(k+1) \rfloor (k+1)/2)) \pmod{m_k'}$ if k is odd or $f_{\mathcal{T}}^k(i, j) = (i + (3k/2 + 1)j) \pmod{m_k'}$ otherwise; $I_{\mathcal{S}}^k(i, a) = i$ if $(a = 1; 3 \wedge G \cong \mathcal{S}) \vee (a = 1; 4 \wedge G \cong \mathcal{T}) \vee (a = 2; 6 \wedge G \cong \mathcal{K})$, $I_{\mathcal{S}}^k(i, a) = i + 1$ $\pmod{\lceil (k+1)^2/2 \rceil}$ if $a = 0$, $I_{\mathcal{S}}^k(i, a) = i - 1 \pmod{\lceil (k+1)^2/2 \rceil}$ if $a = 2$, $I_{\mathcal{T}}^k(i, a) = i + 1 \pmod{\lceil 3(k+1)^2/4 \rceil}$ if $a = 0; 5$, $I_{\mathcal{T}}^k(i, a) = i - 1 \pmod{\lceil 3(k+1)^2/4 \rceil}$ if $a = 2; 3$, $I_{\mathcal{K}}^k(i, a) = i + 1 \pmod{k+1}$ if $a = 0; 1; 7$ and $I_{\mathcal{K}}^k(i, a) = i - 1 \pmod{k+1}$

if $a = 3; 4; 5$; $J_G^k(j, a) = i$ if $(a = 0; 2 \wedge G \cong \mathcal{S}) \vee (a = 0; 6 \wedge G \cong \mathcal{T}) \vee (a = 0; 4 \wedge G \cong \mathcal{K})$, $J_{\mathcal{S}}^k(j, a) = j + 1 \pmod{\lceil (k + 1)^2/2 \rceil}$ if $a = 1$, $J_{\mathcal{S}}^k(j, a) = i - 1 \pmod{\lceil (k + 1)^2/2 \rceil}$ if $a = 3$, $J_{\mathcal{T}}^k(j, a) = i + 1 \pmod{\lceil 3(k + 1)^2/4 \rceil}$ if $a = 1; 2$, $J_{\mathcal{T}}^k(j, a) = i - 1 \pmod{\lceil 3(k + 1)^2/4 \rceil}$ if $a = 4; 5$, $J_{\mathcal{K}}^k(j, a) = i + 1 \pmod{k + 1}$ if $a = 1; 2; 3$ and $J_{\mathcal{K}}^k(j, a) = i - 1 \pmod{k + 1}$ if $a = 5; 6; 7$. Note that the functions I_G^k and J_G^k are used to determine the Cartesian coordinate of a particle using the Cartesian coordinate of a neighbor and the port number of this neighbor.

Since the values of $f_G^k(i, j)$ is bounded by $3(k + 1)^2/4$, if G is isomorphic to one of the three grids, the size of the messages will not exceed $\log_2(3(k + 1)^2/4)$. As for Algorithm 2, the number of sent messages is $|V(P)| - 1$. Figure 7c illustrates the obtained 4-identifiers after the execution of Algorithm 3.

Since the particles can move during the execution of an algorithm, the k-local identifiers may become not valid anymore (i.e., there may be two particles p_1 and p_2 with the same k-local identifier and with $d_G(s(p_1), s(p_2)) \leq k$) if the structure of the particle graph P on G changes. It is possible to keep a valid k-local identifier in case a particle moves in a direction of a port a by setting $id = f_G^k(I_G^k(i, r_G(a)), J_G^k(j, r_G(a)))$ as the new k-local identifier. It corresponds to update the variable id which corresponds to a color in a coloring of the kth power on the grid in function of the new position of the particle. Also, in the case a particle do ℓ movements, by storing the successive directions of movement of the particle during these ℓ movements, it is also possible to update the value of the k-local identifier in order that it remains valid.

Note that for both Algorithms 2 and 3 finish after at most h rounds, h being the height of the spanning tree. Also the number of sent messages in both Algorithms 2 and 3 is $|V(G)| - 1$ (the number of edges in a spanning tree).

5   Conclusion

In this paper, we have presented a new leader election algorithm based on local computation. We have also presented an algorithm which affects a different variable for every two particles p_1 and p_2 at distance at most k. All the presented algorithms only require a $O(1)$-space memory. This complexity makes it possible to use our algorithms for programmable matter. Moreover, in case of movements of particles, there is no need of communication in order to update the k-local identifiers.

As future work, it would be interesting to determine a more general deterministic leader election algorithm in our algorithmic framework that can take into account fault tolerance. Also, it would be interesting to extend the presented results to 3D grids. Another interesting question could be to use our results to clustering the set of particles in several sets which induce subgraphs of small diameter.

Acknowledgments. This work was supported by the French "Investissements d'Avenir" program, project ISITE-BFC (contract ANR-15-IDEX-03).

Appendix A   Proof of Theorem 1 and Bound on the Complexity of the S-Contraction Algorithm

The three lemmas presented in this appendix are used in order to prove Theorem 1.

In the following lemma, we describe how to determine, in the context of programmable matter, if a particles is S-contractible or not.

Lemma 1. *Let G be an infinite grid among $\mathcal{S}$, $\mathcal{T}$ and $\mathcal{K}$ and let $S \subseteq V(P)$, for P the particle graph on G and S the set of vertices occupied by all the particles in the same fixed state. One round is sufficient in order that every particle determine if it is S-contractible or not if G is isomorphic to $\mathcal{S}$. Otherwise, if G is isomorphic to $\mathcal{T}$ or $\mathcal{K}$, no round is necessary.*

Proof. Let $N^+(p)$ be the set of port labels on which p can communicate with particles from its neighborhood. In order to verify that $M_G(p) \cap S$ is connected in the triangular or king grids, it suffices to verify that $N_G^+(p) \cap S$ forms an interval of consecutive integers (by considering that 0 and $\Delta(G) - 1$ are consecutive). For example, $\{0, 4, 5\}$ contains successive integers in the triangular grid but that is not the case for $\{0, 2, 5\}$. Such verification in the triangular and king grids can be done during any local computation. Figure 5 illustrates three possible cases that could happen for a particle in the triangular grid. On the left part of Fig. 5, the particle does not satisfy Property (I) but satisfies Property (II). On the middle part of Fig. 5, the particle satisfies Property (I) and does not satisfy Property (II). Finally, on the right part of Fig. 5, the particle satisfies both Properties (I) and (II).

In the square grid, in order to test if a particle p is such that $M_G(p) \cap S$ is connected, it requires to receive $N_G^+(p') \cap S$, from the particle p' in the neighborhood of p and afterward to test if $N_G^+(p)$ only contains consecutive integers (by considering that 0 and 3 are consecutive) and then to verify, for any two successive particles p' and p'' from the neighborhood, that the vertex which corresponds to the corner adjacent to both p' and p'' is occupied by a particle.

If G is among $\mathcal{T}$ and $\mathcal{K}$, then no round is required to know if a particle is in S or not (since a particle know the state of its neighbors). If G is isomorphic to $\mathcal{S}$, then, in one round, which consists in sending the values of $N^+(p) \cap S$ to the adjacent particles, every particle knows if it is S-contractible or not. $\square$

The following lemma is be useful in order to prove that our leader election algorithm works correctly.

Lemma 2. *Let G be an infinite grid among $\mathcal{S}$, $\mathcal{T}$ and $\mathcal{K}$ and let $S \subseteq V(P)$, for P the particle graph on G. Let p be an S-contractible particle. If S is connected and hole-free, then $S - \{s(p)\}$ is connected and hole-free.*

Proof. First, note that in all three grids, the fact that $|N_G(p) \cap S| < |N_G(p)|$ implies that there is a vertex v in $N_G(p) \setminus S$. By contradiction, suppose we create a hole in $G[S]$ by removing the vertex $s(p)$ from S. This implies, since $G[M_G(p) \cap S]$ is connected, that v was already in a hole from $G[S]$. Second, since $G[M_G(p) \cap S]$ is connected, we are sure that the subgraph $G[S \setminus \{s(p)\}]$ is connected. □

To ensure that our leader election algorithm works correctly, it remains to prove that there always exists an S-contractible particle. That is what we do in the following Lemma.

Lemma 3. *Let $S \subseteq V(P)$, for P the particle graph on G. If $G[S]$ is hole-free, then there always exists an S-contractible particle in S.*

Proof. Note that there exists a particle on the border of $G[S]$ since S is finite. Let A be the set of particles on the border of $G[S]$. For any particle p, the fact that there are at least two connected components B_1 and B_2 in $G[M_G(p) \cap S]$ implies that there is no path in $G[S \setminus \{s(p)\}]$ between any vertex of B_1 and a vertex of B_2, since it would imply the existence of a hole in $G[S]$ containing a vertex from $M_G(p) \setminus S$. Therefore, if $p \in A$ and if p is not an S-contractible particle, then p is an articulation of $G[S]$.

Now suppose, by contradiction, that there is no S-contractible particle in S. By the previous remark, the graph $G[A]$ is connected and all particles of A are articulations of $G[S]$. However, a finite graph containing a cycle contains vertices which are not articulation of $G[S]$. Thus, $G[A]$ contains no cycle ($G[A]$ is a forest). However, by definition, the leaves (the vertices of degree 1) are S-contractible. Thus, we obtain a contradiction with the fact that there is no S-contractible particle in S. □

In the case P is hole-free, note that by Lemmas 2 and 3 there is always a particle which is both on the border of $G[S]$ and not an articulation of $G[S]$.

Proof (Proof of Theorem 1). Note that before the execution of the algorithm, the set S is the set $V(P)$. Since P is hole-free and connected and by Lemma 2, S remains connected and hole-free during the execution of the algorithm. By Lemma 3, there is always a particle in S which is S-contractible (every particle on the border which is not an articulation is S-contractible). Thus, for every round, the number of particles in state **C** strictly decreases. Since $|V(P)|$ is finite, we are sure that at some point, S will only contain one vertex. If at some point, S contains one vertex, there will be at least one elected leader.

Finally, note that the fact that there are two elected leaders contradicts the fact that S remains connected during the execution of the algorithm. □

Let G' be a subgraph of G such that G' is hole-free, the *radius* of G', denoted by $r(G')$, is given by $r(G') = \min_{u \in V(G')} \{\max_{v \in A}(d_{G'}(u, v))\}$, for A the set of the particles on the border of G'. Moreover, let $h(T)$ be the *height* of a tree T, i.e., $h(T) = \min_{u \in V(T)} \{\max_{v \in V(T), |N_T(v)|=1} (d_T(u, v))\}$ and let $mtree(G')$ be the maximum height among all induced subgraphs of G' which are trees, i.e., among the set $\{G'[B] | B \subseteq V(G'), G'[B] \text{ is a tree}\}$.

In the following Proposition, we give a bound on the required number of rounds for the termination of Algorithm 1.

Proposition 1. *Let S be the set of particles in state C and P be the particle graph on G. Moreover, let $b_G = r(P) + mtree(P) + 1$ if G is isomorphic to $\mathcal{T}$ or $\mathcal{K}$ or $b_G = 2(r(P) + mtree(P)) + 2$ if G is isomorphic to $\mathcal{S}$. If P is hole-free, then after $b_G(P)$ rounds of the S-contraction algorithm on P, one particle will be the leader.*

Proof. First, suppose G is isomorphic to $\mathcal{T}$ or $\mathcal{K}$. Let S_t be the set of particles in state C after the first t rounds. Note that after $r(P) + 1$ rounds we are sure that every remaining particle u satisfies $|N_G(u) \cap S_{r(S)+1}| < N_G(u)$. This is due to the fact that each particle u on the border of S_i, for $i \geq 0$, is not in S_{i+1} if $|N_G(u) \cap S_{r(S)+1}| < N_G(u)$. Thus, by Lemma 2, $G[S_{r(P)+1}]$ is either a tree or empty. By definition, we have $h(G[S_{r(P)+1}]) \leq mtree(P)$. Note that in the case $G[S_t]$ is not a trivial tree (a tree containing only one vertex), we have $h(G[S_{t+1}]) = h(G[S_t]) - 1$, for $t \geq r(P) + 1$. Therefore, we obtain that S_t is empty if $t \geq r(P) + mtree(P) + 1$.

Second, suppose G is isomorphic to $\mathcal{S}$. Note that, by Lemma 1, one round is sufficient in order that every particle determines if it is S-contractible. Consequently, it is easy to observe that the required number of rounds in order that the S-contraction algorithm finishes for $\mathcal{S}$ is bounded by two times the required number of rounds in order that the S-contraction algorithm finishes for $\mathcal{T}$ or $\mathcal{K}$. $\qquad\square$

Appendix B   An Example of Algorithm in Order to Construct a Spanning Tree

Our proposed algorithm (Algorithm 4) for constructing a spanning tree consists in setting the particle in state L as the root and, afterward, constructing a spanning tree using a classical distributed spanning tree algorithm.

Algorithm 4. A spanning-tree algorithm for a particle p.

Case 1: State L (leader).
set child$(p) = N_G^+(p)$;
send a message m (which only contains the bit 0) through each port from child(p).
Case 2: State N (not elected).
if the particle receives the message m through the port a **then**
  if the particle has never received the message m before **then**
    set parent$(p) = a$;
    set child$(p) = N_G^+(p) \setminus \{a_1, \ldots, a_\ell\}$, where a_1, ..., a_ℓ are ports on which p has
received the message m;
    send the message m through each port from child(p).
  end if
end if

Appendix C   Combining the S-Contraction Algorithm with a General Leader Election Algorithm

Daymude et al. [5] introduced a leader election algorithm that works on every configuration of P. In this appendix we present a way to reduce the required number of rounds in order that this algorithm finishes its execution (by using the S-contraction algorithm). In the remaining part of this appendix, the leader election algorithm from [5] will be called the *general leader election algorithm* (to the best of our knowledge, it is the only leader election algorithm for programmable matter working on every configuration).

In order to simplify the presentation of the results, we only discuss the results for the triangular grid. However, by modifying the algorithms it is possible to make it work for the square and king grids also. We begin this appendix by describing how the general leader election algorithm works. This description will help the reader to convince itself that, in a lot of cases, combining the S-contraction algorithm with a general leader election algorithm could be a good idea.

For particle p and a port a of p connected to another particle, we denote by $n(a)$ the port number of the first port of p connected to a particle after a in the clockwise order. The general leader election algorithm uses the fact that it is possible to send a message around a boundary. Sending a message around a boundary consists in sending and re-transmitting the message in the following way: for a particle p, if a message is received from the port a, then the particle re-transmits the message to the particle connected to p by the port $n(a)$ of p. Figure 8 represents how the messages are transmitted in this case.

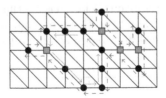

Fig. 8. The way how the messages are re-transmitted in the algorithm of Daymude et al. [5] (square: articulation; dashed arrow: message transmitted along the border; simple arrow: message transmitted along the hole).

For each hole H of the particle graph P on G, we denote by $b(H)$, the set of particles which are adjacent with a vertex of H. Also, when a particle is adjacent to vertices of different holes or when a particle belongs to the border of P, it is possible to decompose particles in agents, each agent corresponding to a different hole or to the border. Thus, an agent will be either adjacent to vertices of at most one hole or belong to the border but never both.

The general leader election algorithm can be summarized as the succession of four phases. A first phase consists in removing the candidacy of each particle having six neighbors. A second phase consists, for each hole H of P, to remove the candidacy of some agents of $b(H)$ using a randomized procedure. Simultaneously, the same process is done for agents in the border of P. A third phase consists in verifying if there is only one candidate in $b(H)$, for each hole H of P and only one candidate in the border of P. They calculate the relative positions of the candidates in order to do such verification. Finally, in the last phase, they verify if the remaining candidates agents are in $b(H)$ or in border of P. The leader will be the candidate particle of the border of P. We can verify if an agent is in $b(H)$ by sending a message around the boundary and verifying when the message comes back to the initial particle if this message has been re-transmitted in the clockwise direction or not. As Fig. 8 illustrates, the messages re-transmitted through the boundary of a hole are re-transmitted in the counterclockwise direction and the messages re-transmitted through the border of P are re-transmitted in the clockwise direction. In all these phases, the messages are re-transmitted around a boundary.

The required number of rounds in order that the general leader election algorithm finishes its execution is $O(\ell)$, where ℓ is the number of particles in the border of P.

We do the following remark about the S-contraction algorithm that comes from the fact that for any S-contractible particle p of S, if $G[S]$ is connected, then $G[S \setminus \{s(p)\}]$ is also connected.

Remark 1. *If the particle graph P has a hole then, after the execution of the S-contraction algorithm on the particle graph P on G there will remain particles in state $\mathbf{C}$. Also, the graph induced by the particles in state $\mathbf{C}$ is connected.*

In particular when P has one hole, the remaining particles in state $\mathbf{C}$ will form a ring in the triangular grid. Thus, it is possible to run the S-contraction algorithm and, afterward, execute the general leader election algorithm on the remaining particle in state $\mathbf{C}$. Let S_c be the set of particles in state $\mathbf{C}$ after the S-contraction algorithm on the particle graph P on G. Depending on the structure of P, it could happen that the number of particles in the border $T[S_c]$ is smaller than in the border of P and that it speeds the execution of the general leader election algorithm. For example, that is always the case when P has at most one hole.

Appendix D   Coloring the kth Power of Graph

The kth power of a graph G is the graph on the same vertex set than G and with edges connecting every two vertices u and v satisfying $d_G(u,v) \leq k$. Note that there is a correlation between this definition and the kth power of the adjacency matrix of G (the adjacency matrix of the kth power of G is easily obtained from this matrix). Our goal in this appendix is to determine an optimal coloring of the kth power of the square, triangular and king grids. We use these colorings in order to propose a distributed algorithm (supposing we have a leader) in order to assign k-local identifiers to the particles (see Sect. 4). A coloring of the kth power of a grid corresponds to assign a value to each vertex of the graph such that every two vertices with the same assigned value are at distance at least

$k + 1$. An example of coloring of the kth power of the square grid is represented by Figs. 9 and 10. In Fig. 9, it is easy to notice that every two vertices with color 0 (or any other color) are at distance at least 4.

More formally, a *k-coloring* of a graph G is a map c from $V(G)$ to $\{0, 1, \ldots, k-1\}$ which satisfies $c(u) \neq c(v)$ for every $uv \in E(G)$. The *chromatic number* $\chi(G)$ of G, is the smallest integer k such that there exists a k-coloring of G. The kth *power* G^k of a graph G is the graph obtained from G by adding an edge between every two vertices satisfying $d_G(u, v) \leq k$. More details about the coloring of the kth power of graphs can be found in the survey from Kramer and Kramer [17]. The results presented in this appendix are inspired by the previous works [12, 16, 22] about the coloring of the kth power of the grids.

D.1   Coloring the kth Power of Square Grids

We give the following result from Fertin et al. [12].

Theorem 2 ([12]). *For any* $k \geq 1$, $\chi(8^k) = \lceil (k+1)^2/2 \rceil$.

Let $m_k = \lceil (k+1)^2/2 \rceil$. In their paper, Fertin et al. define an optimal coloring c of the kth power of the square grid as follows: $c((i,j)) = (i + kj) \pmod{m_k}$. In Figs. 9 and 10, we represent patterns for coloring the *3th* and *4th* powers of the square grid. These patterns have been obtained using the coloring from [12]. Note that since there is a pattern, a vertex (i, j) can determine its color only knowing $i \pmod{m_k}$ and $j \pmod{m_k}$. We recall the definition of the following function $f_8^k(i,j) = (i + kj) \pmod{m_k}$. Note that $f_8^k(i, j) = f_8^k(i', j')$, in the case $i \equiv i' \pmod{m_k}$ and $j \equiv j' \pmod{m_k}$ This function is used in order to assign k-local identifiers to particles.

0	1	2	3	4	5	6	7
3	4	5	6	7	0	1	2
6	7	0	1	2	3	4	5
1	2	3	4	5	6	7	0
4	5	6	7	0	1	2	3
7	0	1	2	3	4	5	6
2	3	4	5	6	7	0	1
5	6	7	0	1	2	3	4

Fig. 9. A pattern for coloring the 3th power of the square grid.

0	1	2	3	4	5	6	7	8	9	10	11	12
5	6	7	8	9	10	11	12	0	1	2	3	4
10	11	12	0	1	2	3	4	5	6	7	8	9
2	3	4	5	6	7	8	9	10	11	12	0	1
7	8	9	10	11	12	0	1	2	3	4	5	6
12	0	1	2	3	4	5	6	7	8	9	10	11
4	5	6	7	8	9	10	11	12	0	1	2	3
9	10	11	12	0	1	2	3	4	5	6	7	8
1	2	3	4	5	6	7	8	9	10	11	12	0
6	7	8	9	10	11	12	0	1	2	3	4	5
11	12	0	1	2	3	4	5	6	7	8	9	10
3	4	5	6	7	8	9	10	11	12	0	1	2
8	9	10	11	12	0	1	2	3	4	5	6	7

Fig. 10. A pattern for coloring the 4th power of the square grid.

D.2   Coloring the kth Power of Triangular Grids

The chromatic number of the kth power of the triangular grid has been determined by Sevcikova [22].

Theorem 3 ([22]). *For any $k \geq 1$, $\chi(\mathcal{T}^k) = \lceil 3(k+1)^2/4 \rceil$.*

Let $m'_k = \lceil 3(k+1)^2/4 \rceil$. We recall the definition of the following function:

$$f_{\mathcal{T}}^k(i,j) = \begin{cases} (i \ _{(\text{mod } 3(k+1)/2)} + j(3(k+1)/2) + \\ \lfloor 2j/(k+1) \rfloor (k+1)/2)) \quad (\text{mod } m'_k) \text{ if } k \text{ is odd}; \\ (i + (3k/2+1)j) \quad (\text{mod } m'_k) \qquad \text{otherwise.} \end{cases}$$

Note that $f_{\mathcal{T}}^k(i,j) = f_{\mathcal{T}}^k(i',j')$, in the case $i \equiv i' \pmod{m'_k}$ and $j \equiv j' \pmod{m'_k}$. This function is used in order to assign k-local identifiers to particles.

D.3   Coloring the kth Power of King Grid

To our knowledge, the chromatic number of the king grid has not been determined yet. However, in contrast with the triangular grid, the chromatic number of the kth power of the king grid is easy to determine. In this subsection, we determine the exact value of the chromatic number of the kth power of the king grid.

Theorem 4. *We have $\chi(\mathcal{K}^k) = (k+1)^2$.*

Proof. Let $\mathcal{K}_k$ be the subgraph of $\mathcal{K}$ induced by the vertices $\{(i,j) \in V(\mathcal{K}) |\ 0 \leq i \leq k,\ 0 \leq j \leq k\}$. Note that $diam(\mathcal{K}_k) = k$ and that $|V(\mathcal{K}_k)| = (k+1)^2$. Thus, since each vertex of $\mathcal{K}_k$ must be colored differently in a coloring of the kth power of the king grid, we obtain that $\chi(\mathcal{K}^k) \geq (k+1)^2$. We define the coloring function $c((i,j)) = i_{\ (\text{mod } k+1)} + (k+1)j_{\ (\text{mod } k+1)}$. Note that we have $d_{\mathcal{K}}(u,v) \geq k+1$, for every two vertices u and v with the same color in $\mathcal{K}$. Therefore, we obtain that $\chi(\mathcal{K}^k) = (k+1)^2$.   □

Note that since there is a pattern, a vertex (i,j) can determine its color only knowing $i \pmod{(k+1)}$ and $j \pmod{(k+1)}$. We recall the definition of the following function $f_{\mathcal{K}}^k(i,j) = i_{\ (\text{mod } k+1)} + (k+1)j_{\ (\text{mod } k+1)}$. Note that $f_{\mathcal{K}}^k(i,j) = f_{\mathcal{K}}^k(i',j')$, in the case $i \equiv i' \pmod{k+1}$ and $j \equiv j' \pmod{k+1}$. This function is used in order to assign k-local identifiers to particles.

References

1. Attiya, H., Welch, J.: Distributed Computing: Fundamentals, Simulations and Advance Topics. Wiley, Hoboken (2004)
2. Bauderon, M., Métivier, Y., Mosbah, M., Sellami, A.: Graph relabelling systems: a tool for encoding, proving, studying and visualizing distributed algorithms. Electron. Notes Theoret. Comput. Sci. **51**, 93–107 (2001)
3. Bourgeois, J., Copen Goldstein, S.: Distributed intelligent MEMS: progresses and perspectives. IEEE Syst. J. **9**(3), 1057–1068 (2015)
4. Butler, Z.J., Kotay, K., Rus, D., Tomita, K.: Generic decentralized control for lattice-based self-reconfigurable robots. Int. J. Rob. Res. **23**(9), 919–937 (2004)
5. Daymude, J.J., Gmyr, R., Richa, A.W., Scheideler, C., Strothmann, T.: Improved leader election for self-organizing programmable matter. In: Fernández Anta, A., Jurdzinski, T., Mosteiro, M.A., Zhang, Y. (eds.) ALGOSENSORS 2017. LNCS, vol. 10718, pp. 127–140. Springer, Cham (2017). https://doi.org/10.1007/978-3-319-72751-6_10
6. Derakhshandeh, Z., Dolev, S., Gmyr, R., Richa, A.W., Scheideler, C., Strothmann, T.: Brief announcement: amoebot - a new model for programmable matter. In: SPAA 2014, pp. 220–222 (2014)
7. Derakhshandeh, Z., Gmyr, R., Richa, A.W., Scheideler, C., Strothmann, T.: An algorithmic framework for shape formation problems in self-organizing particle systems. In: NANOCOM 2015, vol. 21, pp. 1–21 (2015)
8. Derakhshandeh, Z., Gmyr, R., Strothmann, T., Bazzi, R., Richa, A.W., Scheideler, C.: Leader election and shape formation with self-organizing programmable matter. In: Phillips, A., Yin, P. (eds.) DNA 2015. LNCS, vol. 9211, pp. 117–132. Springer, Cham (2015). https://doi.org/10.1007/978-3-319-21999-8_8
9. Derakhshandeh, Z., Gmyr, R., Richa, A.W., Scheideler, C., Strothmann, T.: Universal shape formation for programmable Matter. In: SPAA 2016, pp. 289–299 (2016)
10. Derakhshandeh, Z., Gmyr, R., Porter, A., Richa, A.W., Scheideler, C., Strothmann, T.: On the runtime of universal coating for programmable matter. In: Rondelez, Y., Woods, D. (eds.) DNA 2016. LNCS, vol. 9818, pp. 148–164. Springer, Cham (2016). https://doi.org/10.1007/978-3-319-43994-5_10

11. Derakhshandeh, Z., Gmyr, R., Richa, A.W., Scheideler, C., Strothmann, T.: Universal coating for programmable matter. Theoret. Comput. Sci. **671**, 56–68 (2017)
12. Fertin, G., Godard, E., Raspaud, A.: Acyclic and k-distance coloring of the grid. Inform. Process. Lett. **87**(1), 51–58 (2003)
13. Di Luna, G.A., Flocchini, P., Santoro, N., Viglietta, G., Yamauchi, Y.: Shape formation by programmable particles. In: OPODIS 2017 (2017)
14. Lakhlef, H., Mabed, H., Bourgeois, J.: Optimization of the logical topology for mobile MEMS networks. J. Netw. Comput. Appl. **42**, 163–177 (2014)
15. Itai, A., Rodeh, M.: Symmetry breaking in distributed networks. In: Annual Symposium on Foundations of Computer Science, pp. 150–158 (1981)
16. Jacko, P., Jendrol, S.: Distance coloring of the hexagonal lattice. Discuss. Math. Graph Theory **25**, 151–166 (2005)
17. Kramer, F., Kramer, H.: A survey on the distance-colouring of graphs. Discret. Math. **308**, 422–426 (2008)
18. Naz, A., Piranda, B., Bourgeois, J., Copen Goldstein, S.: A distributed self-reconfiguration algorithm for cylindrical lattice-based modular robots. In: NCA, pp. 254–263 (2016)
19. Mazurkiewicz, A.: Distributed enumeration. Inform. Process. Lett. **61**, 233–239 (1997)
20. Piranda, B., Bourgeois, J.: Geometrical study of a quasi-spherical module for building programmable matter. In: Groß, R., et al. (eds.) Distributed Autonomous Robotic Systems. SPAR, vol. 6, pp. 387–400. Springer, Cham (2018). https://doi.org/10.1007/978-3-319-73008-0_27
21. Rosenstiehl, P., Fiksel, J.-R., Holliger, A.: Intelligent graphs. In: Graph Theory and Computing, pp. 219–265 (1972)
22. Sevcikova, A.: Distant chromatic number of the planar graphs. Safarik University, Manuscript (2001)
23. Tucci, T., Piranda, B., Bourgeois, J.: Efficient scene encoding for programmable matter self-reconfiguration algorithms. In: SAC, pp. 256–261 (2017)
24. Yim, M., et al.: Modular self-reconfigurable robot systems. IEEE Rob. Autom. Mag. **14**(1), 43–52 (2007)

Reaching Consensus in Ad-Hoc Diffusion Networks

Dariusz R. Kowalski[1,2] and Jarosław Mirek[1(✉)]

[1] Department of Computer Science, University of Liverpool, Liverpool, UK
{D.Kowalski,J.Mirek}@liverpool.ac.uk
[2] SWPS University of Social Sciences and Humanities, Warsaw, Poland

Abstract. We consider an algorithmic model of diffusion networks, in which n nodes are distributed in a 2D Euclidean space and communicate by diffusing and sensing molecules. Such a model is interesting on its own right, although from the distributed computing point of view it may be seen as a generalization or even a framework for other wireless communication models, such as the SINR model, radio networks or the beeping model. Additionally, the diffusion networks model formalizes and generalizes recent case studies of simple processes in environment where nodes, often understood as biological cells, communicate by diffusing and sensing simple chemical molecules.

To demonstrate the algorithmic nature of our model, we consider a fundamental problem of reaching consensus by nodes: in the beginning each node has some initial value, e.g., the reading from its sensor, and the goal is that each node outputs the same value.

Our deterministic distributed algorithm runs at every node and outputs the consensus value equal to the sum of inputs divided by the sum of the channel coefficients of each cells. For a node v consensus is reached in $\mathcal{O}\left(\log_\rho\left(\frac{1}{2}\sqrt{\frac{d_{min}}{d_{max}} \cdot \frac{d_v}{b\sum_i d_i}}\right)\right)$ communication rounds, where d_v is the sum of molecule reachability ratios to node v in the medium, $d_{min} = \min_i d_i$, $d_{max} = \max_i d_i$, and b is the sum of the initial values. ρ represents the second largest eigenvalue of a matrix of normalised molecule reachability ratios, that we analyze together with an associated Markov Chain.

Keywords: Ad hoc networks · Cells · Diffusion networks · Molecules · Consensus

1 Introduction

The consensus problem in distributed systems, where multiple processes want to agree on a certain value, is considered to be fundamental for many fault-tolerant and sensors' systems [17]. We consider the consensus problem in an ad-hoc network of cells with a diffusion-based molecular communication model.

Supported by Polish National Science Center (NCN) grant UMO-2017/25/B/ST6/02553.

© Springer Nature Switzerland AG 2019
S. Gilbert et al. (Eds.): ALGOSENSORS 2018, LNCS 11410, pp. 180–192, 2019.
https://doi.org/10.1007/978-3-030-14094-6_12

1.1   Previous Work

The consensus problem has been widely considered in distributed systems in several papers; here we only sample some relevant work. The authors in [1] focused on solving the classic consensus problem in a network of n mobile nodes, where the communication in the environment is considered unpredictable, their main goal was to reach convergence on a common consensus value by all the n nodes. The authors assumed that each node has its own identifier, the communication is synchronous, and when a message is delivered the destination is not certain. The proposed protocols solved the consensus problem with crashes and Byzantine failures. The model in [18] studied systems of autonomous agents with biologically motivated interactions and the evolving of self-ordered motion in these systems. According to a local rule, the agents are updated from time to time. The authors in [18] show that despite the changing neighborhood of each agent and absence of centralized coordination, as the system gradually evolves, all agents eventually move in the same direction.

In [16] a profound study of the present consensus algorithms and convergence was presented accompanied by the performance analysis of these algorithms. It reveals collaboration between various fields of engineering and science like complex networks, distributed computing, graph theory, and Markov chains. The idea of cooperation among dynamic systems is explained through elaborated investigation of formation control for multi-vehicle systems. The connection between spectral and structural properties of complex networks and the speed of information propagation of consensus algorithms is demonstrated. The authors in [15] presented a method for solving the consensus problem in distributed systems. The consensus problem in this method is considered on two levels. The first level represent general methods that can help in solving multi-valued and multi-attributed conflicts. The second level relates to different applications of consensus methods in solving various conflicts that may occur in distributed systems.

1.2   Related Work in Molecular Communication

Few research scopes [7–9] focused on studying the consensus problem in the molecular communication model. In the most relevant work [7], the authors explored consensus-type processes under simplified diffusion based molecular communication. They focused on studying the convergence of a random process of exchanging information about an event (measurement) through a diffusion based network. Through communication, all nodes try to obtain the best estimate. Additionally, they consider two simplified cases of diffusion environments, including distance unification and random deployment of cells.

1.3   Our Results

In this paper, we consider an ad-hoc network where nodes are deployed in a two-dimensional medium and communicate with each other by diffusing and sensing molecule concentrations. This model has been designed as a complete graph with weighted edges. We assume that the edge weight (that we called a channel coefficient up to this point) between two nodes depends on the distance between them, as signal strength in diffusion is inversely proportional to the cube of the distance [14], as well as other diffusion parameters. It is however important to highlight that the weights of the graph are not known a priori.

The most notable novelty of this paper is that the model is very basic and there are few assumptions that the system has to fulfill in order to execute the algorithm. This allows to think that our setup is actually a generalization or a framework for models such as the SINR [11,12]. Precisely, in the SINR model signals transmitted by processors weaken polynomially with respect to distance. Nevertheless the strongest signal contains a legible information and stations are distinguishable by having their unique identifiers. It is worth noticing that in the diffusion-based molecular model signals weaken in a similar manner. On the contrary, however, we do not distinguish nodes with identifiers and the information is actually just the amount of molecules - these similarities remind of the beeping model [5,6].

We assume that each node has its initial value, and these values are supposed to be an estimation of a parameter in the environment. The eventual goal is that nodes share their values between each other until they reach an agreement (consensus) about the average of these values. Nevertheless in biologically-motivated systems many processes are controlled by some overriding processes. Justification is to be found later in this paper. This leads to the notion of an eventual consensus that we introduce - even though the algorithm converges to a consensus, it is controlled by an overriding mechanism. Hence, each node does not know how long it should exchange information, yet will be notified when it is done.

In [7], the authors analyzed the processes of convergence of information-exchange between randomly distributed nodes in a diffusion medium. We generalize that work by proposing a model where nodes are placed arbitrarily. We propose a deterministic distributed algorithm which runs at every node and eventually computes the consensus value equal to the sum of inputs divided by the sum of channel coefficients. For a node v consensus is reached in $O\left(\log_\rho\left(\frac{1}{2}\sqrt{\frac{d_{min}}{d_{max}}} \cdot \frac{d_v}{b\sum_i d_i}\right)\right)$ rounds, where d_v is the sum of the *molecule reachability ratios* i.e. $\sum_i c_{iv}$, where a molecule reachability ratio c_{iv} between nodes i and v is the percentage of molecules sent by i that reaches v in a communication round, $d_{min} = \min_i d_i$, $d_{max} = \max_i d_i$, and b is the sum of the initial values. ρ represents the second largest eigenvalue of a special matrix of normalised reachability ratios that we used for our analysis (c.f. Sect. 3).

Finally, we also elaborate on how the length of a round influences the total time of consensus (defined as the number of rounds multiplied by the round duration) and propose study on local consensus for such systems.

1.4   Structure of the Paper

This paper is organized as follows. In Sect. 2 we give a detailed description of our model. Then we introduce the proposed consensus algorithm and its analysis in Sect. 3. Section 4 discusses time adjustments for optimizing round duration and how the local consensus in a molecular environment is implied by the eventual consensus. Finally, Sect. 5 combines conclusions and prospective future work.

2   Model

2.1   Network Environment

We consider an ad doc diffusion network consisting of n nodes, also called *cells*, deployed arbitrarily in a two dimensional Euclidean space. Nodes are identical and anonymous, in particular, they do not have distinguishing identifiers and have limited computational power. They also form an *ad-hoc* network structure, in the sense that they do not posses any a priori knowledge about network topology. In the remainder of the paper we use letters i, j, k to distinguish nodes for the purpose of notation and analysis only.

2.2   Communication Model

Communication between nodes is by diffusion (also called transmission) and sensing of molecules. More precisely, each node i could decide to diffuse an amount Q of molecules at time τ, and any other node j of distance d from node i can sense altogether

$$c(Q, d, T) = \int_{\tau}^{\tau+T} Q \cdot \frac{1}{(4\pi D t)^{\frac{2}{3}}} \cdot \exp(\frac{-d^2}{4Dt}) \, \delta t \tag{1}$$

of these molecules within time interval $[\tau, \tau + T]$, where D is the diffusion coefficient, resulting from the communication medium. In case when more than one node diffuses molecules, a receiver node j accumulates the sensed molecules through the summation of the values $c(Q, d, T)$ over diffusing nodes i, i.e., in the interval $[\tau, \tau + T]$ node j senses

$$\sum_{i} c(Q_i, \text{dist}(i,j), T_i) - c(Q_i, \text{dist}(i,j), [T_i - T]_+)$$

molecules in total, where $\text{dist}(i,j)$ is the Euclidean distance between i and j, T_i is the time that passed from diffusion of node i up to time $\tau + T$ (i.e., node i diffused at time $\tau + T - T_i$), Q_i is the amount of molecules diffused by i at that time, and $[T_i - T]_+$ equals to $\max\{T_i - T, 0\}$. In other words, the receiver senses the total amount of molecules that have been in its nearest proximity in the time interval $[\tau, \tau + T]$ without being able to distinguish which molecules come from which transmitter. In this sense, molecules are indistinguishable.

2.3   Control Variables

In the bloodstream there is a mechanism, called chemotaxis which gives cells the ability to migrate in a specific direction. During chemotaxis, cells move in response to an external signal, most frequently a small molecule, known as a chemoattractant. When cells sense the concentration of the chemical, they move in the direction where the concentration of that chemical is higher. Such a mechanism of directional movement may be observed while wound healing [3]. Consequently, cells' migration stops when they no longer sense the external signal of the chemoattractant chemical.

Another example of cell movement are the lymphocyte responses to [Ca2+] signals, ranging from short-term cytoskeletal modifications to long-term changes in gene expression [10]. These are only few examples of how biological activities end.

Such biological processes motivated us to introduce *control variables* that allow some *oracle* to provide additional information to the algorithm during the execution. In this work we use it in a very basic form to recognize (e.g., based on some external system behavior) whether the task of consensus has been actually done in the system or not.

2.4   Rounds and Synchronization

We assume that nodes are synchronized and can work in rounds of some predefined length T. We assume that T is a system parameter and discuss its length in Sect. 4. However this parameter strongly implies the topology of the network.

Nodes which want to diffuse molecules do it in the beginning of a round, and all nodes sense the aggregated level of molecules in their close proximities during the whole round. At the end of the round, we assume that the amount of molecules drops to zero; in practice, in biological systems this could be guaranteed by the ability of cells to absorb the molecules remaining at the end of the round in negligible time, or releasing other type of molecules that chemically transform the remaining communication molecules into some other type of molecules without any implications for communication.

Consequently, inspired by a model from [13] we assume that nodes can utilize molecules that were used for communication through the use of a mechanism known as *Destroyer molecules*, which help controlling the communication channel by eliminating remaining molecules from the environment. The authors in [13] assume that destroyer molecules are immobile in the environment and their size is bigger compared to information molecules. Besides, when an information molecule gets close to a destroyer molecule, it binds to it and is removed from the environment, due to the chemical attraction between them.

The conclusion is that treating the system as synchronous is possible because cells perform a repeating sequence of: diffusing molecules, sensing the concentration of molecules and cleaning the environment. Parameter T is therefore strongly connected with sensing concentrations because the longer the round the further broadcasts may be heard. The propagation of molecules across the medium in diffusion based communication is due to spontaneous diffusion, which is a stochastic process and occurs at a much lower speed, i.e. the order of a few millimeters per second. Finally, after the sensing phase, the environment is cleaned and a new round begins.

2.5   (Eventual) Consensus Problem

In the consensus problem, in the beginning each node has its initial value. The goal is that all nodes output the same value. We assume that initial values b_i, for a node i, are in $[0, 1]$, which covers a wide spectrum of potential input types (i.e., many other input ranges could be easily transformed 1–1 into values in $[0, 1]$). Let $I = (b_1, b_2, \ldots, b_n)$ denote the vector of initial values.

The definition of consensus is actually meaningless because it has been already shown in classical distributed systems that if the nodes do not have knowledge about the network topology, then it may be impossible to compute any non-constant function even on a ring, see e.g., [2] for details. Consequently, in such a weak model as the diffusion-based networks some computations may not be exact and are of a different nature. This brings us to the concept of *eventual consensus*, considered before in weak distributed systems: there is a round starting from which candidate values stored at nodes are the same. Note that in some executions nodes may not know this stabilization round, even though it exists.

2.6   Performance Measure

We consider time as the performance measure, defined in two ways. First, as the number of rounds in which all nodes reach consensus, regardless of the initial setting. Second, also called the *total time*, as the number of rounds multiplied by the round duration; we will show that the total time of an algorithm may depend on the actual duration of a round.

3   CONCELLSUS - The Consensus Algorithm in Diffusion Networks

In this section we introduce a deterministic algorithm CONCELLSUS that allows cells to arrive at a consensus in a diffusion-based ad hoc network.

3.1   Technical Preliminaries

The nature of a diffusion-based model allows us to think that nodes/cells deployed in a medium form a distributed system, which can be represented as a complete graph with weighted edges and self loops added to each node. More formally, we denote by $\mathcal{N} = (V, E)$ the graph representing a given ad-hoc network, where $V = \{1, 2, \ldots, n\}$ and $E = V \times V$. Notice that we allow self-loops in the network graph, as when a cell diffuses a certain amount of molecules, it may also sense it from the medium. For an edge $(i, j) \in E$, we associate a weight c_{ij} equal to $c(1, \text{dist}(i, j), T)$, where T is the length of a round. We call this weight the *molecule reachability ratio* between cell i and cell j. Furthermore, it is worth noticing that for any i, j we have that $c_{ij} = c_{ji}$.

Let $C = \{c_{ij}\}$ be the matrix of edges' weights in network $\mathcal{N}$. We denote the sum of weights of edges connected to node i by $d_i = \sum_{j \in V} c_{ji}$, and call it the *impact* of cell i. We associate the stochastic matrix $P = \{p_{ij}\}$, where $p_{ij} = \frac{c_{ij}}{d_i}$, with network $\mathcal{N}$. We call it the *matrix of normalised molecule reachability ratios*.

3.2   The Algorithm

Algorithm 1. CONCELLSUS, code for node i and input b_i; control variable process

---

Input: b_i, process
Output: w_i
1 **diffuse** a unit of molecules to all neighbors of i;
2 **receive** value d_i;
3 initialize $value_i$ to $\frac{b_i}{d_i}$;
4 **while** *process* **do**
5   **diffuse** $value_i$ (to all neighbors of i);
6   **receive** $received_i$ (accumulated from all neighbors j of i or i itself);
7   **update** $value_i := \frac{received_i}{d_i}$;
8 **end**
9 $w_i := \lceil \frac{value_i}{d_i} \rceil$

Nodes start with some initial local measurements resulting from the environment, which could be arbitrary. Recall that we denote the initial sequence of values as $I = (b_1, \ldots, b_n)$.

Having those initial measurements, nodes start exchanging values, stored as a local variable $value_i$ at some node i, in order to arrive at a consensus. They proceed in rounds, starting from round 0, which is slightly different from the other rounds. In round 0, each node diffuses a *unit* of molecules, and during the round, senses/receives some accumulated value (of molecules) from all its neighbors. The node stores it as d_i. We may think that cell i receives a form of a weighted average, which in case of a diffusion of units of molecules corresponds to value d_i. We will call this the *impact* of node i.

In the main part of the algorithm each node diffuses its initial measurement divided by its impact to all the neighbors. During the round, each node i senses the medium to receive some accumulated value (of molecules) from all its neighbors and then updates $value_i$ with $\frac{received_i}{d_i}$. Intuitively, cells that exist farther from cell i have a smaller impact on the actual $value_i$ (in a single round) than cells living closer. Thus, if some cell j with a huge measurement is very far from cell i then this value will more likely "reach" i via multiple other cells (which will be sensing and accumulating a large portion of this value) rather than directly.

Let us take a closer look at the main loop of CONCELLSUS, c.f., Algorithm 1. Motivated by biological processes and the way how they finish, we assume having a control variable **process** that indicates whether cells should continue exchanging information or not. For algorithmic reasons, we treat **process** as a boolean variable, however we may think that this is an overriding mechanism that controls the protocol, as it was motivated in Sect. 2.3.

The main loop in CONCELLSUS represents the molecule exchange between cells in consecutive rounds. Each iteration consists of three steps. (1) Initially a cell i diffuses its own actual value to the medium and then (2) receives the values, modified by the formula of the communication medium, from all the neighbors. The value represents the sum $\sum\{\frac{value_j}{d_j}c_{ij} : j$ is a neighbor of i or $j = i\}$, where value c_{ij} is the weight factor between the cells (the farther the cells, the smaller the value received). Finally, (3) cell i updates and rescales its own value accordingly with its sum of weights d_i that it learned in the initial round.

Cells, or devices of nanometer order size are considered as very simple and of weak computational capabilities. Consequently, the computations may not necessarily need to be of a significant precision. One may imagine that a group of cells needs to decide which direction they should move. It is somehow more natural that the cells will decide on a general direction and repeat the procedure after some movement - especially if the topology of the medium is not known. As a result, we round the consensus value to a natural number when the estimation of each cell is sufficiently close to that number.

Eventually, in the described process, each cell will compute a sort of an average from all the initial measurements, hence make a consensus. The main loop terminates when $value_i$ is sufficiently close to $b\pi_i$, where $b = \sum_{i \in V} b_i$ and π_i is the ith component of the stationary distribution vector. Thus, the value that the algorithm finally computes is $b\frac{d_i}{\sum_j d_j}\frac{1}{d_i}$ that actually is $\frac{value_i}{d_i}$. The most important question, from our perspective is, however, what is the convergence rate of such a process; we will analyze it in the following subsection.

3.3   Analysis of CONCELLSUS

Each cell has an initial measurement as an input, but does not know the measurements of other cells. Cells exchange their measurements, which are implicitly weighted by the medium. This process begins with vector $I = (b_1, \ldots, b_n)$ denoting cells' initial measurements. It changes with time as cells exchange and update their values round by round. Hence the vector of current values I_x in step x corresponds to $I_x = IP^x$, where P is a stochastic matrix defined in Sect. 3.1.

We show first that the Markov chain defined by matrix P converges to a stationary distribution $\pi = (\pi_1, \ldots, \pi_n)$, where $\pi_i = \frac{d_i}{\sum_j d_j}$ for every node i.

Lemma 1. *Consider a Markov chain with transition matrix P defined in Sect. 3.1. Then its stationary distribution is $\pi = (\pi_1, \ldots, \pi_n)$ where $\pi_i = \frac{d_i}{\sum_j d_j}$ for every node i.*

Proof. Firstly, we show that $\pi = (\pi_1, \pi_2, \ldots, \pi_n)$ is a stationary distribution for the process described by P, where $\pi_i = \frac{d_i}{\sum_i d_i}$ for every node i:

As $\pi_i = \frac{d_i}{\sum_j d_j}$, then $\sum_i \pi_i = \sum_i \frac{d_i}{\sum_j d_j} = 1$, and π is a proper stationary distribution.

Let $N(i)$ be the neighborhood of i. The condition $\pi = \pi P$ together with the fact that $p_{ij} = \frac{c_{ij}}{d_i}$ means that

$$\pi_i = \sum_{k \in N(i)} \left(\frac{d_i}{\left(\sum_j d_j \right)} \frac{c_{ik}}{d_i} \right) = \frac{d_i}{\sum_j d_j},$$

what ends the proof.   □

Now we need to show that the process of exchanging and updating values converges and that eventually all the nodes end with the same value. Besides we would like to know what is the rate of convergence. Let us assume that control variable *process*, driven by an external system oracle, does not switch off, or switches off at the same round after (eventual) consensus has been reached in the system.

Theorem 1. *Algorithm* CONCELLSUS *performs an eventual consensus in* $\mathcal{O}\left(\log_\rho \left(\frac{1}{2} \sqrt{\frac{d_{min}}{d_{max}}} \cdot \frac{d_v}{b \sum_i d_i} \right) \right)$ *iterations.*

Proof. We are interested in bounding the number of iterations r after which IP^r will be sufficiently close to the vector of the consensus value. To do this, we will analyze some property of the Markov chain described by P.

Recall that matrix $P = \{p_{ij}\}$ is a stochastic matrix of a primitive, reversible Markov chain, that describes a random walk on $\mathcal{N}$. According to Lemma 1 its stationary distribution is $(\pi_1, \pi_2, \ldots, \pi_N)$. We know that the eigenvalues of P are in the following relation: $1 = \lambda_1 > \lambda_2 \geq \cdots \geq \lambda_k$. Let ρ denote the second largest eigenvalue of P.

From Lemma 2 in [4] we know that

$$p_{ij}^{(r)} = \pi_j + \mathcal{O}\left(\sqrt{\frac{\pi_j}{\pi_i}} \cdot \rho^r \right),$$

where p_{ij} corresponds to the appropriate entry of the rth power of matrix P. Taking $M_P = \max_{i,j}\{\sqrt{\pi_j/\pi_i}\}$, the matrix form of the equation above is

$$P^r = P^\infty + \mathcal{O}(M_P \cdot \rho^r \cdot E),$$

where E is the matrix with all the entries equal 1. The limit $P^\infty = \lim_{r \to \infty} P^r$ of the chain is an $N \times N$ matrix where all the entries in the ith column are equal π_i. In our case we have that $M_p = \sqrt{d_{max}/d_{min}}$. In what follows

$$P^r = P^\infty + \mathcal{O}\left(\sqrt{\frac{d_{max}}{d_{min}}} \cdot \rho^r \cdot E\right).$$

For any vector of the initial measurements of the cells I, we have that $L = IP^\infty$ is the eigenvector of P. The ith entry of L equals $b\pi_i$, where b is the sum of the input vector I (the initial measurements vector). Consequently

$$IP^r = L + \mathcal{O}\left(\sqrt{\frac{d_{max}}{d_{min}}} \cdot \rho^r \cdot b \cdot e\right),$$

where e is the row vector with all its entries equal 1.

In order to be sure that all cells compute the correct value we need to be sure that for every cell v, $c\sqrt{\frac{d_{max}}{d_{min}}} \cdot \rho^r \cdot b < \frac{1}{2}\pi_v$, for some $c > 0$ resulting from the $\mathcal{O}$ notation. The reason being is that the final step of the algorithm computes a value rounded to the closest integer. Hence, as $\pi_v = \frac{d_v}{\sum_i d_i}$ we have that

$$\rho^r < \frac{1}{2}\sqrt{\frac{d_{min}}{d_{max}}} \cdot \frac{d_v}{cb\sum_i d_i},$$

and finally

$$r < \log_\rho\left(\frac{1}{2}\sqrt{\frac{d_{min}}{d_{max}}} \cdot \frac{d_v}{cb\sum_i d_i}\right).$$

We conclude that the whole network of cells will reach a consensus in r iterations that is bounded from above by $\log_\rho\left(\frac{1}{2}\sqrt{\frac{d_{min}}{d_{max}}} \cdot \frac{d_v}{cb\sum_i d_i}\right)$.   □

A final remark is that the result of the theorem concentrates on some values specific for each node. However, one may think that the overall number of iterations (for the whole system to terminate) is bounded by extreme values. Furthermore the number of rounds is defined by the base of the logarithm - value ρ, the second largest eigenvalue of the transition matrix. This leads to an obvious conclusion that the number of iterations of CONCELLSUS differs according to the topology of the network.

4   Extensions of Results

Biologically motivated systems have many subtleties, yet to be discovered. Consequently there are numerous problems that one may encounter, even when establishing the model. In this section we present notions and ideas resulting from our investigations. On one hand they justify, to some extent, our model, but on the other they may be inspirational in the sense of follow-up work.

4.1 Optimizing Round Duration

We stated the notion of rounds in the section regarding the model, but up to this point we treated it as a priori given. One may ask what is the optimal time duration of a round.

It is worth emphasizing that the duration of round a influences the weights in the graph of the network. Clearly, if the round is longer then each transmission of a certain cell reaches further distances, and consequently higher amount of molecules is sensed by nodes during rounds; hence the corresponding weights are also greater. However, the absolute time of the algorithm execution, defined as the number of rounds multiplied by the duration of a round, may be very long. On the other hand, too short rounds may cause that the graph becomes practically "disconnected" in the sense that weights will be too small to guarantee propagation across the network in a reasonable time. Therefore, our goal is to find round duration that provides a trade-off between the above mentioned two extremes.

From the point of view of the analysis from the previous section, we know that CONCELLSUS requires $r = O\left(\log_\rho\left(\frac{1}{2}\sqrt{\frac{d_{min}}{d_{max}}} \cdot \frac{d_v}{cb\sum_i d_i}\right)\right)$ rounds to reach consensus. Hence, the absolute time needed for the algorithm to terminate is bounded by rT there T is the duration of a single round (round a together with round b). The question is what is the optimal value of T, in order to minimize expression rT.

Considering T as a factor that influences the topology of the graph leads us to a conclusion that values d_{min}, d_{max}, d_v and $\sum_i d_i$ from the bound for r, are functions of T.

Thus in order to minimize the function $u(T) = rT$, we solve $\frac{\delta u}{\delta T} = 0$ to find the argument T^* for which $u(T^*)$ is minimal. However, we need to take into consideration, that the found value T^* has to satisfy $T^* \geq T_0$, where T_0 is the minimal duration of a round required to assure connectivity of the network graph. Otherwise reaching consensus would be impossible.

4.2 Local Consensus

The classical consensus problem considered in computer science concerns the matter of achieving a consistent opinion among all nodes in a network. In the world of cells and diffusion-based communication many processes are flexible. This means that even if there are numerous cells in the medium, only a subset of them may be employed for a specified task. If one considers the circulatory system then certainly not every platelet takes part in covering a particular wound. This brought us to a notion of a local consensus.

When considering the consensus problem, cells communicate with each other in order to reach an eventual consensus, where every cell finishes with the same value. What is more, we showed the rate of convergence of such a process for the protocol proposed in the previous section. Nevertheless to the best of our knowledge processes taking place in the human body are often controlled by some overriding mechanisms or organs. This leads to the notion of a local consensus.

If a certain task is considered as performed, by an overriding mechanism, then it is likely to happen that a process will be stopped and some cells may end without reaching the eventual value. Yet such an occurrence could suffice for the considered task, when the consensus has been achieved by a subset of cells. We say that there has been a local consensus.

The local consensus is implied by the eventual consensus. Precisely, if a value is becoming closer to be agreed among the whole network of cells with divergence α, then the divergence between a subset of cells is equal α' and $\alpha' \leq \alpha$.

5   Conclusions and Open Problems

In this work we demonstrated that the presented bio-inspired model of diffusion networks is algorithmically challenging - even on the level of establishing a simple and coherent model - by the design and analysis of a consensus algorithm.

In the consensus problem, it is interesting to see to what extent the external adversary could distract such a simple ad-hoc system from convergence.

Further study includes classical distributed computing and communication problems, mobility, fault tolerance, security, or even robotics.

Acknowledgements. We would like to thank Athraa Juhi Jani for fruitful talks and insight into the literature on the subject.

References

1. Angluin, D., Fischer, M.J., Jiang, H.: Stabilizing consensus in mobile networks. In: Gibbons, P.B., Abdelzaher, T., Aspnes, J., Rao, R. (eds.) DCOSS 2006. LNCS, vol. 4026, pp. 37–50. Springer, Heidelberg (2006). https://doi.org/10.1007/11776178_3
2. Attiya, H., Snir, M., Warmuth, M.K.: Computing on an anonymous ring. J. ACM **35**(4), 845–875 (1988). https://doi.org/10.1145/48014.48247
3. Bray, D.: Cell Movements: From Molecules to Motility. Garland Science (2001)
4. Broder, A., Karlin, A.: Bounds on the cover time. J. Theoret. Probab. **2**(1), 101–120 (1989). https://doi.org/10.1007/BF01048273
5. Cornejo, A., Kuhn, F.: Deploying wireless networks with beeps. In: Lynch, N.A., Shvartsman, A.A. (eds.) DISC 2010. LNCS, vol. 6343, pp. 148–162. Springer, Heidelberg (2010). https://doi.org/10.1007/978-3-642-15763-9_15. http://dl.acm.org/citation.cfm?id=1888781.1888802
6. Du, D.-Z., Hwang, F.K.: Combinatorial Group Testing and Its Applications, 2nd edn. World Scientific (1999). https://doi.org/10.1142/4252. https://www.worldscientific.com/doi/abs/10.1142/4252
7. Einolghozati, A., Sardari, M., Beirami, A., Fekri, F.: Consensus problem under diffusion-based molecular communication. In: 2011 45th Annual Conference on Information Sciences and Systems, pp. 1–6, March 2011. https://doi.org/10.1109/CISS.2011.5766149
8. Einolghozati, A., Sardari, M., Beirami, A., Fekri, F.: Data gathering in networks of bacteria colonies: collective sensing and relaying using molecular communication. In: 2012 Proceedings IEEE INFOCOM Workshops, pp. 256–261, March 2012. https://doi.org/10.1109/INFCOMW.2012.6193501

9. Einolghozati, A., Sardari, M., Fekri, F.: Networks of bacteria colonies: a new framework for reliable molecular communication networking. Nano Commun. Netw. **7**, 17–26 (2016). https://doi.org/10.1016/j.nancom.2015.01.003, http://www.sciencedirect.com/science/article/pii/S1878778915000046

10. Gallo, E.M., Canté-Barrett, K., Crabtree, G.R.: Lymphocyte calcium signaling from membrane to nucleus. Nat. Immunol. **7**(1), 25–32 (2006)

11. Gupta, P., Kumar, P.R.: The capacity of wireless networks. IEEE Trans. Inf. Theoret. **46**(2), 388–404 (2006). https://doi.org/10.1109/18.825799

12. Jurdziński, T., Kowalski, D.R.: Distributed randomized broadcasting in wireless networks under the SINR model. In: Kao, M.Y. (ed.) Encyclopedia of Algorithms, pp. 577–580. Springer, New York (2016). https://doi.org/10.1007/978-1-4939-2864-4_604

13. Kuran, M., Yilmaz, H.B., Tugcu, T.: A tunnel-based approach for signal shaping in molecular communication. In: 2013 IEEE International Conference on Communications Workshops (ICC), pp. 776–781, June 2013. https://doi.org/10.1109/ICCW.2013.6649338

14. Meng, L.S., Yeh, P.C., Chen, K.C., Akyildiz, I.F.: Mimo communications based on molecular diffusion. In: 2012 IEEE Global Communications Conference (GLOBECOM), pp. 5380–5385. IEEE (2012)

15. Nguyen, N.T.: Consensus system for solving conflicts in distributed systems. Inf. Sci. Inform. Comput. Sci. **147**(1–4), 91–122 (2002). https://doi.org/10.1016/S0020-0255(02)00260-8

16. Olfati-Saber, R., Fax, J.A., Murray, R.M.: Consensus and cooperation in networked multi-agent systems. Proc. IEEE **95**(1), 215–233 (2007). https://doi.org/10.1109/JPROC.2006.887293

17. Turek, J., Shasha, D.: The many faces of consensus in distributed systems. Computer **25**(6), 8–17 (1992). https://doi.org/10.1109/2.153253

18. Vicsek, T., Czirók, A., Ben-Jacob, E., Cohen, I., Shochet, O.: Novel type of phase transition in a system of self-driven particles. Phys. Rev. Lett. **75**(6), 1226 (1995)

Filling Arbitrary Connected Areas by Silent Robots with Minimum Visibility Range

Attila Hideg[1]([✉]), Tamás Lukovszki[2], and Bertalan Forstner[1]

[1] Department of Automation and Applied Informatics,
Budapest University of Technology and Economics, Budapest, Hungary
{Attila.Hideg,Bertalan.Forstner}@aut.bme.hu
[2] Faculty of Informatics, Eötvös Loránd University, Budapest, Hungary
lukovszki@inf.elte.hu

Abstract. We study the uniform dispersal problem (also called the filling problem) in arbitrary connected areas. In the filling problem robots are injected one-by-one at $k \geq 1$ Doors into an unknown area, subdivided into cells. The goal is to cover the area, i.e. each cell must be occupied by a robot. The robots are homogeneous, anonymous, autonomous, have limited visibility radius, limited persistent memory, and silent, i.e. do not use explicit communication. A fundamental question is how 'weak' those robots can be in terms of hardware requirements and still be able to solve the problem, which was initiated by Barrameda et al. [4]. In our previous paper [11] we presented an algorithm which solves the filling problem for orthogonal areas with $O(1)$ bits of persistent memory, 1 hop visibility range and without explicit communication. The algorithm utilized the timing of movements and had $O(n)$ runtime, where n is the number of cells in the area. In this paper, we generalize the problem for silent robots for an arbitrary connected area represented by a graph, while maintaining the 1 hop visibility range. The algorithm is collision-free, it terminates in $O(k \cdot \Delta \cdot n)$ rounds, and requires $O(\Delta \cdot \log k)$ bits of persistent memory, where Δ is the maximum degree of the graph.

Keywords: Autonomous robots · Filling · Dispersion

1 Introduction

In swarm robotics a huge number of simple, cheap, tiny robots can perform complex tasks collectively. Advantages of such systems are scalability, reliability, and fault tolerance. In recent years much attention has been paid to the cooperative behavior of simple, tiny robots which have to complete a particular task collectively. Many distributed protocols have been developed for a wide range of problems, like gathering, flocking, pattern formation, dispersing, filling, coverage, exploration (e.g. [1–6,8,12]; see [7,10] for recent surveys). In this paper we study the *uniform dispersal* (or *filling*) of synchronous robots in an unknown, connected area.

© Springer Nature Switzerland AG 2019
S. Gilbert et al. (Eds.): ALGOSENSORS 2018, LNCS 11410, pp. 193–205, 2019.
https://doi.org/10.1007/978-3-030-14094-6_13

The filling problem was introduced by Hsiang et al. [12] for an orthogonal area, where the area is represented by pixels that form a connected subset of integer grid. The robots are placed at the same entry point, called the Door, one-by-one and have to occupy all the pixels. There can be at most one robot per pixel at any given time. When more than one door is present in the area the problem is called *multiple door filling* or *k-door filling*.

Barrameda et al. [4,5] investigated the minimum hardware requirements and the possibilities of solving the filling problem for orthogonal regions by robots with constant visibility radius, constant communication range, and constant number of bits of persistent memory.

In [5], the authors allowed holes to be present in the map. Two methods were proposed: one without communication (**MUTE**) and one with communication (**TALK**). Both methods worked in the asynchronous (ASYNC) model, therefore, it can be used in the fully-synchronous (FSYNC) model, as well. **MUTE** required a visibility range of 6 and was inspired by the dance of bees, where robots implicitly communicated using only their movements. In **TALK** they required a visibility and communication range of 1 and worked strictly in orthogonal areas as they could see diagonally. Both solutions [4,5] only required a constant amount of memory.

In [4], common top-down and left-right directions and externally visible colors were assumed for the multiple door filling. In [9] Das et al. showed that allowing visible colors or lights yields a more powerful computational model than allowing infinite visibility range but no lights.

In our previous paper [11], we have further reduced the hardware requirements of the robots for filling orthogonal regions. The robots do not use any communication and also no lights. The robots have a common sense of North and East directions, but they cannot measure their absolute position and do not share a common coordinate system. They need a $O(1)$ bits of persistent memory and the visibility range of the robots is reduced to 1 hop. We have presented an algorithm solving the single-door and the multiple-door filling problem for orthogonal regions in $O(n)$ time in the synchronous computational model. A key element of the algorithm is to reserve time-slots

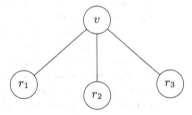

Fig. 1. Robots r_1, r_2 and r_3 would move to vertex v. Only one can move to v at any given time, however, they are not visible by each-other (1 hop visibility means they can see adjacent vertices).

for each possible direction of the movement (labeled as North, East, South, and West) in order to prevent collisions. Unfortunately, this idea could not be extended for the general case, where the area is represented by an arbitrary connected graph (moreover, it relies on notions that are not available in arbitrary graphs: north, south, orientation). A problematic scenario is illustrated in Fig. 1. Using a fixed assignment of time-slots to the edges incident to the occupied vertex of a robot r_1 can lead to collision with other robot(s) r_2, r_3, since another robots can choose the same time slot to target the unvisited vertex v.

Our Contribution: In this paper we present an approach, which is different from [11], to solve the filling problem for any arbitrary connected graphs with robots of visibility range of 1 hop, i.e. the robots can see adjacent vertices. (Note: three-dimensional scenarios and more complex topologies can be modeled.).

First, we present a method, the **Virtual Chain Method** (**VCM**), for the single door filling by a set of autonomous anonymous robots with a visibility radius of 1 hop in $O(\Delta \cdot n)$ time in the synchronous computational model, where n is the number of vertices of the graph with a maximum degree of Δ. The robots require $O(\Delta)$ bits of persistent memory.

Then, we consider the multiple door case, when the robots enter in $k > 1$ doors and we generalize the VCM algorithm for solving this problem. The robots need a visibility range of 1 hop, $O(\Delta \cdot \log k)$ bits of persistent memory, and the algorithm terminates in $O(k \cdot \Delta \cdot n)$ time.

Both algorithms are simple enough to be implemented by a swarm of elementary robots.

Our algorithm is optimal in term of visibility range. This follows from the fact that with a visibility range of less than 1 the robots cannot even distinguish between occupied and unoccupied neighbors. For constant k and constant Δ, our algorithm is asymptotically optimal in the size of the memory. This follows from the result of Barrameda et al. [4]; they proved that oblivious (memoryless) robots cannot deterministically solve the problem. Moreover, for constant k and constant Δ, our algorithm is asymptotically optimal in running time. The asymptotic optimality of the running time $O(n)$ follows from the fact that we can place one robot per round in the single door case and n robots must be placed.

A summary of these previous results and a comparison to our contribution is presented in Table 1.

Table 1. Summary of the requirements of different Filling algorithms. All of these algorithms have $O(n)$ running time in the synchronous model.

Requirements of Filling algorithms				
Method	Visibility range (hops)	Communication range (hops)	Memory (bits)	Graph type
DFLF [12]	2	2	2	Arbitrary
TALK [5]	2	2	4	Orthogonal
MUTE [5]	6	0	9	Orthogonal
k-Door in [4]	2	0	$O(1)$	Orthogonal
Filling in [11]	1	0	13	Orthogonal
k-Door in [11]	1	0	13	Orthogonal
Here: **VCM**	1	0	$O(\Delta)$	Orthogonal
MD-VCM	1	0	$O(\Delta \cdot \log k)$	Arbitrary

Organization: In Sect. 2 we define our model. In Sect. 3 we describe the Virtual Chain Method for filling a connected regions represented by an arbitrary graph. Section 4 extends our algorithm to solve the k-door filling. Finally, Sect. 5 summarizes the paper.

2   Model

We are given an unknown, connected area, represented by a graph. Each vertex of the graph allows only one robot to occupy it at any given time. We assume that for each vertex the adjacent vertices are arranged in a fixed cyclic order, which does not change during the dispersion. The entry points of the graph are called Doors. For simplicity we assume the degree of the Door vertices are 1. Otherwise, we introduce an auxiliary vertex of Degree 1 connected only to the Door, which takes the role of the original Door. This models the two side of a doorstep.

In our model we use common concepts in distributed mobile robotics. For an excellent overview we refer to the book by Flocchini et al. [10].

Each robot has a sensor allowing it to gather information from its vicinity, a computational unit, and locomotion capabilities. They are *autonomous*, i.e. no central coordination is present, *homogeneous*, i.e. all the robots have the same capabilities and behaviors, and *anonymous*, i.e. they cannot distinguish each other. They have a visibility range of 1 hop, i.e. each robot can 'see' only the vertex it occupies and the vertices adjacent to it. The robots are *silent*, i.e. they cannot communicate at all. They have limited bits of persistent memory, which is $O(\Delta)$ bits in the single door case and $O(\Delta \cdot \log k)$ when k-doors are present.

The robots operate corresponding to the Look-Compute-Move (LCM) model. During the Look phase, the robots take a snapshot of their surroundings. In the Compute phase, based on the snapshot they decide whether to stay idle or to move to one of its neighboring vertices, and during the Move phase they move there. The movement is atomic between two vertices, meaning it is either performed and the robot appears at the destination vertex, or does not move at all. We use the FSYNC model, where the robots perform their LCM cycles at the same time, i.e. each robot takes snapshots, computes, and moves at the same time.

The robots are placed on a predefined vertex, which is called the Door. At the beginning of each cycle, if the Door is empty, a new robot is placed there and performs its Look-Compute-Move phases during the same cycle.

3   Virtual Chain Method

We now present the Virtual Chain Method (VCM) for the single door case which is based on the traditional follow-the-leader principle. This principle has also been used in [4,5,11,12]. One robot becomes a leader and the rest of the robots follow it until the leader is blocked, then another robot takes the leadership. During our dispersion algorithm the robots create a virtual chain and move along it. The method mimics a depth first search like exploration of the area. An example for the dispersion can be seen on Fig. 2.

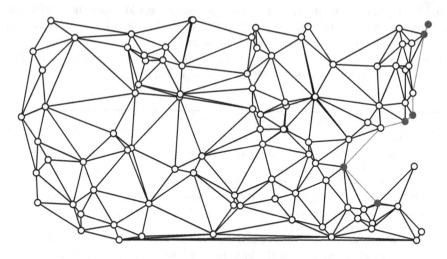

Fig. 2. The robots enter through one vertex called the Door (top right vertex), and follow the leader. The vertices occupied by Followers are blue, while the vertex occupied by the Leader is red. The red line denotes the path of the leader. (Color figure online)

3.1   Concept

In the Virtual Chain Method the following states are permitted to the robots:

- *None*: starting state immediately after the robot is placed at the Door.
- *Leader*: the first robot placed at the Door switches to Leader state. Only the leader moves to previously unoccupied, so called *unvisited* vertices. We will ensure that there can only be at most one leader at a time.
- *Follower*: the robots following their predecessor are in Follower state. A follower can promote itself to leader, when its predecessor is in Finished state.
- *Finished*: the final state of the robots. If a robot detects it cannot move anymore it switches to this state. A Finished robot can never move again. Only the leader can switch to Finished state.

The chain is defined by the path of the current leader from the Door. The followers are following that path, other robots in the area are already in Finished state.

The chain is not visible nor can be detected by the robots; their successor or predecessor might not even be in their visibility range at certain times. To avoid breaking the chain, each robot must follow its predecessor. To ensure the robots can detect their predecessor they are only allowed to move after their successor arrived to their previous vertex. That previous vertex on the chain is called the *entry vertex* of that robot. This way a robot and its predecessor can never be farther than two hops. Only the Leader is allowed to move to vertices, which were never occupied. These vertices are called *unvisited* vertices. When the Leader cannot move anymore, either because its vertex does not have any neighbors (other than its entry) or the movement would result in a collision, the Leader switches to Finished state and the leadership will be taken by its successor. Therefore, there can be at most one Leader at a time during the dispersion. The algorithm terminates when each robot is in Finished state. The rules followed by the robots can be seen on Algorithm 1.

Algorithm 1. (VCM): Rules followed by robot r.

1: If $r.State$ is None:
   r promotes itself to Leader if r has no neighbors, or to Follower otherwise
2: If $r.State$ is Follower:
   If $r.Predecessor$ moved from its place, r follows it.
   If $r.Predecessor$ did not move for two rounds, r switches to Leader state
       and performs the actions of a Leader in the same round.
3: If $r.State$ is Leader:
   If r is an observer, it stores visited neighbors.
   If r is an observed, it finds the first unvisited neighbor in the cyclical order.
   If r has an unvisited neighbor, it moves there, otherwise it switches to Finished.

Round Structure. The algorithm operates in rounds. A *round* is a sequence of Δ consecutive *steps* $S_1 \dots S_\Delta$ (a step is an LCM cycle of the robots). The rounds and the steps are illustrated in Fig. 3. During each round each robot is either an *observer* or an *observed* robot. The observed robot has to schedule its movement, and can move to one of its neighboring vertex v_i, which is the ith vertex in the fixed cyclic order starting from the entry vertex of the robot. It can only perform this movement in step s_i (as in Fig. 4), and not allowed to move more than once in each round. The observer robot counts the number of steps its predecessor has waited before moving. At the end of each round the robots switch their roles (i.e. the observed robots become observers and vice versa). Each robot starts as an observer when it is placed at the Door. The robots also store the occupancy information of the neighboring vertices in order to determine which of them are unvisited vertices. Now we describe the exact behavior of the robots in the different states:

Finished: Robots that are in Finished state have terminated their actions and do not move anymore.

Leader: If robot r is a Leader it leads the chain and moves to unvisited vertices. Thus, the number of unvisited vertices is monotonically decreasing. Robot r can move to vertex v_i in step s_i if and only if v_i is an unvisited vertex. In its observer round r registers each neighbor which is occupied in any step of the round. In its observed round, r moves to only those vertices which had not been registered as occupied in the current or previous round (later we show, it is sufficient to

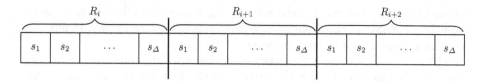

Fig. 3. The structure of the rounds. Three rounds (denoted by R_i, R_{i+1}, and R_{i+2}), each consists of Δ consecutive steps (s_j).

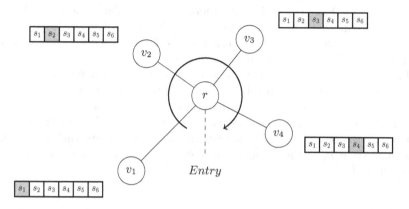

Fig. 4. Robot r, entered from below, with several neighbors. It can only move to v_i in s_i, where v_i is the ith vertex in the cyclic order starting from the entry vertex of r.

determine that v_i is an unvisited vertex). If no such vertices are available, r switches to Finished state.

Follower: Each Follower robot has to know which robot to follow, i.e. which robot is its predecessor. We will keep the invariant that for each Follower robot r in the beginning of its observer round the predecessor r' of r is in a neighboring vertex and r' moves in that round. We will ensure that r is able to determine the next position of r' based on that in which step r' moves away (similarly to the Leader, r' has to move to v_i in s_i). In the next observed round of r, r moves to the previous position of r' in step s_j, where j is the jth neighbor in the cyclic order from the previous position of r.

There are two special cases. First, if r' moves in s_Δ, r will only notice it in s_1 of its next observed round, which will be taken into account. The second special case, if r' cannot move anymore, which can only happen if r' is the leader and do not have any more unvisited neighboring vertices, r will notice it in s_1 of its next observed round. In this case r switches to Leader state and acts accordingly in the same round (i.e. r moves to an unvisited neighbor, or switches to Finished state if there is no unvisited neighboring vertex). For this reason Followers have to maintain unvisited vertices around them.

None: The robot r placed at the Door is initialized with None state. In this state the robot does not move and waits for state transition. This transition can be one of the following: if the neighboring vertex v of r is occupied then r becomes a Follower of the robot on v, otherwise it becomes a Leader. Before this state transition the robot must perform the following tasks.

The first task to solve is how the robots know which step they are in. The first robot is placed in the area in step s_1 of the first round. However, if that robot would move from the Door, the next one would be placed in s_2. Based on its surroundings, the new robot cannot acquire this information and it should not be provided explicitly when it is placed. If the robots are only allowed to move from the Door in s_Δ the new robot is always placed in s_1 of the next round. This can easily be achieved by not letting the robots in None state to switch states before s_Δ. (Note: the robot at the Door does not have a successor yet, therefore, it does not have to schedule its movement.).

The second task for robots at the Door is to ensure their role (observer/observed) in each round. When robot r moves from the Door in round R_i a new robot r' is placed at the Door immediately. The robot r' becomes active in step s_1 of round R_{i+1}, which is the observer round of r. As r' is initialized as an observer they would be both observers in R_{i+1}. To achieve distinct roles, r' stays inactive in R_{i+1} and becomes an observer in R_{i+2} again (and r' observes r moving in that round).

3.2   Analysis

Lemma 1. *The predecessor of the Follower is either in a neighboring vertex v or it was in v in the previous round. In the latter case the Follower moves to the previous position of the predecessor, which is v.*

Proof. Assume r follows its predecessor r'. Let R_i be the round when r observes r' from the Door (r was placed in the previous round). In R_i r' moves to one of its neighbors. In R_{i+1} r moves to the vertex which is left by r'. After this they become neighbors again. This argument can also be repeated for the following rounds. Therefore the claim of the lemma holds.   □

Lemma 2. *Each vertex can be considered an unvisited vertex if it is not occupied in two consecutive rounds.*

Proof. There are three cases regarding vertex v: (i) it is occupied, (ii) it is unoccupied, but has been occupied before, or (iii) it is an unvisited vertex (unoccupied, and never been occupied before). In case (i), any robot observing v knows it is not an unvisited vertex. In case (ii), let r be the last robot that occupied v, and let R_i be the round in which r moved from v. According to Lemma 1 the successor of r moves to v in round R_{i+1}. Consequently, v will not be unoccupied through two consecutive rounds. Therefore, observing v for two consecutive rounds allows any robot to identify cases (i) and (ii). In any other case vertex v is an unvisited vertex, i.e. case (iii) holds.   □

Lemma 3. *Each robot in Follower state always knows where its predecessor is.*

Proof. Using induction, we show that after movement $i \geq 0$ a Follower r knows where its predecessor r' is.

Induction Start: We show that after movement 0 (i.e. before its first movement), r knows where r' is.

Movement 0 means r did not move yet, and is still at the Door. Let v be its only neighboring vertex. Assume a robot is occupying v, as r is in Follower state (if v had not been occupied, r would became a Leader). We show that the robot on v must be predecessor r' of r.

The robots placed at the Door can only move to v. After this movement a new robot is placed at the Door, whose predecessor is the robot which has moved to v. If any other robot had moved to v, a new robot would not been placed at the Door. Therefore, if r is placed at the Door its predecessor is on v.

Inductive Step: We show, if r knows where its predecessor is after movement i, it will know where it is after movement $i + 1$.

After movement i robots r and its predecessor r' are in two neighboring vertices (v and v') as r just moved after r'. Therefore, r knows which robot is to observe. If r' moves, it can only move during an observed round while r is observing it. While observing r counts the steps until r' move to its next vertex v''. As assumed, the neighbors of v' are in a fixed cyclic order, r will know which is the next vertex v'' when it moves to v' (after movement $i + 1$) and the robot occupying v'' will be r'. □

Lemma 4. *Two Followers cannot have the same predecessor.*

Proof. When a robot r moves from the Door, only one robot will be placed at a time, therefore, only one robot can Follow r (and choose it as its predecessor). According to Lemma 3. each Follower knows where its predecessor is after every movement and will not change predecessors. As a result r will never be the predecessor of two (or more) different robots. □

Lemma 5. *The Leader only moves to unvisited vertices.*

Proof. As stated in Lemma 2. it is enough to observe each neighboring vertex for two consecutive rounds to determine whether it is an unvisited vertex or not. Assume R_j is the observer round of r and R_{j+1} is the observed round. In R_{j+1}, in which r moves, it already knows which neighbor is unvisited and – if any – it moves to the first unvisited vertex in the cyclic order of the neighbors in the corresponding step of R_{j+1}. □

Lemma 6. *There can be at most one Leader at a time.*

Proof. The first robot placed at the Door becomes the Leader. Afterwards, the Leader only moves to unvisited vertices. If there is no unvisited neighboring vertex (this can be determined in two rounds), the Leader changes its state to Finished. After changing the state its successor becomes a new Leader, the number of Leaders is still one. The condition of becoming a new Leader is that the old Leader did not move for two consecutive rounds, which implies it has changed to Finished state (otherwise, the old leader would have moved to an unvisited vertex).

Lemma 7. *No collision can occur during the dispersion.*

Proof. The robots in None and Finished states are not allowed to move, therefore, it only has to be shown that robots in Follower and Leader state cannot collide with other robots.

The robots in Follower state are following their predecessor and, according to Lemma 1, if they are not in neighboring vertices, the successor can only move to the previous position of the predecessor. As two robots cannot have the same predecessor (see Lemma 4), it is not possible that multiple Followers would move to the same vertex at the same time.

The Leader can only move to unvisited vertices (see Lemma 5) and Followers can only move to the position of its predecessor (i.e. not unvisited vertices), and as only one Leader is in the system (see Lemma 6), it cannot collide with other robots.

Therefore, no collision can occur.   □

Lemma 8. *Algorithm VCM fills the area (represented by the graph).*

Proof. By contradiction, assume some vertices left unoccupied after the algorithm terminated (meaning each robot is in Finished state). Since the graph is connected and the Door is occupied all of the time, there exists an unoccupied vertex v having at least one occupied neighbor v'. Let r be the robot on v', which is in Finished state. The condition for a robot to switch to Finished state is to be a Leader and not to have unvisited neighbors. However, according to Lemma 2, v will be identified as an unvisited vertex by r and r can occupy it. This contradicts the assumption that v remains unoccupied.   □

Theorem 1. *By algorithm VCM an area (represented by a connected graph) with a single door is filled in $O(\Delta \cdot n)$ rounds without collisions by robots with visibility range of 1 hop and $O(\Delta)$ bits of persistent memory.*

Proof. After placing the robot at the Door it is in None state, in which the robot *skips* one round. In the next round it observes its predecessor moving, then it moves in the third round. In the same round the next robot will be placed at the Door. For this reason, the robots are placed in every third round at the Door. As each round consists of Δ steps it takes $3 \cdot \Delta \cdot n = O(\Delta \cdot n)$ steps to place n robots.

Regarding the memory requirement, the robots require to store the index of the current step within a round, the unvisited neighbors, the direction of their predecessor known from the index in which step it moved away (each requiring at most Δ bits of memory), and some additional information, requiring constant amount of bits: current state, observer/observed role, entry vertex. As a result, $O(\Delta)$ bits of persistent memory is required for the Virtual Chain Method.   □

4   Multiple Doors

In the Multiple Door Filling, the biggest challenge is how the robots entering from different Doors avoid collisions. This is usually defined by some sort of priority order, e.g. in [4], the robots had 2 hops visibility and k different externally visible colors, where $k > 1$ is the number of Doors. In [11] the cells could be entered form different directions in each step.

In the Multiple Door Virtual Chain Method, (MD-VCM) the robots from each distinct Door will form a distinct chain, which are lead by their Leader robot. For each Door (for each chain) we introduce distinct time-slots, in which they can perform their actions. As opposed to the single door case, each step is substituted by k steps. The new round consists of $k \cdot \Delta$ steps. Step $s_{i,j}$ within a round corresponds to the jth step of the chain originating from the ith Door D_i, $1 \leq i \leq k$, $1 \leq j \leq \Delta$. Each robot from D_i only performs their actions in $s_{i,*}$.

Rules for the k-Doors Case: The robots in Finished, Leader, and Follower state follow the same algorithm as in the original VCM. The actions of the robots in None state has to be modified, as they are not necessarily placed in first step of the corresponding chain, e.g., if $i < k$, the robots entering from Door D_i will move from it in $s_{i,\Delta}$, resulting the new robot to be placed at D_i in step $s_{i+1,\Delta}$ (which is the time-slot of the next Door). To make sure the new robot starts their actions in step $s_{1,1}$ of the next round the newly placed robot stays inactive for $k - i$ steps. The only exceptions to this rule are the robots which has been initially placed in the first round R_1. They know they are the first robots in that Door if there are no neighboring robots for two rounds. In this case they become active immediately (not skipping $k - i$ steps) and switch to Leader state in step $s_{i,\Delta}$ of the second round.

Note: on a rare occasion it is possible two Doors are neighbors with eachother. Since we assumed that the degree of each Door vertex is 1 and the graph is connected, it is only possible for $n = 2$. In this very special case, the robots at the Door do not know correctly in which step they were placed there. However, the filling problem is solved right after both robots are placed.

4.1   Analysis

In the Multiple Door case Lemmas 1, 2, and 3 still hold. In case of Lemmas 4 and 5 it still has to be considered that there are multiple chains in the area.

Lemma 9. *In the MD-VCM, each Leader only moves to unvisited vertices and cannot collide with each other.*

Proof. Unlike in Lemma 5 there can be $k > 1$ robots in the system with Leader state, meaning that multiple robots can choose the same unvisited vertex as their destination. However, different leaders (which are from different Doors) are assigned to different steps. As a result, one of them moves there first, after which the vertex cannot be considered an unvisited one and the other Leaders have to choose a new destination.

Lemma 10. *The MD-VCM distinct chains cannot cross each other.*

Proof. The vertices on the paths of the leaders (the chains) are not unvisited vertices, since they were already occupied by the leader. Unvisited vertices are detected by the leaders according to Lemma 2. Because of Lemma 5 the Leaders only move to unvisited vertices. Consequently, the leaders do not cross other chains. Each Follower only moves on the path of the Leader of the chain, chains originating from different Doors are disjoint.

Theorem 2. *An area (represented by a connected graph) with a multiple doors is filled by the MD-VCM in $O(k \cdot \Delta \cdot n)$ rounds without collisions by robots with visibility range of 1 hop and $O(\Delta \cdot \log k)$ bits of persistent memory.*

Proof. Similarly to Theorem 1 the robots are placed in each Door in every three rounds if the chain from that Door is able to move, otherwise no further robots can be placed at that Door anymore. In the worst case, when a single chain blocks all the other Doors, a Door is used to cover the area. Consider a graph which is a simple chain of n vertices whose first k vertices are the Doors $D_1 \ldots D_k$. In this case only the robots placed on D_k can move, the robot in the other Doors are blocked. This is equivalent to the single door case, in which the running time is $3n$ rounds, where each round length consists of $k \cdot \Delta$ steps, yielding $O(k \cdot \Delta \cdot n)$ runtime.

Regarding the hardware requirements, the robots do not need additional visibility, nor other equipment. As the round length increases, and the current step index has to be stored, the memory increases to $O(\Delta \cdot \log k)$. ☐

5   Summary

In this paper, we have presented the Virtual Chain Method for solving the filling problem in unknown region, represented by an arbitrary connected graph, with autonomous robots having a minimum visibility range of 1 hop. The robots are not equipped with communication capabilities, they are working synchronously. We only assumed that the neighbors of each vertex are arranged in a fixed cyclic order, which is the same for each robot stepping into that vertex. For the single door case, the robots need $O(\Delta)$ bits of persistent memory, and $O(\Delta \cdot n)$ time steps, where n is the number of vertices of the graph. We have extended this method to solve the k-door filling problem in $O(k \cdot \Delta \cdot n)$ time steps for robots with memory requirement $O(\Delta \cdot \log k)$. The Multiple Door Virtual Chain Method does not add further hardware requirements for the robots. It remains an open question how the multiplicative factor k can be eliminated in the running time.

Acknowledgment. This work was partly performed in the frame of FIEK_16-1-2016-0007 project, implemented with the support provided from the National Research, Development and Innovation Fund of Hungary, financed under the FIEK_16 funding scheme.

References

1. Albers, S., Kursawe, K., Schuierer, S.: Exploring unknown environments with obstacles. Algorithmica **32**(1), 123–143 (2002)
2. Albers, S., Henzinger, M.R.: Exploring unknown environments. SIAM J. Comput. **29**(4), 1164–1188 (2000)
3. Augustine, J., Moses Jr., W.K.: Dispersion of mobile robots: a study of memory-time trade-offs. In: Proceedings of the 19th International Conference on Distributed Computing and Networking, ICDCN 2018, pp. 1:1–1:10 (2018)
4. Barrameda, E.M., Das, S., Santoro, N.: Deployment of asynchronous robotic sensors in unknown orthogonal environments. In: Fekete, S.P. (ed.) ALGOSENSORS 2008. LNCS, vol. 5389, pp. 125–140. Springer, Heidelberg (2008). https://doi.org/10.1007/978-3-540-92862-1_11
5. Barrameda, E.M., Das, S., Santoro, N.: Uniform dispersal of asynchronous finite-state mobile robots in presence of holes. In: Flocchini, P., Gao, J., Kranakis, E., Meyer auf der Heide, F. (eds.) ALGOSENSORS 2013. LNCS, vol. 8243, pp. 228–243. Springer, Heidelberg (2014). https://doi.org/10.1007/978-3-642-45346-5_17
6. Brass, P., Cabrera-Mora, F., Gasparri, A., Xiao, J.: Multirobot tree and graph exploration. IEEE Trans. Robot. **27**(4), 707–717 (2011)
7. Bullo, F., Cortés, J., Marttınez, S.: Distributed algorithms for robotic networks. In: Applied Mathematics Series. Princeton University Press (2009)
8. Cohen, R., Peleg, D.: Local spreading algorithms for autonomous robot systems. Theoret. Comput. Sci. **399**(1), 71–82 (2008)
9. Das, S., Flocchini, P., Prencipe, G., Santoro, N., Yamashita, M.: Autonomous mobile robots with lights. Theoret. Comput. Sci. **609**, 171–184 (2016)
10. Flocchini, P., Prencipe, G., Santoro, N.: Distributed Computing by Oblivious Mobile Robots. Synthesis Lectures on Distributed Computing Theory. Morgan & Claypool Publishers, San Rafael (2012)
11. Hideg, A., Lukovszki, T.: Uniform dispersal of robots with minimum visibility range. In: Fernández Anta, A., Jurdzinski, T., Mosteiro, M.A., Zhang, Y. (eds.) ALGOSENSORS 2017. LNCS, vol. 10718, pp. 155–167. Springer, Cham (2017). https://doi.org/10.1007/978-3-319-72751-6_12
12. Hsiang, T.-R., Arkin, E.M., Bender, M.A., Fekete, S.P., Mitchell, J.S.B.: Algorithms for rapidly dispersing robot swarms in unknown environments. In: Boissonnat, J.-D., Burdick, J., Goldberg, K., Hutchinson, S. (eds.) Algorithmic Foundations of Robotics V. STAR, vol. 7, pp. 77–93. Springer, Heidelberg (2004). https://doi.org/10.1007/978-3-540-45058-0_6

BSLoc: Base Station ID-Based Telco Outdoor Localization

Jinhua Lv[1], Qinpei Zhao[1], Jiangfeng Li[1], Yige Zhang[1], Xiaolei Di[1],
Weixiong Rao[1(✉)], Mingxuan Yuan[2], and Jia Zeng[2]

[1] Tongji University, Shanghai, People's Republic of China
{jhlv,qinpeizhao,lijf,yigezhang,dixl,wxrao}@tongji.edu.cn
[2] Huawei Noah's Ark Lab, Hong Kong, China
{yuan.mingxuan,zeng.jia}@huawei.com

Abstract. Telecommunication (Telco) localization is an important complementary technique of Global Position System (GPS). Traditional Telco localization approaches requires radio signal strength indicator (RSSI) of mobile devices with the connected base stations (BSs). Unfortunately, many of real-world signal measurement could miss RSSI values, and Telco operators typically will not record RSSI information, e.g., due to the major departure from current operational practices of Telco operators [6]. To address this problem, we design a novel BS ID-based coarse-to-fine Telco localization model, namely BSLoc, which requires only the connected BS IDs, time and speed information of mobile devices. BSLoc consists of two layers: (1) a sequence localization model via Hidden Markov Model (HMM) to localize the mobile devices with coarse-grained locations, and (2) a machine learning regression model with engineered features to acquire the fine-grained locations of mobile devices. Our experiments verify that, on a 2G dataset, BSLoc achieves a median error 26.0 m, which is almost comparable with two state-of-art RSSI-based techniques [9] 17.0 m and [20] 20.3 m.

1 Introduction

Recent years witnessed the popularity of location-based applications such as Google Map, Uber and Wechat on mobile devices. Billions of mobile users make use of such applications in their daily life, which motivates the development of outdoor localization techniques. As the most widely used localization technique, the Global Position System (GPS) still suffers from some shortcomings such as: (1) hungry energy-consuming, (2) easily blocked by high buildings, and (3) usually turned off by users due to privacy leakage consideration.

Meanwhile, telecommunication (Telco) localization has been proposed to localize mobile devices with measurement report (MR) data from Telco networks. The MR data can be collected when mobile devices connect to nearby base stations (BSs). A MR record contains connection information with up to 6 neighboring BSs [3]. Compared with the GPS, telco localization has strong points as: (1) energy-efficient (2) feasible in most mobile devices (3) better network coverage and being available indoors and underground (4) active when

© Springer Nature Switzerland AG 2019
S. Gilbert et al. (Eds.): ALGOSENSORS 2018, LNCS 11410, pp. 206–219, 2019.
https://doi.org/10.1007/978-3-030-14094-6_14

making calls or mobile broadband (MBB) services. Most existing telco local-ization studies involve four categories. *(i)* Measurement-based methods [17] esti-mate the point-to-point distances or angles from a device to its nearby BSs based on a radio propagation model, *(i)* fingerprint-based approaches [4] build a his-togram of received signal strength indicator (RSSI) for each location grid as its fingerprint, *(iii)* machine learning based methods [19,20] learn the relationship between MR features and locations to predict the position of an individual MR record (namely single-point localization), and *(iv)* sequence localization [5,13] uses sequence-to-sequence models to generate a location sequence from a MR sequence.

The majority of the localization methods above assume that MR data con-tains sufficient signal strength information (e.g., RSSI). Nevertheless, a high ratio of real-world MR records collected from mobile users contain such information from at most two BSs [13,20]. In the worst case, MR records contain BS IDs alone even without any signal strength information. Moreover, Telco operators typically will not record the signal strength information due to *(1)* the major departure from current operational practices of Telco operators [6] and *(2)* extra storage and computation overhead caused by logging such information [12].

In this paper, when given such MR records above with BS IDs alone, we design a BS ID-based coarse-to-fine telco localization model, namely BSLoc. BSLoc consists of two layers. In the first layer, we build a sequence localization model via Hidden Markov Model (HMM) encouraged by the good performance of sequence methods [13]. In the second layer, based on the coarse-grained grid locations by the first layer, we employ a machine learning regression model with engineered features to obtain fine-grained locations of mobile devices.

Compared with the state-of-the-art BS ID-based techniques, BSLoc offers three advantages: (1) no need of base station position. The two previous meth-ods [6,12] exploit the infrastructure information of BSs (e.g., precise position of BSs) from Telco providers, which can hardly be obtained by individual users, (2) Map constrained. Perera, et al. [12] computes straight lines by Vironoi as move-ment path which may falsely depart from real road segments. Instead, BSLoc leverages road networks for higher localization accuracy. (3) Good performance. The experimental results verify that our proposed model outperforms the best competitor by 37.3% in median error.

The rest of this paper is organized as follows. Section 2 first introduces the problem statement and then gives an overview of our solution. Section 3 gives the detail of our solution, and Sect. 4 evaluates our solution. Section 5 finally concludes the paper. Table 1 summarizes some symbols and their meanings used in this paper.

2 Overview

2.1 Problem Statement

Problem 1. (BS ID-based Telco Localization): BS ID-based Telco localization problem is to localize a mobile device using its connected BS IDs.

Table 1. Mainly used symbols/names and associated meanings

Symbol	Meaning	Symbol	Meaning
BS	Base Station	RAF	Random Forest
HMM	Hidden Markov Model	RSSI	Received Signal Strength Indicator
MR	Measurement Report	Telco	Telecommunication

When a mobile device connects to nearby BSs in a Telco network, the BSs generate MR records. As shown in Table 2, an example LTE 4G MR record contains a user ID (IMSI: International Mobile Subscriber Identification Number), connecting time (MRTime), up to 6 nearby BSs (eNodeBID and CellID) and radio signal strength indicator (RSSI) if any.

Table 2. An example of an LTE 4G MR record

MRTime	2017/5/31 14:12:06	IMSI	***012	Serving_eNodeBID	99129	Serving_CellID	1
eNodeBID_1	99129	CellID_1	1	RSRP_1	−93.26	RSSI_1	−67.18
eNodeBID_2	99131	CellID_2	4	RSRP_2	−98.44	RSSI_2	−53.65
...	...	...	...	...	...	...	...
eNodeBID_6	99145	CellID_6	5	RSRP_6	−90.02	RSSI_6	−50.92

Problem 1 essentially localizes mobile devices when the given MR records contain empty items of RSRP and RSSI. To solve Problem 1, we have to tackle three challenges. (*i*) The low spatial sensitivity of BS IDs. The coverage radius of a base station is 0.5–5 km [13], and the switch of connected BSs is rather infrequent [9], leading to the low spatial sensitivity of BS IDs. (*ii*) The disparity of MR records. The collected MR data is unevenly distributed on different areas, resulting in the difficulty of localization in rarely visited area. (*iii*) The GPS noise of MR data. The corresponding GPS labels of the MR data can be far away from true locations, leading to the difficulty of accurate localization.

To illustrate the above challenges, Fig. 1 shows an example of collected data from a dataset *Jiading Campus*. Figure 1(a) is a bicycle trajectory around the campus. The part of the trajectory highlighted by a black square is about 3 km but the serving BS did not change. Figure 1(b) dashed by a black square involves plenty of noisy GPS labels of the MR data, and the area dashed by a black circle shows some rarely traveled (by two or three trajectories) roads.

2.2 System Overview

In Fig. 2, the proposed localization model contains two following stages. First, the *offline stage* is to train historical data (i.e., those MR records together with associated GPS positions and speed information of mobile devices) to generate a two-layer machine learning models: Hidden Morkov Model (HMM) and Random

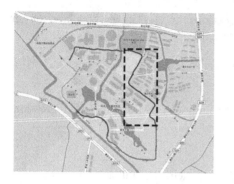 

Fig. 1. Dataset *Jiading Campus*

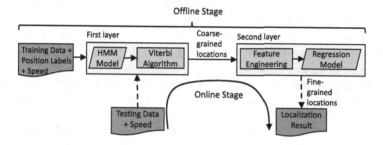

Fig. 2. System overview

Forest (RAF) regression model. In the first layer, we design a sequence localization model via HMM. It maps a sequence of observed BS IDs, timestamp, and speed information to a sequence of coarse-grained grid locations. In the second layer, we employ the RAF regression model to map the features with respect to the coarse-grained locations generated by the first layer to the fine-grained GPS positions.

Next, at the *online stage*, we takes as input a sequence of receiving BS IDs, timestamp and speed information to generate the coarse-grained grid locations as the output, e.g., by using the classic Viterbi algorithm [1]. After that, such grid location is next feed to the second layer RAF regression model which finally generates the fine-grained GPS locations.

To enable the proposed model, we need to perform the preprocessing steps. First, when people are moving along road networks, we adopt a classic map-matching technique such as [8] to project the GPS positions in our data onto the digital road network extracted from OpenStreetMap. The purpose is to mitigate GPS noise. Thus we use the projected GPS points on road networks as the ground truth of moving positions. Second, we divide the ground map area of interest into square grids with width cw. A grid location is a spatial index which refers to an area ($cw \times cw$) in a ground map, and we typically set $cw = 30\,\mathrm{m}$

to represent the road width. The grid locations are used in the first layer of our model to generate coarse-grained locations.

3   System Design

In this section, we give the detail of the two models: HMM and RAF regression.

3.1   Hidden Markov Model

We describe the used HMM $\lambda = (S, V, A, B, \pi)$ with the following variables:

- $S = \{s_1, s_2 \ldots s_N\}$ is the set of states. In our case, each state represents a grid position for each MR record, and N indicates the total amount of divided grid in the area of interest.
- $V = v_1, v_2 \ldots v_M$ is the set of observations. In our case, each observation v_i is the set of up to 6 BS IDs appearing inside MR records. The first ID is the one with the serving base station.
- $A = \{a_{ij}\}$ is the state transition probability distribution, where a_{ij} represents the probability that the grid s_i at time t is transited to the next one s_j at time $t + 1$.
- $B = \{b_j(k)\}$ is the probability distribution of observation k in state j, where $b_j(k)$ is the emission probability of v_k in the grid s_j, i.e., $b_j(k) = P(v_k|s_j)$.
- $\pi = \pi_i$ is the initial state distribution with $\pi_i = P[q_1 = S_i]$.

In the HMM model, the key is to learn the probabilities A and B.

Learn Transition Probability: The transition probability measures the probability of a device moving from a grid location G_j to another G_k with time interval Δt. We learn transition probability from two parts: transition matrices from historical trajectories associated with training MR data, and speed constraint from mobile phone sensor. The detail transition probability computation is described in Algorithm 1.

Transition Matrices. We use the statistics of trajectory data to compute transition probability. We construct transition matrices built by three steps. First, we convert each GPS point into a triplet $\langle TrajID, Time, Location \rangle$. Second, for every two points in the same trajectory, we extract a new triplet $\langle \Delta t, G_j, G_k \rangle$ which indicates the transition from grid G_j to G_k with time interval Δt. Third, triplets with same Δt make a matrix, thus generating multiple matrices with different Δt. Each entry in a matrix denotes the count of movements from grid location G_j to G_k with time interval Δt.

Speed Constraint. Based on the velocity v_t at time t and velocity v_{t+1} at time $t+1$ ($v_t < v_{t+1}$), we heuristically constrain the moving distance inside the interval $d_v = [d_0, d_1]$ where $d_0 = v_t * \Delta t$ and $d_1 = v_{t+1} * \Delta t$. However, the velocity information is noisy due to the common measurement errors. For example, a mobile device pauses for 30 s but the velocity value collected from the accelerometer

---

Algorithm 1. Transition probability calculation algorithm

Input: G^t: candidate locations in time t, G^{t+1}: candidate location in time $t+1$,
    M: offline computed transition matrices, Δt: time interval between t
    and $t+1$, v^t: speed at t, v^{t+1}: speed at $t+1$
Output: $transProb$: the transition probabilities from t to $t+1$

1 $M^- = mat \in M$ with time interval $\Delta t^- = \Delta t$;
2 **for** *each* G^t_j *in* G^t **do**
3     $n_j = \sum_{G^{t+1}_k \in G^{t+1}} M^-[G^t_j][G^{t+1}_k]$;
4   **for** *each* G^{t+1}_k *in* G^{t+1} **do**
5        $transProb[G^t_j][G^{t+1}_k] = \frac{M^-[G^t_j][G^{t+1}_k]}{n_j} * p_{speed}$, where p_{speed}=Eq. 1;

6 **return** $transProb$;

---

sensor might still indicate a moving speed (i.e., $2\,\text{m/s}$). Suppose the velocity noise follow a Gaussian distribution. We set the movement probability as follow:

$$P(G_k|G_j, v_t, v_{t+1}) = \begin{cases} 1, & dist_{j,k} \in [d_0, d1] \\ e^{-\frac{(d-d_0)^2}{2d_0^2}}, & dist_{j,k} \in [d_0 - k * cw, d_0] \\ e^{-\frac{(d-d_1)^2}{2d_1^2}}, & dist_{j,k} \in [d_1, d_1 + k * cw] \end{cases} \quad (1)$$

In the equation above, $dist_{j,k}$ is the distance between two grids G_j and G_k typically computed by the centroid distance of such grids. The parameter k restricts the noisy deviations into a certain range, and we empirically set k as the standard deviation of the trajectory of GPS positions of a given mobile device.

Emission Probability: Given the observed BS IDs as a feature O_k in a grid state s_j, we compute the emission probability $b_j(k)$ by Algorithm 2.

Algorithm 2. Emission probability calculation algorithm

Input: V_k: feature of MR record k, G_j: grid locations
Output: $emissionProb$: the emission probabilities for observation V_k in grid G_j

1 **for** *each* G_j *in* G **do**
2   n_j = amount of BS IDs in grid G_j;
3   n_{ij} = amount of BS IDs equal to V_k in grid G_j;
4     $emissionProb[G_j] = \frac{n_{ij}}{n_j} * w_{ij}$, where w_{ij} = Eq. (2);

5 **return** $emissionProb$;

Bayesian Emission Probability. Providing that we observe feature O_k from a MR record, we first roughly estimate the probability that O_k locates on a grid location G_j. First of all, the training samples indicate the empirical distribution of the BS IDs in the area of interest. Then according to Bayes' rule, we formulate the emission problem as $P(O_k|G_j) = \frac{P(O_k, G_j)}{P(G_j)}$. For each grid location G_j, we count the number of training samples as n_j. Similarly we count the number n_{kj} of training samples with feature O_k in grid G_j. The emission probability $P(O_k|G_j)$ is then estimated as $\frac{n_{kj}}{n_j}$, which is proportional to $\frac{P(O_k, G_j)}{P(G_j)}$.

Grid Weight Enhancement. Beyond the rough estimation $\frac{n_{kj}}{n_j}$ above, we are interested in the reliability of the empirical distribution in that grid, and thus define the weight of grid G_j with observation O_k as w_{kj}:

$$w_{kj} = \frac{n_j}{\sum_{G_k \in G(O_k)} n_k} \tag{2}$$

where n_j denotes the number of training samples in grid G_j, and $G(O_k)$ indicates all those grids containing feature O_k and thus $\sum_{G_k \in G(O_k)} n_k$ computes the total amount of training samples in all such grids.

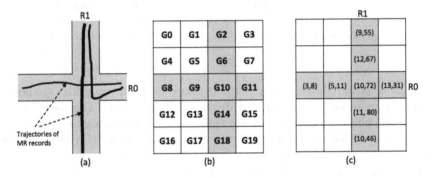

Fig. 3. Example emission probability calculation (from left to right): (a) Example trajectories on two roads R_0, R_1; (b) Map division by grids $G_0 \cdots G_{19}$; (c) Empirical distribution on the divided grids

Figure 3 shows an example to compute the emission probability. In this figure, mobile devices are moving on two intersected roads R_0 and R_1, and we divide the map into $4 \times 5 = 20$ grids $G_0 \cdots G_{19}$. Assume that all the 20 gray grids are with a certain feature (e.g., two BS IDs), i.e., O_k are all inside such 20 grids. In the right-most figure, each pair in the gray grids indicates n_{ij} and n_j, respectively. We then emission probability of O_1 locating on G_8 is computed as $P(O_1|G_8) * w_{1,8} = \frac{3}{8} * \frac{8}{8+11+55+67+72+80+46+31} \approx 0.0081$.

3.2   Regression Model

Based on the predicted grid locations of HMM model, we use the center of grids with size of $cw \times cw$ as its coarse locations. After that, based on the coarse

Table 3. Engineered features

Features	Description
pred_loc	p_i's longitude/latitude from the output of the first layer
obsv	p_i's observed BS IDs
speed	p_i's speed
last_loc	p_{i-1}'s longitude/latitude from the output of the first layer
last_obsv	p_{i-1}'s observed BS IDs
last_speed	p_{i-1}'s speed
next_loc	p_{i+1}'s longitude/latitude from the output of the first layer
next_obsv	p_{i+1}'s observed BS IDs
next_speed	p_{i+1}'s speed

locations generated by the first layer, we compute the contextual information such as observed BS IDs, speed and predicted grid locations into feature vectors in Table 3. The RAF regression model trains the features with the fine-grained GPS coordinates. Our experiment will validate that the two-layer design performs much better than the approach using the first layer HMM model alone.

We use standard RAF regression model to build the mapping from engineered features to GPS locations (longitude/latitude pairs). The regression target is to minimize the total error in the leaves of trees in RAF. We formulate the regression objective as

$$S = \sum_{t=1}^{T} \sum_{i \in L^t} D(i) \tag{3}$$

where T is the number of trees in the forest, L^t is the leaves of a tree in RAF and $D(i)$ is the squared error of samples in the leave i. During the offline training stage, the regression target S leads to the minimization of the training error. Then as the online stage, the trained RAF model predicts GPS locations by engineered inputs from the first layer.

4   Evaluation

Table 4. Statistics of two data sets

	Jiading Campus	Siping Campus
Number of samples (4G)	19542	2650
Number of 4G BSs	39	23
Number of samples (2G)	13416	3585
Number of 2G BSs	91	53

4.1 Datasets

Our experiments use two data sets collected in Shanghai city: (1): *Jiading* dataset is collected from a university campus located in a rural area of the North-west Shanghai (2): *Siping* dataset is collected from another university campus in an urban area of the North-east Shanghai. We developed an Android app to collect MR records, speed information and associated GPS position to collected the two data sets above. Specifically, when collecting MR data, we meanwhile turn on GPS sensors to acquire GPS coordinates.

Table 4 summarizes the two data sets. A piece of sample of the two data sets is an MR record with a GPS location. Both two data sets are collected with sampling rate of three seconds. Although the amount of samples and coverage area of Siping data set are smaller than that in Jiading data set, Siping data set includes more BSs per unit area due to dense BS deployment in urban Siping campus than the one in rural Jiading campus.

4.2 Counterparts and Evaluation Metrics

We implement three state-of-art algorithms (the detail refers to Sect. 5): (1) BS ID-based algorithm Cell* by Leontiadis et al. [6], (2) RSSI-based algorithm NBL by Margolies et al. [9], and (3) RSSI-based algorithm CCR by Zhu et al. [20]. Our evaluation objectives include:

- How BSLoc can outperform the BS ID-based algorithm Cell*.
- How BSLoc is comparable to the existing RSSI-based algorithms.

We evaluate BSLoc against the three algorithms by the metrics including Mean, Median and 67% error. The three algorithms compute localization errors by the distance between predicted locations and true locations except Cell*. Cell* predicts the path of mobile device, which consists of several road segments. For a MR record, the localization error is computed as the minimum distance between its true location and predicted road segment.

4.3 Baseline Study

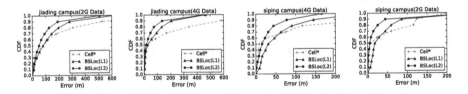

Fig. 4. Comparison with best competitor for BS ID-based techniques.

Figure 4 shows the comparison of our solution including one layer (L1) and two layer (L2) design with Cell*. Here, BSLoc only uses the serving BS ID as observation. From the result, we can see BSLoc(L2) achieves much better accuracy than

Cell* and BSLoc(L1) in general. For example, on *Jiading* 2G dataset, the median errors (and mean errors) of three algorithms BSLoc(L1), BSLoc(L2) and Cell* are {80.8 m, 50.3 m, 80.2 m} ({116.0 m, 79.0 m, 171.3 m}), respectively. BSLoc(L2) achieves 37.3% improvement than Cell* in median error and 53.9% improvement in mean error.

Specially, we find that the former 50% errors of Cell* are often lower than our BSLoc. The reason is the difference of evaluation metrics. The localization error of Cell* is computed by the distance between true location and predicted road segment. In general, both BSLoc(L1) and BSLoc(L2) behave better than Cell* on the four datasets.

4.4   Comparison with RSSI-Based Methods

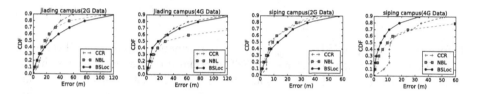

Fig. 5. Comparison with latest RSSI-based algorithms.

Figure 5 shows the comparison of BSLoc with the two RSSI-based algorithms (CCR and NBL). Here, BSLoc uses first two BS IDs as observation. From the result, we can conclude that BSLoc has comparable performance with NBL and CCR in general. For example, the median errors (and mean errors) of three algorithms (NBL, CCR, BSLoc) are {17.0 m, 20.3 m, 26.0 m} ({49.7 m, 30.0 m, 47.7 m}) on Jiading 2G dataset, and {9.4 m, 6.5 m, 13.1 m} ({20.1 m, 26.7 m, 26.3 m}) on Siping 2G dataset respectively. Considering the missing of the RSSI information, the proposed method has a good performance on both of the datasets. Since both the other two algorithms depend on the connection with neighboring BSs. Such neighboring BS IDs can hardly be obtained by mobile apps in LTE network. The results show our new BS ID-based method could have comparable performance with RSSI-based techniques with missing RSSI information.

4.5   Sensitivity Study

In this section, we vary the values of the parameters, which are the grid size and the extent of missing RSSI information, and study the sensitivity of BSLoc.

Effect of Grid Size: Figure 6(a) gives the experimental results when changing the grid size cw. A smaller grid size means more precise locations. In this experiment, we test five grid sizes: {15 m, 20 m, 30 m, 50 m, 100 m}. The result shows

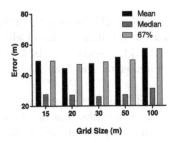
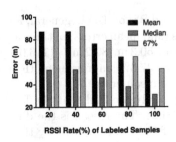

Fig. 6. Sensitivity study on the dataset *Jiading 2G Campus*: (a) Effect of grid size (b) Effect of missing RSSI

that the best grid size for *Jiading* dataset is between 20 m and 30 m. This is because a smaller grid size also means less training samples in each grid, leading to more inaccurate emission probabilities. Thus, the errors become higher when grid size is smaller than 20 m.

Effect of Missing RSSI: In Fig. 6(b), the performance of our method changes when adding different percentage of RSSI information. The figure shows that a higher percentage of RSSI information brings lower localization errors in general. The experiment validates the enhancement of signal strength information from base stations.

5   Related Work

In this section, we review the literature work of Telco localization, including RSSI-based and BS ID-based techniques. The RSSI-based techniques can be classified into four categories: measurement-based methods [7,11,15,17], fingerprinting methods [4,9,18], machine learning based methods [19,20] and sequence methods [5,13,16]. The BS ID-based techniques are all BS infrastructure based methods [6,10,12].

Measurement Based Methods: Measurement based methods employ signal measurement to estimate the location distance and the angle for telco localization. These methods suppose the signal information follows signal propagation model and estimate the distance/angle from neighboring base stations. Then the location of mobile device is computed via trilateration. There are a variety of measurements such as AOA, TOA and RSS [7,11,15,17]. However, the signal measurements are often noisy in urban areas due to multi-path propagation, non-line-of-sight propagation and multiple access interference, leading to large localization errors.

Fingerprinting Methods: Fingerprinting methods locate devices by comparing an input MR record against a fingerprint database which is constructed during an offline phase. The representative work CellSense [4] first divides the map area into square grids and then builds a fingerprint database that stores the

RSSI histogram for each grid at offline stage. Then at online stage it searches the K nearest grid neighbors for a given MR record via empirical distribution and returns the weighted average location. Moreover, a very recent work NBL [9] builds a Gaussian distribution in each grid for each base station, and achieve improvement than CellSense. Compared with measurement-based methods, fingerprinting methods lead to much lower localization error.

Machine Learning Based Approaches: Machine learning based methods build a representative feature on MR records and learn the mapping function from the built feature to actual location through well-trained models, such as Random Forest (RaF) and artificial neural network (ANN) [3,20]. For instance, Zhu et al. [20] first propose a two-layer random forest regression model to learn the location from the RSSI based features, and achieve good performance with high accuracy. Huang et al. [3] implement a variety of machine learning based methods for localization including Random Forest, MLP (Multilayer perceptron), XGBoost and etc., which has verified the effectiveness of machine learning models. In addition, Zhang et al. [19] propose a confidence level-based data repair method to optimize Telco localization.

Sequence Methods: Sequence methods map a sequence of MR records to a trajectory of locations. The sequence methods consider the contextual information, i.e., time and speed context, yielding more accurate estimations than single point methods. Mohamed et al. [5] propose a HMM model and employ Viterbi algorithm to map a sequence of MR records to a trajectory. Ray et al. [13] employ HMM and particle filtering algorithm to localize a sequence of MR records. These methods have demonstrated better localization accuracy.

BS ID-Based Methods: BS ID-based methods just take as input the ID of connected base stations (one or two) without signal strength information to locate mobile devices. Paek et al. [10] matches cell-id sequence with location sequence by Smith-Waterman algorithm. Leontiadis et al. [6] exploits the location and azimuth information of connected base station and builds a coverage map for each base station. The applied A* algorithm searches a path with maximum likelihood on the generated weighted road network. Perera et al. [12] aims to provide a realtime localization approach by mathematical computations based on the connected base station's location and coverage shape. Nevertheless, mobile users can hardly obtain such base station information from commercial Telco providers. It is not hard to find that the performance of both works is not as good as RSSI-based methods.

6   Conclusions

In this paper, we propose a BS ID-based coarse-to-fine telco localization approach without signal strength information or position of BSs. The two-layer localization framework first locates mobile devices in square grids, and next predict a precise GPS location. Our experiments on two data sets have successfully validated the advantages of our method over the state-of-art BS ID-based methods, and almost comparable to RSSI-based approaches. As the future, we plan to employ sequence-to-sequence learning framework [2,14] for more precise localization.

Acknowledgment. This work is partially supported by National Natural Science Foundation of China (Grant No. 61572365, 61503286, 61702372) and sponsored by The Fundamental Research Funds for the Central Universities.

References

1. Forney, G.D.: The viterbi algorithm. Proc. IEEE **61**(3), 268–278 (1973)
2. Hochreiter, S., Schmidhuber, J.: Long short-term memory. Neural Comput. **9**(8), 1735–1780 (1997)
3. Huang, Y., et al.: Experimental study of telco localization methods. In: 2017 18th IEEE International Conference on Mobile Data Management, MDM, pp. 299–306. IEEE (2017)
4. Ibrahim, M., Youssef, M.: CellSense: a probabilistic RSSI-based GSM positioning system. In: 2010 IEEE Global Telecommunications Conference, GLOBECOM 2010, pp. 1–5. IEEE (2010)
5. Ibrahim, M., Youssef, M.: A hidden Markov model for localization using low-end GSM cell phones. In: 2011 IEEE International Conference on Communications, ICC, pp. 1–5. IEEE (2011)
6. Leontiadis, I., Lima, A., Kwak, H., Stanojevic, R., Wetherall, D., Papagiannaki, K.: From cells to streets: estimating mobile paths with cellular-side data. In: Proceedings of the 10th ACM International on Conference on Emerging Networking Experiments and Technologies, pp. 121–132. ACM (2014)
7. Lopes, L., Viller, E., Ludden, B.: GSM standards activity on location (1999)
8. Lou, Y., Zhang, C., Zheng, Y., Xie, X., Wang, W., Huang, Y.: Map-matching for low-sampling-rate GPS trajectories. In: Proceedings of the 17th ACM SIGSPATIAL International Conference on Advances in Geographic Information Systems, pp. 352–361. ACM (2009)
9. Margolies, R., et al.: Can you find me now? Evaluation of network-based localization in a 4G LTE network. In: IEEE Conference on Computer Communications, INFOCOM 2017, pp. 1–9. IEEE (2017)
10. Paek, J., Kim, K.H., Singh, J.P., Govindan, R.: Energy-efficient positioning for smartphones using Cell-ID sequence matching. In: Proceedings of the 9th International Conference on Mobile Systems, Applications, and Services, pp. 293–306. ACM (2011)
11. Patwari, N., Ash, J.N., Kyperountas, S., Hero, A.O., Moses, R.L., Correal, N.S.: Locating the nodes: cooperative localization in wireless sensor networks. IEEE Signal Process. Mag. **22**(4), 54–69 (2005)
12. Perera, K., Bhattacharya, T., Kulik, L., Bailey, J.: Trajectory inference for mobile devices using connected cell towers. In: Proceedings of the 23rd SIGSPATIAL International Conference on Advances in Geographic Information Systems, p. 23. ACM (2015)
13. Ray, A., Deb, S., Monogioudis, P.: Localization of LTE measurement records with missing information. In: The 35th Annual IEEE International Conference on Computer Communications, IEEE INFOCOM 2016, pp. 1–9. IEEE (2016)
14. Sutskever, I., Vinyals, O., Le, Q.V.: Sequence to sequence learning with neural networks. In: Advances in Neural Information Processing Systems, pp. 3104–3112 (2014)

15. Swales, S., Maloney, J., Stevenson, J.: Locating mobile phones and the US wireless E-911 mandate (1999)
16. Thiagarajan, A., Ravindranath, L., Balakrishnan, H., Madden, S., Girod, L.: Accurate, low-energy trajectory mapping for mobile devices (2011)
17. Vaghefi, R.M., Gholami, M.R., Ström, E.G.: RSS-based sensor localization with unknown transmit power. In: 2011 IEEE International Conference on Acoustics, Speech and Signal Processing, ICASSP, pp. 2480–2483. IEEE (2011)
18. Vo, Q.D., De, P.: A survey of fingerprint-based outdoor localization. IEEE Commun. Surv. Tutor. **18**(1), 491–506 (2016)
19. Zhang, Y., Rao, W., Yuan, M., Zeng, J., Yang, H.: Confidence model-based data repair for telco localization. In: 2017 18th IEEE International Conference on Mobile Data Management, MDM, pp. 186–195. IEEE (2017)
20. Zhu, F., et al.: City-scale localization with telco big data. In: Proceedings of the 25th ACM International on Conference on Information and Knowledge Management, pp. 439–448. ACM (2016)

Orientation Estimation Using Filter-Based Inertial Data Fusion for Posture Recognition

David Segarra[1(✉)], Jessica Caballeros[1(✉)], Wilbert G. Aguilar[1,2(✉)], Albert Samà[3(✉)], and Daniel Rodríguez-Martín[3(✉)]

[1] CICTE Research Center, Universidad de las Fuerzas Armadas ESPE, Sangolquí, Ecuador
{desegarra,jacaballeros,wgaguilar}@espe.edu.ec
[2] GREC Research Group, Universitat Politècnica de Catalunya, Barcelona, Spain
[3] CETPD Research Group, Universitat Politècnica de Catalunya, Vilanova i la Geltrú, Spain
{albert.sama,daniel.rodriguez-martin}@upc.edu

Abstract. In this article, the Kalman filter, Mahony filter and Madgwick filter are implemented to estimate the orientation from inertial data, using an IMU called 9×3 of the MoMoPa3 project which contain various sensors including a gyroscope, an accelerometer and a magnetometer, each one of them, equipped with three perpendicular axes, in the magnetometer the measurement was modified to correct the distortions by hard metals, demonstrating improvements in the accuracy of the orientation estimates. In addition, the Kinovea video analyzer software is used as reference and gold standard to calculate the Root-Mean-Square Error (RMSE) with each filter. When comparing the angles estimated by the filters with those obtained from Kinovea, it was observed that one of the filters was better in performance. The information obtained in this article can be involved in several fields of science, one of the most important in the field of medicine, helping to control Parkinson's disease since it allows to evaluate and recognize when a patient suffers a fall or presents Freezing of the gait (FOG).

Keywords: Orientation estimation · IMU · Kalman filter · Madgwick filter · Mahony filter · Magnetometer distortion

1 Introduction

The life expectancy throughout the world has increased continuously for more than half a century [1], diseases caused by age or longevity in people has increased, being Parkinson's disease considered of special importance. Parkinson's disease (PD) is a complex degenerative neurological condition that appears in adulthood and is the second most common neurodegenerative disease, mainly affecting the motor system [2–4]. According to the Global Declaration for Parkinson's Disease, PD affects up to 6.3 million people worldwide [5], Parkinson's disease is increasingly a public health challenge in our progressively aging societies.

© Springer Nature Switzerland AG 2019
S. Gilbert et al. (Eds.): ALGOSENSORS 2018, LNCS 11410, pp. 220–233, 2019.
https://doi.org/10.1007/978-3-030-14094-6_15

The main clinical features of Parkinson's disease are related to body movement including tremor, spontaneous shaking (mainly in the upper limbs), muscle stiffness and bradykinesia (slow-moving physical movements) [6], in the advanced phase of the disease freezing of gait (FOG) is present, which is a disabling symptom and a movement disorder, and becomes a problem that could cause falls, this can be prevented by identifying when the FOG happens, with that, being able to control and help the patient by means of auditory signals to not lose coordination and freezing of the motor system, for this, it is necessary being able to recognize the orientation of the patient.

When observing the high index affectation of Parkinson's Disease, the developed project tries to improve the quality of life of the people who suffer it, raising the objective of recognizing the freezing of gait (FOG) of the patients, to avoid falls during this stage, in order to achieve this, first it needs to be able to estimate the orientation of the individual, for this, inertial sensors, such as, accelerometer, gyroscope and magnetometer are used, which together form an Inertial Measurement Unit (IMU). The device with these characteristics used in this article, was originally created for the MoMoPa3 project [7], which provides raw data, later this information from the sensors will be processed and filtered through the use of Kalman filter, Madgwick filter and Mahony filter to evaluate which filter best estimates the position compared to the angles obtained by a video analyzer software called Kinovea, each of the filters was implemented in Matlab.

The remaining of this article is organized as follows Sect. 2 describes briefly related works on orientation estimation with inertial sensors in different fields of science, Sect. 3 shows each of the filters used, and mathematically what corrections had to be made before inputting the measured data to the filters as well as how the angles were calculated from the coordinates given by Kinovea, on Sect. 4 we show the results obtained with each filter compared to our gold standard and finally in Sect. 5 conclusions and future works that could be done are shown.

2 Related Works

The orientation estimation is useful in several areas of research, one of which is to control unmanned aerial vehicles (UAV) or aerial robots, which allow the autonomous or semi-autonomous development of missions that cover the defense and security sectors [8]. In order to guarantee the permanent availability of the control system, it is essential to have certain instruments and sensors to control the stability of an UAV, such as the inertial measurement unit or IMU that provides data from the course followed by the UAV [9].

There are several methods for the estimation of the orientation as in [10] a real-time system was developed to monitor tremors and detection of falls, because of freezing of gait (FOG) which is a symptom present in the late stages of Parkinson's disease. The system consists of a 3D camera sensor based on the Microsoft Kinect architecture, which is able to recognize episodes of freezing (a state of inactivity), tremors and fall incidents. In case of an event, an alert is sent automatically to relatives and health care providers. This project was developed on an ideal environment, so its operation is influenced, depending on the vision range of the Kinect so in a real environment the algorithm can fail to find obstacles in the range of the visual sensor, where the test subject is placed, that's why it is more convenient to use sensors mounted directly on the patient, as proposed in this article.

Another development of great importance are exoskeletons for rehabilitation, which are used to assist movements and/or increase the capabilities of the human body [11]. In the article called "Lower-limb wearable exoskeleton" [12], it shows a system to compensate and evaluate the pathological gait, for applications, in real conditions, as a methodology of assistance of the problems that affect the mobility of individuals with neuromotor disorders. The implementation of sensors consists of an inertial measurement unit (IMU) on the foot (below the ankle joint in the orthosis), and a second unit in the lower bar of the exoskeleton, to achieve the control system of the exoskeleton, through the information acquired from the estimation of the orientation of each articulation. In this case, the work developed to obtain the estimation of the orientation of each articulation, is not robust enough, since it does not present an alternative system to be able to verify and compare that the estimated orientation is correct, in our project it is used both invasive and non-invasive sensors to be able to obtain the sensor orientation.

The determination of the orientation of objects in motion is involved in several fields of science [13]. In order to obtain relevant data for the estimation of the orientation it is necessary to process the signals received from the inertial measurement unit (IMU), the advanced methods of signal processing are strongly researched to optimize the performance of the existing detection hardware [13, 14]. When faced with a dynamic system, the magnitudes of the vector of parameters to be estimated will vary with time, which makes the problem more complex, in these cases, recursive estimation methods are used [15]. Currently there are several methods to manipulate the data, among the most important are the extended Kalman filter, Madgwick filter and Mahony filter, in this article each of the filters is implemented and then compared to obtain which one gives a lower error in the estimation of the orientation. In addition, it is intended to perform an exact system by using two different processes for the estimation of the orientation, using Kinovea as a visual sensor and using the IMU as inertial physical sensors, once the process of acquisition and processing of the data is completed, it is sought to get an autonomous system to incorporate the algorithm of orientation estimation, to be used in several applications within the field of health to detect FOG in Parkinson's disease.

3   Our Approach

3.1   Sensor

For the development of this research, an inertial measurement unit (IMU) was used. It is a device used in several fields, since they are effective, small and light. The IMU used is the same as the work done in [7] called 9×3 that has the objective of evaluating the symptoms of Parkinson's disease (PD). The 9×3 Unit can be used 2 or 3 days working continuously and it independently registers the signals of each of its sensors. This device has several elements such as: three accelerometers, a gyroscope, a magnetometer and a barometer to detect small changes in altitude [7].

The signals used for the estimation of the orientation in this work are obtained only from the inertial sensors integrated in LSM9DS0 [16], which is a 9-axis system composed of a magnetometer module, a triaxial accelerometer and a triaxial gyroscope module, these signals are stored on a microSD card, with a sampling rate of 400 Hz.

3.2 Mahony Filter

The notation used to mathematize the filters that we are going to follow is the one described by Madgwick in [17]. For example: the symbol ^ denotes a normalized vector, the symbol * means the conjugate of the vector, the operator ⊗ denotes a quaternion product, $^A_B\hat{q}$ means the orientation of frame B with respect to frame A, $^A\hat{v}$ is a vector in frame A.

$^E\hat{h}$ represents the direction of the magnetic field in the frame of the earth, calculated by means of the quaternary product of the previous estimation of orientation with the normalized magnetometer measurement and with the conjugate quaternion of the previous estimation of orientation, as can be seen in Eq. (1), m is the measure delivered by the magnetometer.

$$^E\hat{h}_t = {}^S_E\hat{q} \otimes \left(\hat{m} \otimes {}^S_E\hat{q}^* \right) \tag{1}$$

A quaternion is defined by Eq. (2).

$$^S_E\hat{q} = [q_1 \quad q_2 \quad q_3 \quad q_4] \tag{2}$$

Using the vector $^E\hat{b}$, which is the normalization of h, the effect of an erroneous inclination of the measured direction of the earth's magnetic field can be corrected, obtaining only components on the x and z axes of the earth, as shown in Eq. (3).

$$^E\hat{b}_t = \left[0 \quad \sqrt{h_x^2 + h_y^2} \quad 0 \quad h_z \right] \tag{3}$$

From Eqs. (4) and (5) the estimated direction of gravity and magnetic field, can be calculated respectively.

$$^E\hat{v} = \left[2 \times (q_2 \times q_4 - q_1 \times q_3); \quad 2 \times (q_1 \times q_2 - q_3 \times q_4); \quad q_1^2 - q_2^2 - q_3^2 + q_4^2 \right] \tag{4}$$

$$^E\hat{w} = \begin{bmatrix} 2 \times b_2 \times \left(0.5 - q_3^2 - q_4^2\right) + 2 \times b_4 \times (q_2 \times q_4 - q_1 \times q_3) \\ 2 \times b_2 \times (q_2 \times q_3 - q_1 \times q_4) + 2 \times b_4 \times (q_1 \times q_2 - q_3 \times q_4) \\ 2 \times b_2 \times (q_1 \times q_3 - q_2 \times q_4) + 2 \times b_4 \times \left(0.5 - q_2^2 - q_3^2\right) \end{bmatrix} \tag{5}$$

After that the Error is calculated with Eq. (6), which is the result of the sum of the cross product between the estimated direction and the measured direction of the field vectors, a is the measure by the accelerometer.

$$^E\hat{e} = a \times {}^E\hat{v} + m \times {}^E\hat{w} \tag{6}$$

The Mahony filter is used particularly because it allows to correct the bias of the gyroscope, applying an integral and proportional controller with Eqs. (7) and (8), where sp is the sample frequency (400 Hz for this work) and $\hat{g}_{k-1}$ is the measurement given by the gyroscope.

$$i_k = i_{k-1} + {}^E\hat{e} \times sp \tag{7}$$

$$\hat{g}_k = \hat{g}_{k-1} + k_p \times {}^E\hat{e} + k_i \times i_k \tag{8}$$

At last, the quaternion change rate is calculated with Eq. (9) and integrated to produce the quaternion with Eq. (10), later converted to Euler angles to compare these measurements with the measurements of angles found in the analysis of Kinovea.

$$\dot{q} = 0.5 \times {}^S_E\hat{q} \otimes {}^E\hat{g}_t \tag{9}$$

$$^S_E\hat{q}_k = {}^S_E\hat{q}_{k-1} + \dot{q} \times sp \tag{10}$$

3.3 Madgwick Filter

The algorithm of this filter starts by normalizing the values obtained from the accelerometer and magnetometer, calculating the direction of the magnetic field in the frame of the earth and the effect of a wrong inclination measurement of the direction of the magnetic field, to find these values the same equations described in Sect. 3.2 are used.

This filter is characterized by adding a corrective stage using the gradient de-census algorithm, as described in [17]. Where the Jacobian is defined by Eqs. (11) and (12).

$$f_b\left({}^S_E\hat{q}, {}^E\hat{b}, {}^S\hat{m}\right) = \begin{bmatrix} 2b_x\left(0.5 - q_3^2 - q_4^2\right) + 2b_z(q_2q_4 - q_1q_3) - m_x \\ 2b_x(q_2q_3 - q_1q_4) + 2b_z(q_1q_2 - q_3q_4) - m_y \\ 2b_x(q_1q_3 - q_2q_4) + 2b_z\left(0.5 - q_2^2 - q_3^2\right) - m_z \end{bmatrix} \tag{11}$$

$$J_b\left({}^S_E\hat{q}, {}^E\hat{b}\right) = \begin{bmatrix} -2b_zq_3 & 2b_zq_4 & -4b_xq_3 - 2b_zq_1 & -4b_xq_4 + 2b_zq_2 \\ -2b_xq_4 + 2b_zq_2 & 2b_xq_3 + 2b_zq_1 & 2b_xq_2 + 2b_zq_4 & 2b_xq_1 + 2b_zq_3 \\ 2b_xq_3 & 2b_xq_4 - 4b_zq_2 & 2b_xq_1 - 4b_zq_3 & 2b_xq_2 \end{bmatrix} \tag{12}$$

Equations (13) and (14) combine the measurements of gravity and the magnetic field of the Earth, to provide a unique orientation.

$$f_{g,b}\left({}^S_E\hat{q}, {}^S\hat{a}, {}^E\hat{b}, {}^S\hat{m}\right) = \begin{bmatrix} f_g\left({}^S_E\hat{q}, {}^S\hat{a}\right) \\ f_b\left({}^S_E\hat{q}, {}^E\hat{b}, {}^S\hat{m}\right) \end{bmatrix} \tag{13}$$

$$f_{g,b}\left({}_E^S\hat{q}, {}^E b\right) = \begin{bmatrix} f_g^T\left({}_E^S\hat{q}\right) \\ f_g^T\left({}_E^S\hat{q}, {}^E b\right) \end{bmatrix} \tag{14}$$

After that, compute rate of change with Eq. (15).

$$\dot{q} = 0.5 \times \left({}_E^S\hat{q}_{k-1} \otimes {}^E\hat{g}_k\right) - \beta \times S^T \tag{15}$$

Where S^T is the transpose matrix of the multiplication and normalization of Eqs. (14) and (15) and β is the proportional controller gain. Finally, the rate of change of quaternion is integrated as in Sect. 3.2 with Eq. (10), and this result later converted to Euler angles.

3.4   Kalman Filter

The Kalman filter is widely used to make data combinations with a lot of noise, in this case, the filter integrates the data given by the gyroscope (with Drift) and with the combination of the accelerometer and magnetometer, which tend to be quite noisy, an estimation is performed. and puts each one an appropriate weight from their models to be able to perform the estimation of the orientation.

The filter implemented in Matlab for this work is the same described in [18], where the algorithm consists of two important parts: the predictive part and the corrective part.

In the predictive part, the projection of the forward state must be carried out using the Eq. (16), and then project the error covariance forward with the Eq. (17).

$$\hat{x}_k^- = F\hat{x}_{k-1} \tag{16}$$

$$P_k^- = FP_{k-1}F^T + Q \tag{17}$$

Then, in the corrective part, the calculation of the Kalman gain with the Eq. (18) is performed, the state update is performed with the measurement z_k using the Eq. (19), and finally the covariance error was updated with Eq. (20), this corrective part we have to do twice, once to correct by measuring the magnetometer and another to correct by the accelerometer.

$$K_k = P_k^- H^T \left(HP_k^- H^T + R\right)^{-1} \tag{18}$$

$$\hat{x}_k = \hat{x}_k^- + K_k\left(z_k - H\hat{x}_k^-\right) \tag{19}$$

$$P_k = (I - K_kH)P_k^- \tag{20}$$

Where k is the step time; x, state vector, and has the quaternion values; z, input vector; F, state matrix; B input matrix, H output matrix; state and measurement noise; Q and R are covariance matrix from the state and the measurement noise, respectively. All this process is done repetitively.

3.5   Magnetic Field Measurement with Magnetometer

The magnetic field that the magnetometer measures can be affected by metals that are around the sensor, this can be by components that are located on the same Printed Circuit Board (PCB) that the sensor is mounted, by metals found in the battery, or external metals, for example, the metal structure of the building where the measurements are made.

There's a way to prove that there are distortions in the measurement, this is done by moving around the sensor in space in a three-dimensional way, rotating it in angles from 0 to 360° in each of its axes and combinations, and the data in x, y, and z of the magnetic field are plotted as if they were point coordinates, a perfect sphere centered on the point (0, 0, 0) is obtained if there is no any type of interference, however, if there are distortions by metals that are affecting the sensor, this sphere is no longer centered.

In Fig. 1 we can observe data of how the measurements of the magnetometer used in this paper are affected by metals, where the sphere is not centered at the origin (distortion by hard metals).

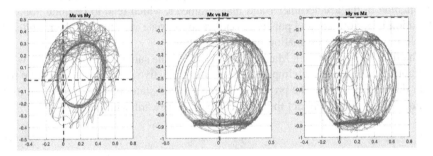

Fig. 1. Magnetic field with distortions due to metals

To correct this problem of hard metals, the following formulas can be used, which calculate offsets that can be added to the measurements on each axis of the Magnetometer measurement.

$$Offset_x = \frac{max(M_x) + min(M_x)}{2} \tag{21}$$

$$Offset_y = \frac{max(M_y) + min(M_y)}{2} \tag{22}$$

$$Offset_z = \frac{max(M_z) + min(M_z)}{2} \tag{23}$$

If we add to the data in Fig. 1 the offsets obtained by Eqs. (21), (22) and (23), we obtain Fig. 2 where we can see that the sphere is centered on the point (0, 0, 0).

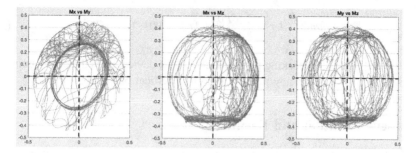

Fig. 2. Magnetic field with distortions due to metals with offsets applied

Another way to check if the measurements obtained are correct is with the module of each of the components of the magnetic field, this module should be a constant value and in the range between 0.25 and 0.65 Gauss, with an average of 0.5 Gauss [19]. With the on-line tool shown in [20] you can enter coordinates, altitude, a date and the software will give a value of the earth's magnetic field with those parameters entered, for example, in Vilanova i la Geltrú, where the tests were performed, when entering the coordinates of 41.223238 N, 1.733494 E, altitude of 10 m, and the date of October 10, 2017, a magnetic field value of 45.471.7 nT or 0.4547 Gauss is obtained, this is the value that will be used later to verify that the magnetometer reading is correct.

3.6 Kinovea Angle Measurement

With the use of Kinovea, the tracking of points in a video is possible, from these points we can obtain data such as acceleration, speed, position and other parameters, this data is stored in a file that can then be read from Matlab, we are interested in examining the angle θ shown in Fig. 3 but the software does not record angles in the aforementioned file, because of this, from the coordinates of the points P1 and P2, and use basic trigonometry and the required angle θ can be calculated.

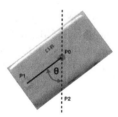

Fig. 3. Tracking points of 9 × 3 sensor for angle measurement

To calculate this angle, the function atan2 is used, this function is the four-quadrant inverse tangent between the intervals of [−pi, pi] as is shown in Eq. (24).

$$\theta = atan2\left(\left|det\begin{bmatrix} P2 - P0 \\ P1 - P0 \end{bmatrix}\right|, (P2 - P0) \cdot (P1 - P0)\right) \tag{24}$$

Where $P0$ and $P1$ are the tracking points placed on the device, $P0$ is the center point, $P1$ is an outer point and $P2$ is any point in the same vertical of the point $P0$.

4   Experimentation and Results

To evaluate the performance of each algorithm, the mean square error (RMSE) is used, where each Euler angle obtained by the filters is compared to a gold standard which is obtained from Kinovea.

The experiment consists of taking the sensor used in the MoMoPa3 project in which the signals of the magnetic field, acceleration and angular velocity are recorded, these signals are then passed through each one of the filters and the Orientation estimation is achieved.

At the same time that the signals of the IMU are being recorded, a video of the device is taken through a camera placed with a zenith perspective, this video is then analyzed image by image through Kinovea and coordinates of tracking points strategically placed in the sensor (green marks and center, see Fig. 4) are saved. Finally, through basic trigonometry (as show in Sect. 3.6) the angles that the device has traveled are obtained.

Fig. 4. Device tracking points (Color figure online)

In order to be able to synchronize the data obtained from the IMU, with the data obtained from Kinovea to calculate the RMSE, the sensor was placed in a vertical position and then was dropped horizontally, with this, the accelerometer records a peak of acceleration when the device touches the surface on which it is resting (see Fig. 5).

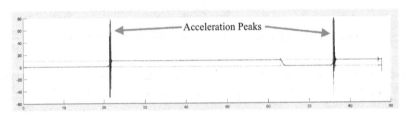

Fig. 5. Peaks of acceleration for synchronization

4.1   Measurement Correction of Magnetic Field

As seen in Sect. 3.5, a problem regarding the measurement of the magnetic field due to metal distortions that surround the sensor is present, but this can be corrected applying the previously explained formulas to obtain offsets and apply them to the original signal.

Some tests were made to find the average value of these offsets that are to be applied to each measurement on each axis of the Magnetometer and later to be able to use them with our orientation estimation algorithms. The values found in each test and an average of all of them can be observed in Table 1.

Table 1. Magnetometer offsets (Gauss)

Test	Mx offset	My offset	Mz offset
1	0.1861	0.0474	−0.5453
2	0.1864	0.0473	−0.5419
3	0.1802	0.0506	−0.5454
4	0.2056	0.0342	−0.5456
5	0.1765	0.0539	−0.5389
Average	0.18696	0.04668	−0.54342

These average value offsets are added to the raw data signals given by the magnetometer: 0.18696 to Mx, 0.04668 to My and −0.54342 to Mz. These values should not vary much if the sensor is used in the same place as the tests, but if it is moved to a different location, the new location may contain metals in the structure of the building that affect the measured magnetic field, or if it is placed next to nearby electronic devices, the offset values must be recalculated.

In order to verify that these offsets are appropriate, the module of the measurements of the magnetometer with and without these offset values is obtained, as shown in Fig. 6(a), the raw data of the magnetometer is outside the range of normal values of the

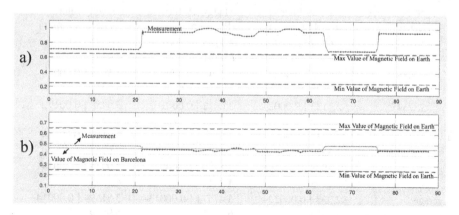

Fig. 6. (a) Raw magnetometer measurements (b) Magnetometer measurements with added offsets

earth's magnetic field (between 0.25 and 0.65 Gauss) and after applying the offset values (Fig. 6(b)) it is seen that, the measured values, besides being within the range, are close enough to the magnitude of the magnetic field in Barcelona, which is approximately 0.4547 Gauss.

4.2 Device Performance Compared to Gold Standard (Kinovea)

The signals obtained from Kinovea and the IMU have different sample frequencies (Fs), the data from Kinovea has a Fs of 30 Hz, that is 1 sample per image, each second contains 30 images (30 frames per second), the sampling frequency of IMU is 400 Hz, so in order to obtain the RMSE an upsampling of the Kinovea data from 30 Hz to 400 Hz had to be performed in order to match the sensor data.

In Table 2 the results obtained for RMSE are summarized for each filter for 5 different tests. The RMSE was calculated from the moment in which the sensor was dropped to be synchronized with the video, until it is back in a vertical position to the plane of movement. Then, between the tests, we calculate the average RMSE value for each filter and we can determine which one works best for our sensor (The MoMoPa3 project sensor). For each test, around 17000 angles were calculated, so, there's enough data to prove that the filters work properly.

Table 2 shows that the Kalman filter is the one that has a lower RMSE value (measured in degrees) with just 2.333° variation from the gold standard, followed up by the Madgwick filter with 3.47574° and then Mahony with 3.51594°, therefore Mahony and Madgwick get very similar results, these variations are acceptable for medical use, but they wouldn't be suitable in other purposes where an exact estimation is needed, like military application.

The maximum error that was obtained in the Kalman filter was 5.9523° in test number 2, while in Madgwick's filter was 7.6186°, and in Mahony's it was 7.9391°, i.e. the Kalman filter was the one with less error in all estimated angles.

Table 2. Root mean square in degrees

Test	N° of angles tested	Kalman filter (RMS | min-max)	Madgwick filter (RMS | min-max)	Mahony filter (RMS | min-max)
1	16360	1.6704 | 0.0001–4.6665	3.1136 | 0.0001–6.4581	3.5683 | 0.0002–7.9391
2	20356	2.8683 | 0.0007–5.9523	3.5744 | 0.00009–7.6186	3.5573 | 0.0001–6.3006
3	13784	2.7123 | 0.0002–5.4585	3.6612 | 0.0001–5.2534	3.6458 | 0.0019–4.1583
4	18469	2.5317 | 0.00005–4.3502	3.9126 | 0.0002–7.3543	3.8167 | 0.0001–6.6734
5	16696	1.8825 | 0.00006–4.4998	3.1169 | 0.0002–4.8514	2.9916 | 0.0001–4.8533
Average		2.333 | 0.00022–4.98546	3.47574 | 0.00013–6.3071	3.51594 | 0.00014–5.9849

Figure 7 provides an example of a recorded signal, in red the angle estimation obtained by the Kalman filter is shown and in blue our gold standard, i.e. the angles computed from the analysis given by Kinovea, one thing to point out is that the red line seems smoother than the blue one, that's because of errors of tracking made by Kinovea, so, the result obtained by the Kalman filter are, in some way, better than the ones on Kinovea.

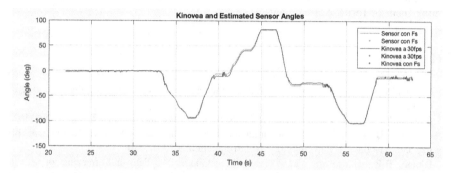

Fig. 7. Comparison between angles obtained with Kalman filter and Kinovea (Color figure online)

Video results are provided on https://www.youtube.com/watch?v=X0SCZ33uQX4.

5   Conclusions and Future Works

There are some ways to get the orientation of a device, robot, vehicle, etc., this could be done using encoders or inclinometers but these solutions are not feasible for certain applications such as wearables, as is the sensor used in the Project MoMoPa3, for these cases, inertial sensors, such as accelerometer, gyroscope and magnetometer are of better use, but with these, it is not possible to obtain directly an orientation measurement in degrees, for this it is necessary to use certain filters that allow the fusion of the data and as a result get an estimate of which direction the sensor is oriented.

In this work, a comparison was made between three filters, namely the Kalman filter, the Madgwick filter and the Mahony filter. The orientation estimation obtained by them, that is, the angles in which the sensor was positioned, was compared to a gold standard, the angles calculated from a video analysis performed by the Kinovea software, where it was obtained that the filter that gave the best results for the sensor used was the Kalman filter, followed by the Mahony filter and finally the Madgwick filter.

Before entering the raw data obtained from the inertial sensors of the IMU, a certain processing must be carried out to each one of them so that, in this way, better results can be achieved, the magnetometer must correct offsets which are produced due to soft and hard metals, in addition to certain interferences where the sensor is taking a measure, the gyroscope must add certain bias due to the problem of drift they have, and the accelerometer must ensure that as long as no force is applied to it, values of 1G or 9.8 m/s corresponding to the measure of the acceleration of gravity that exists in the earth are to be obtained.

As a future work, genetic algorithms can be implemented in order to determine an adequate gain value for each filter, with this one could find an optimal measure that gives better results. In addition, with the angles obtained, future research should consider more the context of the medical field, that is, within the MoMoPa3 project or other applications that involve diseases that affect the human being, the ways of preventing them or forms of treatment.

References

1. Brynjolfsson, E., Mcafee, A.: Race Against the Machine. Digital Frontier Press, Lexington (2011)
2. Singh, M., Murthy, V., Ramassamy, C.: Neuroprotective mechanisms of the standardized extract of Bacopa monniera in a paraquat/diquat-mediated acute toxicity. Neurochem. Int. **62**, 530–539 (2013). https://doi.org/10.1016/j.neuint.2013.01.030
3. Bloem, B.R., Hausdorff, J.M., Visser, J.E., Giladi, N.: Falls and freezing of gait in Parkinson's disease: a review of two interconnected, episodic phenomena. Mov. Disord. **19** (8), 871–884 (2004)
4. Okuma, Y.: Freezing of gait in Parkinson's disease. J. Neurol. **253**(Supplement 7), vii27–vii32 (2006)
5. European Parkinson's Disease Association. http://www.epda.eu.com/about-parkinsons/symptoms/motor-symptoms/rigidity/
6. Rahmatian, S., Torija, G.: Enfermedad de Parkinson, Últimos Avances en el Tratamiento (2017)
7. Rodríguez-Martín, D., et al.: A waist-worn inertial measurement unit for long-term monitoring of Parkinson's disease patients. Sensors **17**, 827 (2017). https://doi.org/10.3390/s17040827
8. Barrientos, A., Del Cerro, J., Gutiérrez, P., San Martín, R., Martínez, A., Rossi, C.: Vehículos aéreos no tripulados para uso civil. Tecnología y aplicaciones, pp. 1–29. Grup. Robótica y Cibernética, Univ. Politécnica Madrid (2009)
9. Benini, A., Mancini, A., Longhi, S.: An IMU/UWB/vision-based extended Kalman filter for mini-UAV localization in indoor environment using 802.15.4a wireless sensor network. J. Intell. Robot. Syst. Theory Appl. **70**, 461–476 (2013). https://doi.org/10.1007/s10846-012-9742-1
10. Bigy, A.A.M., Banitsas, K., Badii, A., Cosmas, J.: Recognition of postures and Freezing of Gait in Parkinson's disease patients using Microsoft Kinect sensor. In: International IEEE/EMBS Conference on Neural Engineering, NER, pp. 731–734. IEEE (2015)
11. Chávez Cardona, A.M., Rodríguez Spitia, F., Baradica López, A.: Exoskeletons to enhance human capabilities and support rehabilitation: a state of the art. Revista Ingeniería Biomédica. **4**, 63–73 (2010)
12. Pons, J., Moreno, J., Brunetti, F., Rocon, E.: Lower-limb wearable exoskeleton. Rehabil. Robot. **3**, 471–498 (2007). https://doi.org/10.5772/5176
13. Sabatini, A.M., Member, S.: Quaternion-based extended Kalman filter for determining orientation by inertial and magnetic sensing. IEEE Trans. Biomed. Eng. **53**, 1346–1356 (2006)

14. Sabatini, A.M.: Inertiel sensing in biomechanics: a survey of computational techniques bridging motion analysis and personal navigation. Comput. Intell. Mov. Sci. 70–100 (2006). https://doi.org/10.4018/978-1-59140-836-9
15. González Jiménez, J., Baturone, A.O.: Estimación de la Posición de un Robot Móvil. Automatica **29**, 3–18 (1996)
16. ST-Microelectronics: LSM9DS1 iNEMO inertial module, pp. 1–74. STMicroelectronics, Ginebra (2013). DocID02476
17. Madgwick, S.O.H.: An efficient orientation filter for inertial and inertial/magnetic sensor arrays, p. 32. Report x-io and University of Bristol, Bristol, UK (2010). https://doi.org/10.1109/icorr.2011.5975346
18. Brigante, C.M.N., Abbate, N., Basile, A., Faulisi, A.C., Sessa, S.: Towards miniaturization of a MEMS-based wearable motion capture system. IEEE Trans. Ind. Electron. **58**, 3234–3241 (2011). https://doi.org/10.1109/TIE.2011.2148671
19. Macmillan, S.: Earth's magnetic field. Geophys. Geochem. (2013)
20. National Geophysical Data Center: Magnetic Field Calculators. https://www.ngdc.noaa.gov/geomag-web/#igrfwmm

Author Index

Printed in the United States
By Bookmasters

Printed in the United States
By Bookmasters